SPECIFICITY IN BIOLOGICAL INTERACTIONS

PONTIFICIAE ACADEMIAE SCIENTIARUM SCRIPTA VARIA

55

WORKING GROUP

ON

SPECIFICITY IN BIOLOGICAL INTERACTIONS

November 9-11, 1983

EDITED BY

C. CHAGAS AND B. PULLMAN

PONTIFICIA ACADEMIA SCIENTIARVM

EX AEDIBUS ACADEMICIS IN CIVITATE VATICANA

MCMLXXXIV

SPECIFICITY IN BIOLOGICAL INTERACTIONS

PROCEEDINGS OF A WORKING GROUP
AT THE
PONTIFICAL ACADEMY OF SCIENCES

November 9-11, 1983

EDITED BY
C. CHAGAS AND B. PULLMAN

D. REIDEL PUBLISHING COMPANY
A MEMBER OF THE KLUWER ACADEMIC PUBLISHERS GROUP
DORDRECHT / BOSTON / LANCASTER
1984

Library of Congress Cataloging in Publication Data

CIP

DATA APPEAR ON SEPARATE CARD
ISBN 90-277-1813-X

Published by D. Reidel Publishing Company,
P.O. Box 17, 3300 AA Dordrecht, Holland.

Sold and distributed in the U.S.A. and Canada
by Kluwer Academic Publishers,
190 Old Derby Street, Hingham, MA 02043, U.S.A.

In all other countries, sold and distributed
by Kluwer Academic Publishers Group,
P.O. Box 322, 3300 AH Dordrecht, Holland.

Printed in Italy

INDEX

SCIENTIFIC PAPERS

FOREWORD

" The specificity of biological interactions " was the subject of a Study Group that met at the Pontifical Academy of Sciences from November 9-11, 1983. This theme is one of the most advanced fields in modern biology.

The knowledge that biological reactions present a specificity which can be sometimes extremely strict has existed for a long, long time and its recognition grew up during the development of many biological disciplines. Molecular basis for this specificity became quite obvious with the study of the action of enzymes and the molecular cause for this specificity became clear after our recognition of the secondary and tertiary structures of proteins. Many techniques have been responsible for these findings, but it seems to me fair to cite X-ray diffraction. Measures obtained between different chains made it possible to envisage how a certain number of reactions occur.

These techniques also allowed for the construction of models in the field of nucleic acids. They were particularly useful in the field of nucleic acid structure determination. Their results have changed biology.

But the idea of image models for biological interactions was also an important contribution in immunology since the days of Erlich and Landensteiner. These were models based on a geometric configuration. They still inspire up to now our graphic concepts of the immune reaction.

These were data from studies made in vitro. *In the field of pharmacology, where experimental animals are used, the problem was also approached, and its interest increased on account of the rapid growth of a quantitative pharmacology. The concept of a receptor was early introduced in pharmacology bv A. J. Clark. For Clark a receptor is the molecular complex able to interact with a pharmacon. This interaction is responsible for the biological response obtained. Nevertheless, it became quite clear in a very short time that a pharmacon injected in an experimental animal would attach itself to a series of molecular complexes different from the " Specific Receptor ". In certain of the cases, more drug was found in the " Non-Specific Receptors " than in the " Specific ones! " The classic experiment to prove it is an old one undertaken by Orchita, who prepared Digitalis Glycosides by breeding the digitalis plants in an atmosphere of radioactive* CO_2. *The radioactive glycoside thus obtained could hardly, if at all, be found in the heart. The same was shown with curares. Thus it became clear that chemical nature and the spatial distribution of the receptors should be known and* in vitro *experiments became again necessary. The development of this endeavour began with a search of the structure of one receptor, the acetylcholine, and is now extended to many other transmitter receptors. What is exciting about the study of the specificity of biological interactions where at least two partners interact, is that their molecular structure in terms of atomic constitution and quantum forces must be known. This not only requires the use of a variety of techniques, but also a high-sophisticated level of theoretical approach.*

This Study Group held at the Pontifical Academy of Sciences has certainly not resolved all problems but has, in my opinion, made progress in the knowledge of one of the most important phenomena of the natural world.

It is for me a great honour to present my profound gratitude to Professor Bernard Pullman for the excellent organization given to the Study Group. I want also to express my deepest thanks to all participants for their good will in coming to the meeting and for their efforts to present their papers thus making possible the publication of this document. Another element for the success of the meeting is the interest which every participant took in the discussion. And last, but not least, I want to express my thanks for the help I received from Fatber Di Rovasenda, Michelle Porcelli, Gilda Massa, Simonetta Ulisse and Silvio Devoto.

CARLOS CHAGAS
President of the
Pontifical Academy of Sciences

LIST OF THE PARTICIPANTS

Prof. David R. DAVIES, National Institute of Arthritis and Metabolic Disease, Laboratory of Molecular Biology, Bethesda, Maryland 20205, U.S.A.

Prof. Jean-Pierre EBEL, C.N.R.S., Institut de Biologie Moléculaire et Cellulaire, 15, rue René Descartes, 67084 Strasbourg Cedex, France.

Prof. A.R. FERSHT, Imperial College of Science and Technology, Department of Chemistry, South Kensington, London SW7 2AY, Great Britain.

Prof. Michael LEVITT, The Weizmann Institute of Science Chemical Physics Department, Rehovoth, Israel

Prof. Brian W. MATTHEWS, University of Oregon, Institute of Molecular Biology, Eugene, Oregon 97403, U.S.A.

Prof. Yuri OVCHINNIKOV, U.S.S.R. Academy of Sciences, Shemyakin Institute of Bioorganic Chemistry, Ul. Vavilova 32, Moscow 117312, U.S.S.R.

Prof. Dinshaw PATEL, Bell Laboratories, 6000 Mountain Avenue, Murray Hill, New Jersey 07974, U.S.A.

Prof. Alberte PULLMAN, Institut de Biologie Physico-Chimique, Fondation Edmond de Rothschild, 13, rue Pierre et Marie Curie, 75005 Paris, France.

Prof. Bernard PULLMAN, Institut de Biologie Physico-Chimique, Fondation Edmond de Rothschild, 13 rue Pierre et Marie Curie, 75005 Paris, France.

Prof. Wolfram SAENGER, Freie Universität Berlin, Institut für Kristallographie, Takustrasse 6, 1000 Berlin 33, Federal Republic of Germany.

Prof. Harold SCHERAGA, Cornell University, Department of Chemistry, Baker Laboratory, Ithaca, New York 14853, U.S.A.

Prof. Arieh WARSHEL, University of Southern California, Department of Chemistry, University Park, Los Angeles, Calif. 90007, USA.

Prof. Ch. ZIMMER, Akademie der Wissenschaften der DDR, Zentralinstitut für Mikrobiologie und Experimentelle Therapie Jena, Bautenbergstrasse 11, 69-Jena, German Democratic Republic.

AUDIENCE OF THE HOLY FATHER

On November 12, 1983, His Holiness John Paul II granted an Audience in the " Sala Regia " of the Apostolic Palace in the Vatican to 45 Members of the Pontifical Academy of Sciences gathered in Plenary Session to discuss the subject " Science in the Service of Peace ", to 20 scientists participating in the Study Week organized by the Academy on " Chemical Events in the Atmosphere and their Impact on the Environment " and to 15 scientists of a Working Group convened by the Academy to discuss " Specificity in Biological Interactions ".

The group, introduced in the Apostolic Palace by the President of the Pontifical Academy of Sciences, His Excellency Prof. Carlos Chagas, accompanied by the Director of the Chancellery, Rev. Father Enrico di Rovasenda, was paternally received by His Holiness, who at the end of the Audience wished to greet personally all the participants.

On the occasion of the Plenary Session the Academy awarded to Prof. Gérard t'Hooft, Institute for Theoretical Physics, Utrecht, the Pius XI Gold Medal, which the Holy Father personally bestowed upon him.

The President of the Academy, Prof. Carlos Chagas, delivered the following address:

Holy Father,

I have the honour to present you my most devout homage, together with members of the Pontifical Academy of Sciences and scientists who have participated in the Study Week and the Study Group which have just ended.

It was exactly four years ago that you received our Academy in this Sala Regia, *whose beauty has so impressed us. The Pontifical Academy of Sciences had met at that time and you so kindly consented to preside over our meeting for the Centenary of the birth of Albert Einstein, who had devoted a great part of his life to the cause of Peace. It is therefore not pure coincidence that you have decided to receive in this magnificent hall your Academicians, who for three days will discuss the subject: " Science in the Service of Peace ".*

Peace, which the Academy has taken as the theme of its Plenary Session, is of great concern to you, as it was to all your predecessors. Since Benedict VI, the Popes, including Your Holiness, have often deplored the useless massacre, " useless slaughter " of human lives and the destruction of cultural treasures which are the dire consequences of war.

We will therefore endeavour during our discussions to examine the various aspects of this problem. One of them will be the possibilities that scientific research offers to impede the many technological developments used for war. I shall mention two examples.

The defense of the civil population and of the ecology against toxic substances, which can contaminate the atmosphere, the rivers, the reservoirs, plants and animals, requires special attention. This can be achieved by the discovery and development of antidotes against the harmful chemical or biological substances released into the biosphere, by the production of new medications against burns, and of products that can neutralize the action of ionizing radiations.

Science can act by making toxic products inactive and — hopefully — by immunizing the toxins extensively and rapidly. With the horizons opened up by modern biology, this is not at all a wild dream. Biology, biochemistry, biophysics, pharmacology, genetics, plant physiology, biomedicine in general and pedology, among other disciplines, must contribute to this effort. The increasing development of grafting and reparatory surgery is necessary to rehabilitate human beings who have been stricken by war.

The continuous perfection of systems of alert in the event of aggression is necessary, even if these will not serve to consign to the curse of history those who have been the first to attack with atomic bombs.

I believe that the very important anti-nuclear declarations which we have published and which have been subscribed to by many scientists, must be followed by action on the part of our Academy, together with Men of Culture and Citizens of the World, in order to renew the appeal to all humanity to avoid the danger of a nuclear holocaust, to banish all fears of war, so that a just and lasting peace can prevail in the world. I believe that a proposal along these lines will be presented to Your Holiness.

Science in the service of peace must also be based on the relationship between individual and collective behaviour; the complexity of this relationship has not yet been resolved and it would be important to know its biological and social mechanisms, inasmuch as they can be a threat to peace. We could thus perhaps find out whether or not individual aggressiveness is independent of social and institutional aggressiveness, and vice versa; this would help us to understand something of the dramas of the past and would shed light on the psychological means which could help to consolidate peace.

Another subject of our discussions will be: development of the underdeveloped countries. How wisely your predecessor Paul VI explained the importance of social development for peace! The words of Paul VI: "The new name of peace is development", which are found in his Encyclical "Populorum progressio" have had a great resonance among us.

The profound differences which exist between poor countries and rich or industrial countries are a constant threat to world peace. The countries of the Third World are becoming more and more the prey of economic interests, which are always egoistic, and of geopolitical interests, which are always imperialistic; and they are subjected to a technological colonialism which is extremely destructive of their national culture.

Your Academy, Holy Father, believes it is its duty to furnish to the governments of these countries as well as to your humanitarian and evangelizing missionaries — who exhaust themselves in work in the forests, in the swamps, the marshes, the bush and the bidonvilles — scientific information which can be useful to them, and to propose to those responsible the norms of a scientific policy which will permit them to face the challenge which they meet at every moment in the fields of education, health, nutrition, agriculture, energy, the protection of nature and renewable resources. The Academy has done so in the past and will do so even more in the future.

However, if development requires wisdom, it also requires economic efforts to be made by all the countries of the world, but especially by those that are rich or imperialistic, which until now have known only how to take over the natural resources of the under-developed countries for their own profit and power, without due compensation. If one compares the sums spent on the 157 wars which have taken place since 1945, with the sums necessary to improve human life and dignity for more than two-thirds of the world's population, one is indeed amazed — I would even say indignant.

I will cite for you only one statistic: the School of Public Health at Harvard University has published a report which proves that the stabilization of the Third World would cost over ten years 7% of what is spent each year on preparations for war.

I am convinced that during our discussion a proposal of great scope will be made in order to have the under-developed countries share in the advantages of science and technology. We will propose action which, above all political rivalries, national passions, geopolitical ambitions and egoism of the governing classes, will bring hope to this immense part of humanity which is beginning to despair because of the wretchedness in the hearts of the rest of the world. This project will be only for the building of a better world.

Humanity, after the devastation of 1939-45, has had some benefit from reconstruction projects; but today only You, Holy Father, have sufficient authority to guide a universal plan for solidarity and peace.

I have presented to Your Holiness the principal lines of the work which we will undertake and for which we need your guidance and advice. The distress which is revealed in my words is due to the terrible reality in which we are living today. But I must confess to you that we are all convinced that our action, no matter how vigorous, will not produce any results so long as the spiritual and moral values which you proclaim are unable to conquer man's egoism and the distress of which he is prisoner.

I therefore take the liberty, Holy Father, to recall the words of Albert Einstein, which I quoted on November 10, 1979, in this same hall: " Humanity has every right to place the propagators of moral values above the discoverers of objective reality. What humanity owes to Buddha, to Moses, and to Jesus Christ is much more important than the success of the research carried on by scientists".

To this I would like to add this testimony. On our television screens in Brazil we saw a few days ago a terrible scene: a man, overcome with grief, buried his wife 27 years old and his child of nine months, both assassinated in the course of a hold-up of a bank. Questioned by a journalist, he replied: "This will continue to happen so long as men and women do not heed the message of Christ".

This message of love is one that you preach continually. It is only by bearing it in our hearts that we can approach our task with fervor, for we believe that in the end, ideals, good will and material detachment will win out and that our labor will not be in vain.

Most Holy Father, our Study Week on "Chemical Events in the Atmosphere and their Impact on the Environment" has just ended. It has studied one of the most serious threats confronting our society, the creation by our civilization of unfavorable conditions for life in our atmosphere. The Working Group which has just met was devoted to "Specificity in Biological Interactions". These discussions have dealt with the complexities of the reactions produced in cells, which determine vital phenomena.

The third group which the Academy will convene in a few days will study the "Contribution of Modern Biology to Agriculture". How much hope is raised in us by the revolution which has taken place in biology and which has taken form in biotechnology! The discussions in this group will certainly give new momentum to the battle against hunger, which is one of the most serious aspects of the pollution caused by the poverty which affects so many of our contemporaries.

Permit me, Holy Father, to express to you my admiration for the eminent specialists who are not members of the Academy and who have interrupted their daily life to come from all parts of the world to help us in our work, without even stopping to admire the beauties of the Vatican or to bask in the historic and artistic climate of Rome. Without them we could not accomplish our task.

And now I wish to announce the names of the new Academicians whom you, Holy Father, have appointed upon their nomination by the

Council of the Academy:

Daniel Adzei Bekoe, of Ghana
Stanislaw Łojasicwicz, of Poland
Wa Kalenga Malu, of Zaire
Charles Townes, of the United States
Hamao Umezawa, of Japan
Pietro Salviucci, of Rome, Honorary Academician.

I also ask you kindly to present to Professor Gérard t'Hooft of the Netherlands the Pius XI Gold Medal, which the Academy has awarded him for his excellent work on elementary particles.

Holiness, in thanking you from the bottom of my heart for the support which you give to the Academy, I wish to assure you that it will always place itself at the service of man and not of economic or political interests. Science increases our knowledge of the natural world, and its correct applications help men, women and children to live as human beings and not just exist. A knowledge of nature and the material improvement of the human condition are and always will be, I believe, for the perfection of Man, who is the glory of God.

The Holy Father answered with the following discourse:

Mr. President, Ladies and Gentlemen,

1. *In this prestigious Assembly of Scientists, honoured by your presence, Cardinals and brother Bishops, and by the Diplomatic Corps accredited to the Holy See, as well as numerous representatives and leaders of the world of culture, I wish to extend heartfelt greetings and an expression of my highest consideration to the distinguished members of the Pontifical Academy of Sciences as they prepare to address in their Plenary Session the theme: " Science in the Service of Peace ".*

With the same cordial sentiments I greet the distinguished scientists who have come from every part of the world for a week of study on the theme: " Chemical Events in the Atmosphere and their Impact on Environment ", and, in a working session, another equally important theme: " Specificity in Biological Interactions ".

In a few days' time another working-group will meet to deal with " Modern Biology applied to Agriculture ".

I congratulate you, Mr. President, Professor Carlos Chagas, for the wisdom and dedication with which you have contributed new and important developments to the life of the Academy. I congratulate you for the planning and promotion of this present series of meetings of personages who devote their energies to the search for truth at the service of humanity.

2. *All knowledge takes its nobility and dignity from the truth that it expresses. Only in the unbiased pursuit of truth do culture and especially science preserve their freedom and are able to defend it from any attempt at manipulation by ideologies or powers.*

" The truth will set you free ": these words from the Gospel enjoy perennial validity and illumine with divine light the endeavours of the scientist who refuses to subordinate his commitment and his research to anything but the truth.

Truth is the goal of the whole universe: finis totius Universi est veritas, *as one of the greatest thinkers of all time, Thomas Aquinas, wrote* (1). *The truth of all beings, their forms and their laws are hidden in the bosom of the Universe, which yearns for its truth to be discovered by the human intellect. You, Men of Science, who welcome the world into your minds, work upon it in your laboratories, and investigate its most secret byways in your dedicated efforts, what are you seeking if not the truth?*

Have courage and the boldness of reason that untiringly seeks the truth and you will find in the Church and in this Apostolic See your most convinced allies. Of course, the conquests of science are at times provisional, subject to review and rethinking, and they will never succeed in expressing the whole truth hidden in the Universe. The sense of mystery is part of your intellectual heritage and reminds you that what you do not know is much vaster than what you do know. In the search for truth, the boldness of reason is linked with the humility of its own limits, the joy of knowing goes hand-in-hand with wonder at the unknown.

Furthermore the sense of mystery also envelops those truths which science cannot discover, but which question the mind of the scientist in the innermost part of his being, where he experiences an irresistible longing and yearning for the divine. The goal of the Universe is not only to reveal the truths it holds within itself, but to manifest the First Truth which gave the world its origin and form.

(1) *Contra Gentiles*, I.I-C.I.

3. *Whatever the paths of your scientific research, may you always be assisted by the sense of the divine. How can one fail to mention here Isaac Newton? He in no way thought, as Auguste Comte was to claim later, that science must rise on the ruins of religion and metaphysics; but he saw in the Universe the presence of God, a presence not immanent but transcending nature.*

In the Scholium Generale *added to the second edition of his* Philosophiae Naturalis Principia Mathematica, *Newton wrote: " This most elegant order of the sun, planets and comets could not have been born without the design and power of an intelligent and powerful being. He rules all things, not as the soul of the world, but as Lord of the Universe... From blind metaphysical necessity that is absolutely identical always and everywhere, no variety in things can be born. The whole truth of things, including places and times, could only have arisen from the ideas and the will of a necessarily existing being "* (2).

The message addressed by the Second Vatican Council to the " men and women of thought and science " agrees with Newton in his conviction that scientific thought and religious thought are inseparable: " Never perhaps, thank God, has there been so clear a possibility as today of a deep understanding between real science and real faith, both in the service of the one truth. Do not lose the occasion of this important encounter: have confidence in faith, this great friend of intelligence ".

Gentlemen, the scientific truth which ennobles your intellect and lifts your research to the level of contemplation of the world and of its Creator, must be transmitted to the whole of humanity for the

(2) Cf. L. Geymonat, *Storia del pensiero filosofico e scientifico*, Milano, Garzanti 1970, vol. II, p. 646.

integral development of each human being and of all nations, for the service of peace, which is the object of your reflections and projects.

4. *There are different ways for men and women of culture to live the precious value of knowledge. Bernard of Clairvaux, one of the strongest personalitites in history, who came down from the loftiest peaks of mysticism to share divine and human truth with the ecclesiastic and civil society of his time, as a true master of love and knowledge, described the different types of men and women of culture always found in history. According to Saint Bernard there are five motives that lead human beings to study: " There are people who only wish to know for the sake of knowing: this is base curiosity. Others wish to know in order that they themselves may be known: this is shameful vanity, and such people cannot escape the mockery of the satyrical poet who said about their likes: ' For you, knowing is nothing unless someone else knows that you know '. Then there are those who acquire knowledge in order to re-sell it, and for example to make money or gain honours from it: their motive is distasteful. But some wish to know in order to edify: this is charity. Others in order to be edified: this is wisdom. Only those who belong to these last two categories do not misuse knowledge, since they only seek to understand in order to do good "* (³).

The words of Saint Bernard the mystic indicate a profound grasp of what motivates those who engage in culture, and they are more than ever relevant in order to remind both the teachers of thought and their disciples of the true purpose of knowledge. In my address of 15 November 1980 in Cologne to the scholars and students of the German Universities I pointed out that " all areas of

(³) SAINT BERNARD, *Thirty-sixth Sermon on the Canticle of Canticles - Oeuvres Mystiques de Saint Bernard*, Editions du Seuil, 1953, pp. 429-430.

our culture are impregnated by a science that proceeds in a mostly functional way ". And I warned that " a purely functional science, deprived of values and alienated from truth, can be completely enslaved by one or other ideology ".

I gladly recall here what an illustrious and now deceased member of the Pontifical Academy of Sciences had to say some forty years ago in a conference to young university students at Lausanne: " One has come to replace the search for truth by the search for what is useful. The young people who previously turned to the masters of thought in order to enlighten their minds, began to ask them for the secrets of nature from which would spring material goods in such abundance. Of the various fields of knowledge, one little by little came to value not those that seek the highest reaches of thought but those that seem more fertile in practical applications " (4).

Saint Bernard of Clairvaux raised knowledge to the level of love, to the level of charity and understanding: " Sunt qui scire volunt ut aedificent et charitas est ".

5. *Members of the Academy, Men of Science, at this very grave moment of history, I ask from you the love of knowledge that builds peace.*

Peace is a gift of God offered to people of good will. I speak now to all men and women of good will, whatever faith they belong to, and especially to you who are listening to me now.

The science which brings together those engaged in research, technicians and workers, which mobilizes political and economic powers, which transforms society at all levels and in all its institu-

(4) G. Colonnetti, *Pensieri e fatti dall'esilio. Conferenza del 12 giugno 1944*, Accademia Nazionale dei Lincei, Roma 1973, p. 31.

tions, has a task today which is proving more urgent and indispensable than ever, namely the task of cooperating in saving and building up peace.

From the depths of centuries past there rises the voice of an unarmed prophet, Isaiah: " *They shall beat their swords into ploughshares and their spears into pruning hooks* " (*Is. 2:4*).

In recent times, at a moment when war was imminent, there rose with biblical force the prophetic voice of an unarmed Pope, Pius XI, who quoted the Psalm: " Dissipa gentes quae bella volunt " (*Ps. 67:31*).

Unarmed prophets have been the object of derision in every age, especially on the part of shrewd politicians, the supporters of power. But today must not our civilization recognize that humanity has need of them? Should not they alone be heard by the whole of the world's scientific community, so that the laboratories and factories of death may give place to laboratories of life? The scientist can exercise his freedom to choose the field of his own research. When, in a particular historical situation, it is all but inevitable that a certain form of scientific research will be used for purposes of aggression, he must make a choice that will enable him to work for the good of people, for the building up of peace. By refusing certain fields of research, inevitably destined, in the concrete historical circumstances, for deadly purposes, the scientists of the whole world ought to be united in a common readiness to disarm science and to form a providential force for peace.

Faced with this great patient in danger of death which is humanity, scientists, in collaboration with all the other members of the world of culture and with the social institutions, must carry out a work of salvation analogous to that of the doctor who has sworn to use all his powers to heal the sick.

6. *Peace is born not only from the elimination of hotbeds of war. Even if all these latter were eliminated others would inevitably appear, if injustice and oppression continue to govern the world. Peace is born of justice:* Opus iustitiae pax (*Is. 32:7*). *Now science, which seeks the truth and is free from all ideologies, can and must promote justice in the world; while not remaining a slave of the economically privileged peoples, it can and must spread everywhere, in order to ensure, through appropriate technical means, that all peoples and all individuals are given their due. The modern world awaits the liberation of science that is a result of the liberation of the mind. Gentlemen, be united in the defence of your liberties in order to build up peace in justice throughout the world.*

This is a relentless work that will never cease, for because of sin, both individual and social, sources of injustice continually arise in the world. With an acute sense of history, the Second Vatican Council warned us of this: " The common good of people is in its basic sense determined by the eternal law. Still the concrete demands of this common good are constantly changing as time goes on. Hence peace is never attained once and for all, but must be built up ceaselessly " (Gaudium et Spes, *48*).

Pax perpetuo aedificanda: *peace has to be ceaselessly built up. Peace is a continuous effort which, in so far as it is up to you, is entrusted to your research, to the technical applications that you must direct, through your authority, to the promotion of justice, with that freedom, that freedom of thought which enables you to make other choices when efforts are made to do you violence, in order to exploit your research and discoveries against justice and peace.*

7. *More than any other, the scientific community is a community of peace, for your rigorous search for the truth in the field of*

nature is independent of ideologies and therefore of the conflicts that result from them. Your activity demands sincere collaboration, and the frank communication of the results of your research.

The scientific community, a community of peace, must be extended to all nations, through the foundation everywhere of institutes for research and sound technological application. It is not enough that political colonialism has ceased; every form of scientific and technological colonialism must cease as well. I cannot fail to note with satisfaction that the Pontifical Academy of Sciences includes an ever greater number of scientists from all the nations of the world, with no racial or religious discrimination. This is a form of cultural ecumenism which the Church, as the promoter of true religious ecumenism, cannot but regard with a sense of lively satisfaction.

8. *From the scientific community, especially when it extends to all the regions of the world, there have come discoveries which have helped the development of humanity in every field: diseases and epidemics have been conquered, new food resources have been found, communications between people have been intensified, the peoples of all the continents have come closer together, natural disasters have been foreseen and overcome. Who can list the benefits brought by science? And cannot one say that these benefits would have been much more important if the techniques resulting from science had not been manipulated by evil powers? Who can deny that science and its applications can be placed at the service of man and of a greater justice?*

It is an irreplaceable task of the scientific community to ensure, as is your intention, Mr. President of the Pontifical Academy of Sciences, that the discoveries of science are not placed at the service of war, tyranny and terror.

The intention to direct science to the promotion of justice and peace demands a great love for humanity. Every human virtue is a form of love. This is the case, in particular, of justice, which is love of neighbour, of individuals, and of peoples. Only the person who loves wants justice for the other person. The person who does not love seeks only to obtain justice for himself.

9. *Truth, freedom, justice and love: such, Gentlemen, must be the cornerstones of the generous choice of a science that builds up peace. These four values, the cornerstones of science and of the life of civilized society, must be at the basis of that universal call of scientists, of the world of culture, of the citizens of the world, which the Pontifical Academy of Sciences, with my full and convinced approval, desires to address to the world for the reconciliation of peoples, for the success of the only war that must be fought: the war against hunger, disease and the death of millions of human beings whose quality and dignity of life could be helped and promoted with seven per cent of the amount spent each year for the incessant and threatening rearmament of the richest nations.*

Permit me at this point to recall with you, in the name of science and in the name of your personal moral authority, the need for a universal conversion to the true goods of man. Peace cannot be invoked, as it is too often, in order to guarantee ethical permissiveness and consumerism. The universal call for peace must be marked by profound reflection on the destiny of man, on the meaning and quality of life. When conversion to truth, freedom, justice and love does not become a necessity widely recognized and put into practice everywhere, social peace is unstable, because it lacks its deepest root, which is found in the heart of man.

10. *It is from God that peace comes for those who are in communion with Him and also for those who, even though they have not found Him, are seeking Him with a sincere heart, in a spirit which, far from stifling the divine, seeks to liberate it within itself. Mr. President, Members of the Academy, and distinguished scientists, I renew the expression of my confidence in you, and as I bring this speech to a close I would like to make my own the words which my predecessor Paul VI addressed in 1966 to the Pontifical Academy of Sciences: " More than anyone else the Church rejoices in every true conquest of the human spirit in whatever field it may be. She recognizes and values highly the importance of scientific discoveries... she not only sees in them a magnificent use of the intelligence, but she discovers in them also the exercise of high moral virtues, which confer on the scientist the character and the merit of an ascetic, sometimes of a hero, to whom humanity must repay a generous tribute of praise and gratitude "* (5).

Gentlemen, as men of thought and science, as pilgrims of the truth, as explorers in the different branches of science and knowledge, about man and the universe, who submit yourselves to the labour of observing, thinking, searching, so that man may be ever more man and may find in nature the proper environment for his development: I ask you to work for justice, love and peace, and to believe that today more than ever the Catholic Church is your ally, this Church which loves true science and right thinking, this Church which prays for you and which in my person, respecting your beliefs, invokes upon each one of you the blessing of God.

(5) Speech of 23.4.1966; Pontificiae Academiae Scientiarum: *Scripta Varia* n. 31, p. XLV.

SCIENTIFIC PAPERS

INTRODUCTORY REMARKS. ELECTROSTATICS AND SPECIFICITY IN NUCLEIC ACID REACTIONS

BERNARD PULLMAN

Institut de Biologie Physico-Chimique
Fondation Edmond de Rothschild
13, rue Pierre et Marie Curie - 75005 Paris
France

ABSTRACT

The increased reactivity of DNA towards electrophiles, with respect to that of its constituents or fragments, which occurs in spite of the decrease of the accessibility to its reactive sites, is attributed to the parallel increase of the molecular electrostatic potential. The role of the electrostatic factor in the specificity or at least selectivity in interactions involving such systems is demonstrated on three examples pertaining to 1) the polymorphism and microheterogeneity of DNA, 2) the interaction of non intercalating ligands in the grooves of DNA and 3) the displacement of reactive sites in oligonucleotides as a function of their length and base sequence.

1. Introduction

The determination of the factors and mechanisms which govern the specificity of intermolecular interactions is certainly one of the main problems of molecular biology. It could possibly also be the basis for a substantial improvement of the chemotherapeutical procedures as employed nowadays. These two reasons are responsible for the selection of this topic for the discussions of the present Study Group because of the hope that, in the general spirit of the Study Groups and Symposia organized under the auspices of the Pontifical Academy of Sciences, such discussions will not only contribute to the development of our understanding of nature but also strengthen our fight against human sufferings.

Among the most challenging but also the most fascinating aspects of the problem that we are going to deal with are, in my opinion, its diversity and its complexity. The diversity refers to the numerous different types and degrees of specificity that are frequently considered. *Stricto senso* specificity implies the exclusive interaction of two selective partners in a selective way; in practice its meaning is frequently more relaxed, referring to more or less exclusive associations such as, for example, the preferential binding of, say, a number of drugs to, say, the minor groove of A-T rich sequences of DNA or the preferred intercalation of other drugs between G-C base pairs of DNA. As to the complexity, it resides in the large number of factors which contribute to intermolecular interactions in general and whose different relative combinations may produce different types of association. If the steric accessibility or better adaptability or even complementarity is frequently considered as a prerequisite for specific associations, few modern biophysicists would pretend, I believe, that these geometrical factors are the only ones involved. Due to a large number of recent studies, some of which carried out by scientists present here, it becomes evident today that what broadly may be referred to as *electronic factors* play also a very important role in specificity. This situation signifies among others that specificity, even though it frequently takes up the appearance of a *local* phenomenon, may in fact be determined, and sometimes in a quite important if not decisive way, by distant influences springing from the overall, global structure of the reacting species. The problem takes, of course, a particular dimension in the field of macromolecules both for interactions involving two or more such species or the interaction of a macromolecule with a relatively small ligand.

This Study Group was organized in view of exchanging and comparing knowledge and opinions of the nature and role of biological specificity, with the aim of arriving, if possible, at a synthetic, overall representation of the present status of the problem, each of the distinguished participants contributing his share to this general construction following his particular expert achievements.

Because of the orientation of my work and the work carried out in my laboratory, I would like to concentrate in my talk on some aspects of *the role of the overall structure and in particular of the overall electronic structure of a macromolecule for reactions with an apparent local specificity.* For the same reasons, I shall discuss this problem essentially in relation *to nucleic acids* and limit myself to their interactions with ligands of relatively small or moderate dimensions.

2. Molecular electrostatic potential in DNA and its constituents

In order to determine the possible role of the overall electronic structure of a macromolecule on the specificity of its local interactions with external ligand, it is essential to choose an appropriate electronic characteristic of the polymer which is liable to translate the effect of its distant parts on the region of space involved. Such a suitable characteristic seems to be available today in the form of *the molecular electrostatic potential* generated by the charge distribution of the polymeric species. It is the role of this potential, and when possible its contribution to the overall interaction scheme, that I wish to examine in some details.

We recall that the molecular electrostatic potential (V) is defined by (Scrocco and Tomasi 1973, 1978):

$$V\ (P) = \sum_{\alpha} \frac{Z_\alpha}{|\underline{r}_{\alpha P}|} - \int \frac{\varrho\,(i)}{|\underline{r}_{1P}|}\, d\tau_1$$

where Z_α is the charge of nucleus α, distant by $r_{\alpha P}$ from point P and $\varrho\,(i)$ is the electronic distribution whose volume element $d\tau_1$ is at a distance of r_{1P}.

The methods of evaluating the potentials in biomolecules and biopolymers have been described in (Pullman, A. and Pullman, B., 1981; Pullman, B. *et al.*, 1982) and will not be repeated here. May we just recall that for simple biomolecules, of limited dimensions, such as the fundamental constituents of the nucleic acids (phosphates, sugars, purine and pyrimidine bases), they are computed from the electronic distributions of these subunits, obtained from *ab initio SCF* computations. The potentials of the macromolecules are constructed by the superposition of the potentials of all the subunits forming the nucleic acid, appropriately positioned in space. In order to facilitate the calculation of the macromolecular electrostatic properties, the electronic distribution of each subunit is replaced by a multicenter, multipole expansion which is capable of reproducing the exact electrostatic properties of the subunit down to a short distance from its constituent atoms.

Before investigating the problem of how such, at first sight a relatively non-discriminatory property, as the electrostatic potential may contribute to the specificity of interactions, a more simple question may be considered of whether the electrostatic properties can be shown altogether to contribute significantly to the reactivity of the nucleic acids. The answer to this question is yes and can be reached in particular by studying the evolution of the *potential minima* associated with the *reactive sites of the purine and pyrimidine bases* of the nucleic acids in the series free bases, nucleosides, nucleotides, single helices and finally double helices, comparing it which the parallel evolution of the *accessibilities* to these sites and, finally, with their known *appropriate* reactivities.

Thus, table I presents the results of computations on the most prominent electrostatic potential minima associated with the bases in the different structures indicated above. (For details of the computations see Pullman, A., and Pullman, B., 1981). At this stage, we limit our considerations concerning the double helix to the B-form of DNA, the computatuons referring to a helical turn of this model for poly (dG.dC) and poly (dA.dT) sequences. The results for the single helices were derived simply by separating the two strands, without alteration of their internal geometry. They should thus not be confused with the usual representations of homopolynucleotides which correspond to different geometrical and spatial configurations and in some of their forms involve inter-base interactions. It seems that they may nevertheless be considered as a first approximation to the representation of partially denatured DNA or of a " breathing " double helix and translate to a large extent the main aspects of the differences

TABLE I. *Electrostatic molecular potential minima (in kcal/mol) at the nucleophilic sites of the four nucleic acid bases, either isolated or incorporated into different compounds*

Site	Single bases	Nucleosides	Nucleotides	Single helices	Double helices
N7 (G)	— 88.2	— 88.0	—146.8	—420.6	—682.6
N3 (G)	— 55.6	— 58.9	—100.9	—373.3	—670.0
N3 (A)	— 62.6	— 65.8	—107.1	—390.5	—668.8
O2 (T)	— 56.2	— 75.4	—110.5	—367.9	—662.8
O6 (G)	— 75.0	— 75.4	—115.2	—387.7	—654.4
N7 (A)	— 67.1	— 66.9	—127.2	—385.2	—650.2
O2 (C)	— 82.3	— 84.7	—122.6	—410.4	—645.1
C8 (G) 5′	— 8.0	— 9.6	— 73.3	—350.7	—630.3
C8 (G) 3′	— 8.0	— 5.5	— 68.2	—366.5	—623.4
N2 (G) 3′	3.3	2.7	— 29.5	—302.5	—623.4
N2 (G) 5′	3.3	0.8	— 31.6	—260.1	—622.6
O4 (T)	— 55.0	— 55.3	— 97.2	—353.2	—611.7
C8 (A) 5′	0.9	— 5.1	— 69.2	—341.9	—610.5
N4 (C) 3′	— 8.3	— 8.2	— 51.4	—299.6	—601.9
N6 (A) 5′	— 12.9	— 14.0	— 52.2	—313.3	—600.4
N6 (A) 3′	— 12.9	— 13.0	— 50.9	—302.7	—597.7
C8 (A) 3′	9.0	1.3	— 60.0	—337.3	—597.2
N4 (C) 5′	— 8.3	— 9.3	— 52.9	—322.2	—593.1
N3 (C)	— 85.7	— 85.2	—126.7	—404.4	—
N1 (A)	— 70.4	— 71.3	—104.0	—367.4	—

between essentially single stranded and essentially double stranded polynucleotides.

Now, what centainly appears as one of the most striking results of the table is *the general, constant, progressive increase in the absolute value (depth) of the potential minima associated with the different reactive sites when we follow the series of substrates in order of increasing complexity.* Particularly striking are the strong increases in these values in the last three columns of table I. They are due to the penetration of the strong potential generated by the phosphate(s) into the vicinity of the bases and its (their) superposition upon the potential inherent in the bases themselves. The only noticeable exceptions to this rule occur upon the transition from a single helix to a double helix for the ring nitrogens such as N1 of adenine and N3 of cytosine, which are exposed in the former and involved in hydrogen bonding in the latter. Such a hydrogen bonding produces a strong depletion of the associated molecular potential minimum. On the other hand, the minima associated with hydrogen bonded carbonyl oxygens and amino groups follow the general trend.

On the other hand, Table II presents thc parallel evolution, for the two extremes of the series, the isolated bases and the bases

TABLE II. *Accessibilities (in Å²) of the bases (towards a model sphere of* 1.2 *Å radius)*

	Isolated	In B-DNA
G	O6 > N7 > N3 > C8 > N2 > N1 11.1 7.7 5.6 4.4 3.5 1.9	N7 > O6 > C8 > N3 4.1 2.7 1.0 0.05
A	N3 > N7 > N1 > C8 > N6 7.1 6.5 5.9 4.5 3.4	N7 > C8 > N3 2.6 1.0 0.7
C	O2 > N3 > C6 > C5 > N4 11.6 5.4 3.6 3.5 3.4	C5 > C6 ∾ N4 ∾ O2 0.3 0.2 0.2 0.2
T	O2 > O4 > C6 > N3 ∾ C5 11.2 9.3 3.3 1.7 1.7	O4 > O2 2.2 0.9
	O2 (C) > O2 (T) > O4 (T) > N7 (G) > N3 (A) > N7 (A)	N7 (G) > O6 (G) > N7 (A) > O4 (T) > C8 (G) = C8 (A) > O2 (T) > N3 (A)

in DNA, of the accessibilities to these same reactive sites (computed following the procedure described in Lavery *et al.*, 1981). A not surprising feature of this table is the *general decrease in the accessibilities of the sites involved when going over from the isolated bases to the bases in the macromolecule.*

There is thus an antagonism between the evolution of the molecular electrostatic potential at the reactive centers of the bases and their accessibility when the bases are incorporated into more complex structures, in particular into the double helix.

Turning now to experiment, one finds frequently stated that the duplex DNA is relatively unreactive with respect to its constituents. However, when looking closer into the data, one observes that this statement refers generally either to reactants which attack the base ring nitrogens implicated in DNA in hydrogen bonding between the complementary pairs (e.g. N1 of adenine, N3 of cytosine etc.) or concerns nucleophilic reagents (e.g. hydroxylamine, semicarbazide etc.). On the other hand, recent literature indicates quite clearly that an overwhelming number of *electrophilic reagents* reacts, on the contrary, much better with DNA than with its constituents. We have investigated from that point of view a large variety of such reagents in our study on the mechanism of chemical carcinogenesis (Pullman, A. and Pullman, B., 1980; Pullman, B. and Pullman, A., 1980). It is well established today that the active forms of the great majority of carcinogens (the so-called proximate or ultimate carcinogens) are electrophilic agents, generally positive ions. This is true both for carcinogens considered as being active *per se* (a number of relatively simple alkylating agents) and those which necessitate a metabolic activation (aromatic hydrocarbons, aromatic amines, aflatoxin B_1 etc.). Now practically all these substances, whose list is presented in table III (which contains

TABLE III. *Electrophilic carcinogens showing increased reactivity in DNA with respect to constituents or single stranded polynucleotides*

Diethyl mustard $CH_3N(CH_2CH_2CL)_2$	Aflatoxin
N-Methyl-N'-Nitrosoguadinine	Benz [a] Pyrene
N-Methyl-N'-Nitrosourea	7-Bromomethylbenz [a] Anthracene
Dimethylsulfate	cis-Dichlorodiammine Platinium (II)
Diethylsulfate	Cis-Dichloro (Ethylenediamine) Platinium (II)
Ethylmethansulfonate	Pyrrolo (1,4) Benzodiazepine Antibiotics e.g. Anthramycin or Neothramycin
N-Hydroxy-1-Naphtylamine	
N-Hydroxy-2-Naphtylamine	

also at its end similarly behaving antitumor drugs), react much better with double stranded DNA than with its constituents or any simpler systems. (For detailed references see Pullman, A. and Pullman, B., 1980; Pullman, B. and Pullman, A., 1980). Because of the reduced target accessibility in DNA, it seems logic, to attribute this increase in reactivity to the effect of the increased electrostatic potential. In fact, this proposition introduces a *unifying principle* for an important aspect of the interaction of carcinogens with one of their probable major receptors.

A related example of a similar effect, which I would like to quote here in relation to a problem that I am going to discuss later, may be found in the interaction of netropsin with the series poly $(dA)_n$·poly $(dT)_n$ (Zimmer *et al.*, 1976) in which this antibiotic induces the formation of a base-paired duplex by binding to its minor groove. While netropsin spans, upon interaction, about four adjacent base pairs, it is nevertheless observed that the efficiency of the interaction of this antibiotic with various oligomers is considerably lower than with the polymer and increases from the hexamer to the decamer. This evolution of the efficiency of interaction can be accounted for by the parallel increase of the molecular electrostatic potential in the groove involved.

3. Electrostatic factors and specificity of interaction

A) *Polymorphism and microheterogeneity of DNA*

I would now like to investigate closer the possible significance of the electrostatic factors for the *specificity* of interactions of DNA. We shall consider a few examples.

The first one continues to refer to the base active sites. It involves, however, the notion of the *polymorphism* and *microheterogeneity* of DNA, a subject among the most actively investigated in todays molecular biology. Polymorphism is a well established feature of DNA structure. It is known that it may be brought about not only by the effect of external conditions (humidity, salt concentration etc.) but also by the variation of the sequence of the base pairs along the helical axis, to the point that it seems universally admitted that this macromolecule may exhibit important regional heterogeneities (microheterogeneity) as a function of its base composition and sequence. It is obvious that such a microheterogeneity may be of particular significance for the efficiency or selectivity of interaction of, say, carcinogens with different conformers or conformational zones of DNA and may thus play a most significant role in the specificity of its interactions.

The problem acquired recently a particular acuteness in conjunction with the discovery of the left-handed variety of DNA, called Z-DNA, by Rich and colleagues, and I would like to consider this case as an example.

The analysis of the crystallographic data and the construction of the corresponding models have suggested to the authors of this discovery (Wang *et al.*, 1979, 1980) that a number of reactive sites on the bases should be more accessible in Z-DNA than in its B-DNA counterpart. The observation concerns in particular a number of atoms on the guanine base, known to be the receptor sites for covalent bond formation with a series of carcinogenic compounds. This situation led these authors to hypothesize that Z fragments interspersed within B-DNA could form particularly sensitive targets for the action of such carcinogens. They paid particular attention to carbon 8 of guanine, C8 (G), known to be the preferred site of action of the carcinogenic N-2-acetoxy-N-2-acetylaminofluorene (AAAF).

Calculations in our laboratory (Zakrzewska *et al.*, 1980, 1981) confirm and quantify the increased accessibility of a number of important acceptor sites of Z-DNA, in particular of C8 (G), with respect to the same sites in the related B-DNA. The comparison of the molecular electrostatic potentials of the same sites in the same species leads, however, to the conclusion that this potential decreases for the C8 (G) receptor site in Z-DNA with respect to B-DNA. This comparison is presented, both for the accessibility and the potential, for C8 of guanine, in fig. 1, in which, in fact, it is not limited to the B and Z forms of DNA but englobes all the main known forms of this biopolymer.

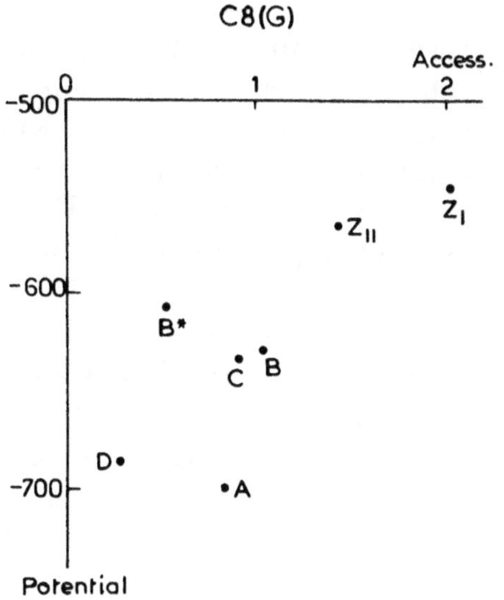

FIG. 1. Potential (kcal/mole) *versus* accessibility (Å^2) for C8 of guanine in the principal conformational forms of DNA.

Such comparative diagrams are available for all the main reactive sites of the nucleic acids (Pullman, B. *et al.*, 1982) and represent thus a basis for the classification of the different allomorphs of DNA with respect to their relative reactivity at these sites.

In our example we are faced, as concerns position 8 of guanine, with the dilemma of which of the two factors, accessibility or molecular electrostatic potential, will dominate its behaviour toward and attacking electrophilic species.

The experimental situation in this respect is, unfortunately, not clear. Thus, in their first attempt, Santella *et al.*, (1981), using as substrate poly (dG-dC) · poly (dG-dC) in the appropriate conditions, were unable to demonstrate that Z-DNA was more reactive than B-DNA towards AAAF. Spodheim-Maurizot *et al.* (1982) reached a more negative conclusion, showing, through kinetic binding studies involving poly (dG-dC). poly (dG-dC) in the B-form and poly (dG-m^5 dC) · poly (dG-m^5 dC) in the Z form, that the reactivity of AAAF is substantially smaller toward the latter than toward the former, a result considered as an indication of the preponderance of the effect of the decrease of the electrostatic potential at C8 (G) in Z-DNA with respect to B-DNA, a decrease which seemed thus to dominate over the greater accessibility of that position in Z-DNA. Recently, however, Santella *et al.*, (1982) claim to obtain a reverse result by studying the equilibrium binding levels of the same polymer as a function of the concentration of the carcinogen. Still more recently, Rio and Leng (1983) find that N-acetoxy-N-2-acetylaminofluorene reacts to about the same extent with B-DNA as with Z-DNA and that N-hydroxy-2-aminofluorene binds to B-DNA and almost not to Z-DNA.

The problem obviously cannot thus be considered as settled, but this rare available example shows the general importance of the microheterogeneity of DNA, in relation to its reactional specificities or preferences, a problem for which important development may be expected to occur, I suppose, in the near future.

B) *Interaction of non-intercalating ligands in the grooves of DNA*

Our second example will be concerned with a specificity no more characteristic of atomic reactive sites on DNA but of its more *extended reactive regions.* In our example these regions will be the grooves of DNA. Much attention has been drawn recently to a large series of non-intercalating drugs which show a remarkable specificity for interaction with one of the grooves of B-DNA, combined with a specificity for a given base sequence. A striking and very much investigated case of such a preferential binding is represented by netropsin (Fig. 2), an antibiotic with antitumor and antiviral activity about which you will hear much more in this meeting from Prof. Zimmer. Netropsin binds in a highly specific way to the minor groove of A-T rich sequences of B-DNA. Following the most widely accepted model for this interaction (see e.g. Berman *et al.*, 1979, Patel, 1982), the concave side of the

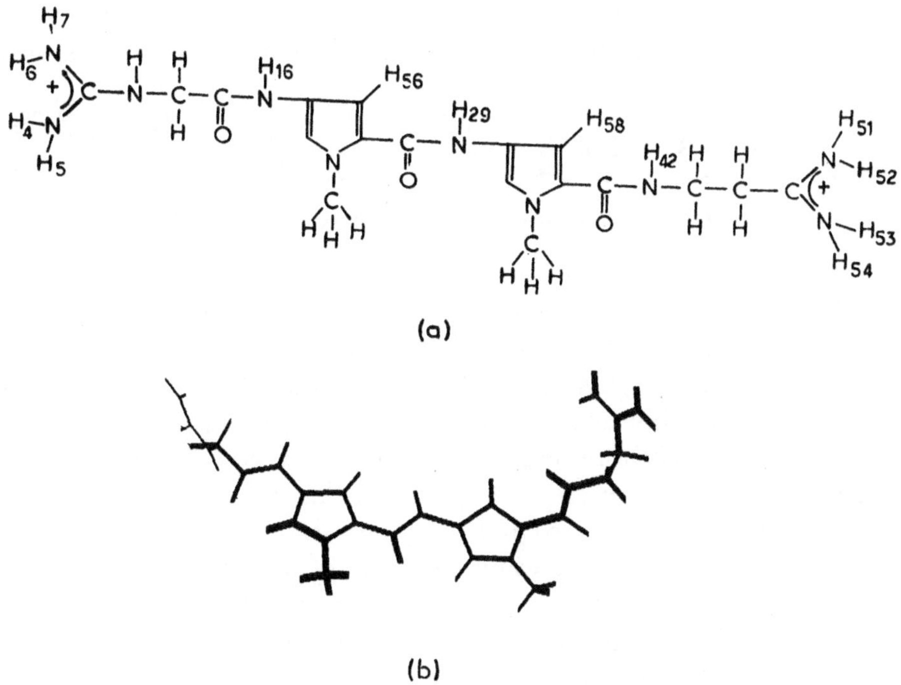

FIG. 2. Netropsin. (a) chemical formula and atom numbering as referred to in the text. b) Minimum energy conformation in the complex with the minor groove of poly (dA) · poly (dT) screened by Mg^{2+}.

antibiotic carrying its hydrogen donors lies in the minor groove of DNA, approximately parallel to the sugar-phosphate backbone and pans four base pairs. The amide nitrogen atoms (numbers 16, 29, 42 in fig. 2 a) can form hydrogen bonds with thymine 02's and/or adenine N3's of one strand. The propionamidininium group is associated with the opposing strand of DNA, whereas the guanidinium residue is not involved in any specific association.

This model was developed on the basis of physicochemical studies and CPK model building, but the exact roles played by the charged ends of netropsin and by the hydrogen bonds remain unclear. The charged ends appear to be important since they are a common feature of many groove-binding ligands. Nevertheless, a netropsin derivative with both ends removed, although showing a decrease in binding efficiency, still complexes to poly (dA). poly (dT) (Zimmer 1975). The ability to form hydrogen bonds has also been thought to be essential. In fact, it is not, as it has been shown that a bisquarternary ammonium heterocycle, SN 18071 (Fig. 3), which cannot form such bonds, binds also

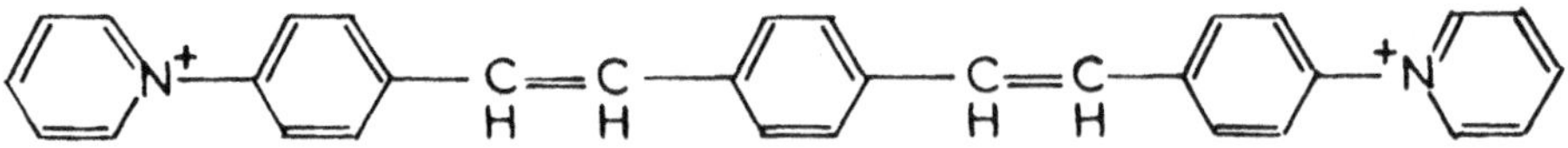

FIG. 3. Bisquarternary ammonium heterocycle SN 18071.

to DNA and shows a similar AT minor gorove specificity (Braithwaite and Baguley, 1980; Baguley 1982).

We have originally suggested (Pullman and Pullman 1981) that this preferential binding could be related to the presence of the most negative surface potential in the minor groove of the nucleic acids involved. More recently (Zakrzewska, Lavery and Pullman 1983) we have gone a step further by performing explicit comparative calculations of the complexing energy of netropsin and SN 18071 with model poly (dA). poly (dT) and poly (dG). poly (dC) duplexes, taking into account the electrostatic and Lennard-Jones type components and allowing for the conformational adaptability of the drugs. Because netropsin has been shown to be specific for the B form of DNA and when interacting with A-DNA converts it to the B form (Ivanov *et al.*, 1974), we have also explored this aspect of the specificity by considering the polynucleotides in both conformations. Moreover, calculations were carried out for free DNAs, in view of understanding their intrinsic preferences, and also for two different models of counterion-screened DNA, so as to be closer to experimental conditions.

The main results for netropsin binding are shown in tables IV and V. It is seen that in both circumstances the lowest energy

TABLE IV. *Netropsin binding to free B-DNA (kcal/mol)*

Sequence	groove	Energy of the complex	Interaction energy			Ligand conformational energy
			Total	Electrostatic	Lennard-Jones	
AT	Minor	—1190.6	—1198.8	—1158.5	—40.3	8.3
	Major	—1079.4	—1096.5	—1077.9	—18.6	17.2
GC	Minor	—1145.6	—1154.8	—1121.0	—33.8	9.3
	Major	—1108.9	—1126.1	—1111.9	—14.2	17.2

TABLE V. *Netropsin binding to B-DNA screened by Mg^{2+} (kcal/mol)*

Sequence	groove	Energy of the complex	Interaction energy			Ligand conformational energy
			Total	Electrostatic	Lennard-Jones	
AT	Minor	— 153.8	— 161.4	— 118.8	—42.6	7.6
	Major	— 37.3	— 52.3	— 32.9	—19.4	15.0
GC	Minor	— 102.5	— 110.8	— 75.1	—35.7	8.3
	Major	— 66.9	— 84.1	— 68.5	—15.6	17.2

complex is found in the minor groove of the AT sequence. It is the most favoured one by both the electrostatic and Lennard-Jones energy terms. This last term may be considered as a measure of the quality of steric fit between the ligand and the macromolecule, because of its strong distance dependence which causes it to become favourable only for very close interactions, but also to rapidly become extremely repulsive if any close contacts are produced. Its significant value in the case of netropsin signifies a good fit. From the detailed results on the interaction we learned that when complexing in the minor groove of the AT sequence the molecule is deeply imbedded, forming six hydrogen bonds between its terminal and amide hydrogens and the Adn (N3)'s, Thy (02)'s and a Sug (01'). In particular, both charged ends of netropsin are well in the groove, forming hydrogen bonds with Thy (02)'s. We were unable to produce any complex which combined the in-groove location of the ligand with a direct interaction of its charged ends with the phosphates (in agreement with conclusions reached by Marky *et al.*, 1983).

A much looser arrangement is found for the GC minor groove, due apparently to the steric and electrostatic influence of the amino groups of guanines. Only one hydrogen bond was found between the guanidinium end and a Sug (01').

Table VI and VII indicate the results obtained for the compound SN 18071.

Once again the best binding position is obtained for the minor groove of the AT sequence, stabilized by both electrostatic and Lennard-Jones terms. (This molecule which is conjugated through-

TABLE VI. *SN* 18071 *binding to free B-DNA (kcal/mol)*

Sequence groove		Energy of the complex		
		Total	Electrostatic	Lennard-Jones
AT	Minor	—1057.4	—1031.1	—26.3
	Major	—1008.3	— 989.2	—19.1
GC	Minor	—1012.3	— 933.2	—19.0
	Major	—1031.5	—1009.5	—22.0

TABLE VII. *SN* 18071 *binding to B-DNA screened by* Mg^{2+} *(kcal/mol)*

Sequence groove		Energy of the complex		
		Total	Electrostatic	Lennard-Jones
AT	Minor	—42.1	—14.3	—27.8
	Major	20.4	40.6	—20.2
GC	Minor	— 7.9	13.1	—21.0
	Major	— 6.6	14.7	—21.3

out its length is much more rigid than netropsin and hence its conformational energy may be neglected). Its complexation energy, although smaller than that for netropsin, remains of a comparable order of magnitude. This result demonstrates thus that *the formation of hydrogen bonds is not necessary neither for binding nor for the preference for the minor groove of the AT sequences of B-DNA.* It seems that if a relatively good steric fit can be obtained in the minor groove the substrate will be sufficiently stabilized by the favourable potentials generated by the AT sequences. This result verifies thus our initial proposition about the dominant role of these potentials in such interactions.

At this stage of the calculations netropsin would be expected to bind better to DNA than SN 18071. Quantitative experimental results (in the form of C_{50} values, defined as micromolar drug concentration required to halve the observed fluorescence due to DNA-bound ethidium) (Baguley 1982) indicate, however, rather comparable affinities, in fact a somewhat greater affinity for SN 18071. An improved agreement with experiment may be obtained by taking into consideration the energies necessary to dehydrate the drugs prior to their binding to DNA.

For this sake, a study of the hydration of netropsin and of SN 18071 was performed with the same program which was used for studying the interaction of these molecules with model DNAs. First, one water molecule was added to netropsin and its optimum position was determined by minimizing the netropsin-water interaction energy. Next, a second water molecule was added to the system and the energy minimization was performed taking into account the influence of the first water molecule. In total, 20 water molecules were added one by one in this way to the system. The same procedure was employed for the hydration of SN 18071, for which 16 water molecules were successively positioned.

The first hydration shell of the solute is then defined as consisting of all the water molecules located for which the interaction energy with the solute exceeds the optimal interaction energy for two water molecules in the same systematics, -7.2 kcal/mole. This definition leads to a first hydration shell of netropsin containing 14 molecules and an associated estimate of the hydration energy of -165 kcal/mole. The first hydration shell for SN 18071 contains only 4 water molecules with a corresponding energy of 32 kcal/mole. Fig. 4 presents the first hydration shell of netropsin and of SN 18071.

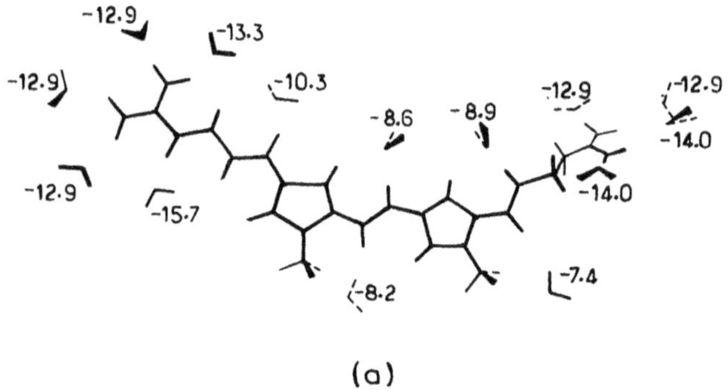

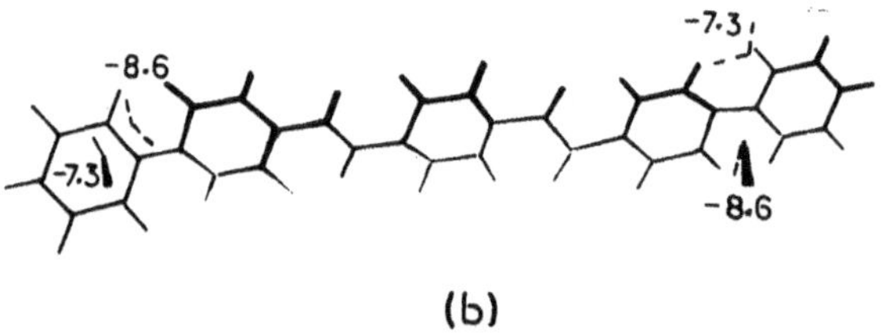

FIG. 4. First hydration shell of a) netropsin and b) SN 18071.

The hydration energy of netropsin is consequently 133 kcal/mole stronger than that of SN 18071. Since the complexation energies of these two ligands with DNA (unscreened) are respectively -1191 kcal/mole for netropsin and -1057 kcal/mole for SN 18071, representing a favouring of netropsin by 134 kcal/mole, we see that by taking into consideration a total dehydration of both ligands before they bind to DNA we obtain approximately equal stabilities for both complexes. In the case of DNA screened with magnesium cations the complex with netropsin is favoured over that with SN 18071 by 112 kcal/mole and dehydration would thus push the balance slightly in favour of binding the latter ligand. These results are thus in very satisfactory agreement with the experimental data noted above. It must, however, be remarked that these calculations are clearly a simplified view of the effects of water and that the ligands studied are probably not totally dehydrated when bound to DNA. Moreover, a more detailed study would have to take into consideration the hydration changes around DNA itself. The results presented indicate nevertheless the qualitative effects of water on such DNA-ligand interactions. This points to the probable overestimation of the importance of hydrogen bonds in such complexes because, generally, a compound capable of strong hydrogen bonding to a substrate will also manifest a similar tendency to bind water. Its affinity for binding to DNA will imply a delicate balance between these two tendencies. We have discussed recently (Gresh and B. Pullman, 1984) a closely related case in the example of the relative affinities of an aliphatic and an aromatic bisguanylhydrazones for similar non-intercalative binding to DNA. (Other examples on the significance of dehydration in biological interactions will be presented by Mme Pullman in her talk on ionophores-cations interactions; see also A. Pullman 1982, 1983).

As announced above, in order to verify the selectivity of netropsin for B-DNA and its inability to form a stable complex with A-DNA, we have investigated also a hypothetical association of both ligands with this latter form. The most significant results refer certainly to the Mg^{2+} screened acid. It is found that in this case too, the best binding site is the minor groove of the AT sequences. The value of the binding energy is, however, extremely small (- 11.1 kcal/mol), much smaller than the similar value for the binding with the B-form of DNA (- 153.8 kcal/mol), a situation which makes it understandable that when put in the presence of A-DNA netropsin converts into the B form.

C) *Reactions of oligonucleotides*

Finally, my last example will be concerned with oligonucleotide duplexes. Due to recent progress in the preparation, crystallization and exploration, in particular by X-ray and NMR procedures, of the structure of these types of compound, both in the solid state and in solution, their study has gained great prominence in the last few years, with the natural attitude of considering these short duplexes as models of nucleic acids. My aim here is to draw attention to the possible dangers of such an extrapolation when referring to the specificity of interaction with external ligands.

We have investigated this problem by studying the interaction of tetramethylammonium ions (TMA) with nucleic acid oligomers d $(Pu)_n$.d $(Py)_n$ of differing sequences and varying chain lengths (n = 2 to 5) (Gresh and Pullman 1983). The available experimental results concern only the interaction of TMA with DNA, in which it occurs preferentially at the A-T base pairs. We wanted to know whether this preference was an inherent one of A-T pairs or the result of the overall structure of the helix. For this purpose we have, as stated above, studied a series of nucleic acid oligomers of varying sequence and chain length. Moreover, in this case, as in that of netropsin, we have computed explicitly the interaction energies of the association, including in the computations, besides the electrostatic term, also the polarization, repulsion and dispersion like terms (for details see Gresh and Pullman, 1983).

The results are summarized in Table VIII, in which the symbols M and m represent the major and minor groove of the oligonucleotide, respectively. It is seen that in the $(dG)_n$. $(dC)_n$ systems, the preferred location of TMA is always in the major groove. With $(dA)_n$. $(dT)_n$, the location of TMA is energetically

TABLE VIII. *The interaction of TMA with oligonucleotide duplexes $d(Pu)_n.(dPy)_n$. Total interaction energies and their components (in kcal/mol) for the most stable associations. M = major groove, m = minor groove.*

Oligonucleotide		n = 1	n = 2	n = 3	$p(dPu)_5 \cdot d(Py)_5p$
$(dG)_n \cdot (dC)_n \cdot$	E_{Tot}	—125.1 (M)	—209.9 (M)	—323.5 (M)	—414.6 (M)
	E_{el}	—113.1	—196.0	—306.0	—396.8
	E_{pol}	— 8.4	— 9.2	— 11.9	— 12.2
	E_{ds}	— 25.6	— 28.4	— 27.5	— 27.5
	E_{rep}	+ 21.2	+ 23.8	+ 21.9	+ 21.9
$(dA)_n \cdot (dT)_n \cdot$	E_{Tot}	—119.2 (M)	—203.5 (M)	—336.7 (m)	—443.8 (m)
	E_{el}	—106.4	—189.0	—308.8	—415.5
	E_{pol}	— 8.2	— 9.1	— 17.9	— 18.1
	E_{ds}	— 24.3	— 27.	— 50.8	— 51.0
	E_{rep}	+ 19.7	+ 22.2	+ 40.8	+ 40.8

favoured in the major groove over the minor groove up to n = 3, but a reversal in the relative affinities for these grooves occurs for n = 5. Moreover, whereas TMA is shown to interact more favourably with $(dG)_n.(dC)_n$ than with $(dA)_n.(dT)_n$ up to n = 3, a marked reversal in the base sequence preferences occurs for n = 5: the minor groove of the adenine-thymine double helix is preferred over the major groove of the guanine-cytosine double helix in $(dPu)_5.(dPy)_5$, and, more markedly, in $p(dPu)_5(dPy)_5p$. This last preference is in agreement with the experimental situation in DNA (Shapiro *et al.*, 1969; Melchior and Von Hippel 1973; De Murcia *et al.*, 1978; Privat *et al.*, 1972; Granados and Bellos 1980).

From the evolution of the components of the interaction energy it may be seen that the electrostatic factor contributes importantly, although not exclusively, to the shaping of the specificity towards the minor groove of the A-T sequences in the longer oligonucleotides.

4. Conclusion

I hope that it is clear from the few examples quoted above that specificity or selectivity in nucleic acid reactions, however local it may sometimes appear, is dependent to a large extent on the overall macromolecular structure of these polymers. Among the features playing an important role in this global effect, the molecular electrostatic potential seems to be particularly prominent. I am sure that the following lectures in this Study Group will point to the important significance of a number of other factors and I hope that before the end of this meeting a synthetic view of the components of specificity and perhaps a better appreciation of their relative responsibilities will become attainable.

REFERENCES

Baguley, B.C., 1982, Mol. Cell. Biochem. *43*, 167.
Berman, H.M., Neidle, S., Zimmer, Ch. and Trum, H., 1979, Biochem. Biophys. Acta *561*, 124.
Braithwaite, A.W. and Baguley, B.C., 1980, Biochem. *19*, 1101.
Granados, E. and Bello, J., 1981, Biochem. *20*, 4761.
Gresh, N. and Pullman, B., 1984, Theoret. Chim. Acta *64*, 383.

Gresh, N. and Pullman, B., 1983, Intern. J. Quantum. Chem., *24*, 491.

Ivanov, V.I., Minchenkova, L.R., Minyat, E.E., Frank-Kamenetskii, M.D. and Schyolkina, A.K., 1974, J. Mol. Biol. *87*, 817.

Lavery, R., Pullman, A. and Pullman, B., 1981, Intern. J. Quantum. Chem. *20*, 49.

Marky, L.A., Blumfeld, K.S. and Breslauer, K.J., 1983, Nucl. Acids Res. *11*, 2857.

Melchior, W. and Von Hippel, P., 1973, Proc. Natl. Acad. Sci. USA *70*, 298.

De Murcia, G., Wilhelm, B., Wilhelm, F. and Daune, M., 1978, Biophys. Chem. *8*, 377.

Patel, D.J., 1982, Proc. Natl. Acad. Sci. USA *79*, 6424.

Privat, M., Spach, G. and Leng, M., 1972, Eur. J. Biochem. *26*, 90.

Pullman, A., 1982, in Ions and Molecules in Solution, Tanaka, N., Ohtaki, H. and Tamamuh R. eds., Elsevier, Amsterdam, 373.

Pullman, A., 1983, in Physical Chemistry of Transmembrane Ion Motions, Spach, G., Ed., Elsevier, Amsterdam, 153.

Pullman, A. and Pullman, B., 1981, Quart. Rev. Biophys. *14*, 289.

Pullman, B., and Pullman, A., 1980, in Carcinogenesis: Fundamental Aspects and Environmental Effects. Proceedings of the 13th Jerusalem Symposium in Quantum Chemistry and Biochemistry. Pullman, B., Ts'o, P.O.P. and Gelboin, H. eds, Reidel Publishing Co. Dordrecht, Holland, 55.

Pullman, B. and Pullman, A., 1981, Studia Biophys, *86*, 95.

Pullman, B., Lavery, R. and Pullman, A., 1982, Eur. J. Biochem. *124*, 229.

Rio, P. and Leng, M., Nucl. Acids. Res. *11*, 4947.

Santella, R.M., Grunberger, D., Weinstein, J.B. and Rich, A., 1981, Proc. Natl. Acad. Sci. USA *78*, 1451.

Santella, R.M., Grunberger, D., Nordheim, A. and Rich, A., 1982, Biochem. Biophys. Res. Comm. *106*, 1226.

Scrocco, E., and Tomasi, J., 1973, Topics Current Chem., *42*, 95.

Scrocco, E., and Tomasi, J., 1978, Adv. Quantum Chem. *11*, 115.

Shapiro, J., Stannard, B. and Felsenfeld, G., 1969, Biochemistry, *8*, 3233.

Spodheim-Maurizot, M., Malfoy, B. and Saint-Ruf, G., 1982, Nucl. Acids. Res. *10*, 4423.

Wang, A.H.-J., Quigley, G.J., Kolpak, F.J., Crawford, J.L., van Boom, J.H., van der Marel, G. and Rich, A., 1979, Nature *282*, 680.

Wang, A.H.-J., Quigley, G.J., Kolpak, F.J., van der Marel, G., van Boom, J.H. and Rich, A., 1980, Science *211*, 171.

Zakrzewska, K., Lavery, R., Pullman, A. and Pullman, B., 1980, Nucl. Acids Res. *8*, 3917.

Zakrzewska, K., Lavery, R. and Pullman, B., 1981, in Biomolecular Stereodynamics, R.H. Sarma Ed., Adenine Press, N.Y., 163.

Zakrzewska, K., Lavery, R. and Pullman, B., 1983, Nucl. Acids Res. *11*, 8825.

Zimmer, Ch. 1975, Progress Nucl. Acids. Res. and Mol. Biol. *15*, 285.

Zimmer, Ch., Luck, G. and Fric, I., 1976, Nucl. Acids Res. *3*, 1521.

THEORETICAL STUDIES OF MOLECULAR RECOGNITION AND CATALYSIS BY ENZYMES *

HAROLD A. SCHERAGA

Baker Laboratory of Chemistry
Cornell University
Ithaca, New York 14853 (USA)

Abstract

The procedures used in conformational energy calculations, and some typical results of such computations on model systems and on polypeptides and proteins, are summarized. It is then shown how application of this methodology to two enzyme-substrate systems (α-chymotrypsin plus oligopeptides and hen egg white lysozyme plus oligosaccharides) leads to computed binding modes and energies that agree with experiment. This agreement attests to the validity of the potential functions and methodology used in the computations, and demonstrates the utility of this methodology in providing an understanding of the interactions that lead to molecular recognition.

I. Introduction

Theoretical conformational analysis is being used to provide a molecular basis of recognition, or specificity in biological systems, in particular of: (a) the interactions between parts of a polypeptide chain so that they can recognize each other and fold into the biologically active form of the native protein, (b) the interactions between two biological molecules, e.g. the recognition of a substrate by an enzyme or of an antigen by an antibody, and (c)

* This work was supported by research grants from the National Institutes of Health (GM-14312), the National Science Foundation (PCM79-20279), and the National Foundation for Cancer Research.

This paper will also appear in the Proceedings of the XVIIIth International Solvay Conference on Chemistry.

the interactions that enable biological molecules to recognize cell surfaces. We are approaching this problem by a combined experimental and theoretical approach, but this paper is concerned primarily with our theoretical studies.

We provide here a brief summary of the computational methodology and some examples of calculations on model systems. We also show the results of computations on the conformations of natural polypeptides, fibrous proteins, globular proteins, and binary complexes of enzymes and substrates.

II. Methodology

Empirical potential energy functions are used in the theoretical calculations (Momany *et al.*, 1975; Dunfield *et al.*, 1978; Pottle *et al.*, 1980; Némethy *et al.*, 1983). They are represented as interactions between pairs of atoms, and the total energy of a molecular system is taken as a sum over all pair interactions. The potential functions were parameterized by applying them to compute experimental data on gas-phase and crystal structures, and thermodynamic properties, of small oligopeptides and related molecules. While originally derived for polypeptides, they were found to be applicable also to polysaccharides without any change of parameterization (Pincus *et al.*, 1976). Further, they apply equally well to both intramolecular and intermolecular interactions.

The total potential energy is expressed in terms of contributions from non-bonded, electrostatic, and hydrogen-bonding interactions, and intrinsic torsional potentials, and is embodied in an algorithm referred to as ECEPP (Empirical Conformational Energy Program for Peptides). The details are given by Pottle *et al.* (1980), and the most recent set of parameters is described by Némethy *et al.* (1983). Procedures are also available for including entropy effects (Gibson and Scheraga, 1969a,b, Gō and Scheraga, 1969, 1976; Gō *et al.*, 1970) and hydration effects (Hodes *et al.*, 1979; Meirovitch and Scheraga, 1980, 1981a) in the computations. [As shown in Figure 1 of a paper by Scheraga (1981), the entropy arising from flexibility can stabilize a conformation that does not have the lowest energy. While examples of such situations have been observed (Gibson and Scheraga, 1969b; Gō *et al.*, 1970), inversion of stability by the entropy contribution generally occurs only if the minima are quite close in energy.]

In order to find the most stable state of a molecular system, the "conformational energy" (including entropy effects) is minimized. The multiple-minima problem is circumvented by incorporating various constraints into the minimization procedure, e.g. short-, medium- and long-range information, and distance constraints (Meirovitch and Scheraga, 1981b,c; Wako and Scheraga, 1982).

The variables in the minimization procedure are the dihedral angles for rotation about the single bonds of the molecules. When dealing with intermolecular systems, such as enzyme-substrate complexes, additional degrees of freedom are involved, viz. overall translation and rotation of the enzyme and substrate as they approach each other in a "docking" process. In such a case, partner in the complex influences the conformation of the other because of intermolecular interactions.

III. Some Calculations with Model Systems

It has been known for some time that the α-helices and β-sheets of proteins have a right-handed twist but the reason for such a twist was unknown. Several years ago, we used conformational energy calculations to predict the handedness of various homopolymer α-helices, and identified the interatomic interactions that determine the right-handed twist of these structures (Ooi *et al.*, 1967; Yan *et al.*, 1970). Recently, we applied the same methodology to the more complicated β-sheets and found that the low-energy structures of such sheets are indeed right-handed, as originally observed for β-sheets in globular proteins by Chothia (1973). Moreover, we identified the interatomic interactions that lead to a right-handed twist (Chou *et al.*, 1982; Chou and Scheraga, 1982; Scheraga *et al.*, 1982a; Chou *et al.*, 1983a,b). In both α-helices and β-sheets, local interactions within these structures (rather than longer-range interactions with the rest of the globular protein) appear to play the dominant role in determining their twist.

We next considered the problem as to how α-helices and β-sheets interact with each other in forming a globular protein. For example, the most favorable arrangement of two α-helices is an anti-parallel one (see Fig. 1) dictated by the most favorable arrangement of the dipoles of the α-helices (Chou *et al.*, 1983c).

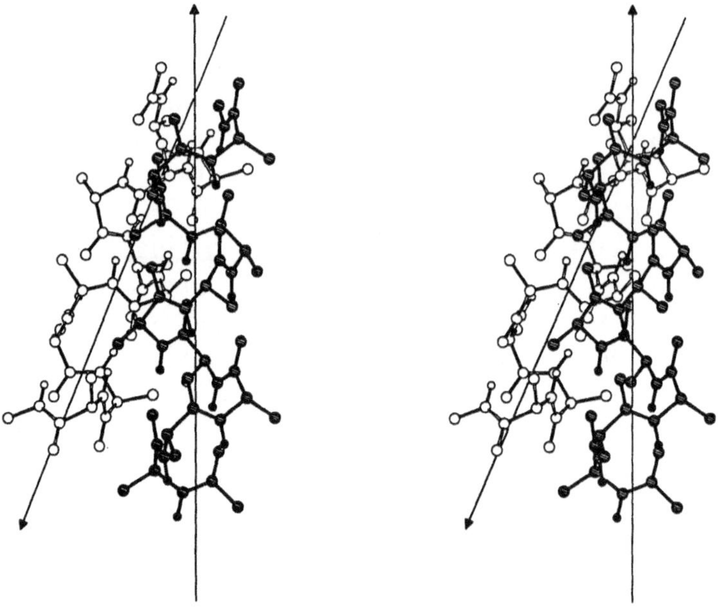

FIG. 1. Stereoscopic picture of two CH_3CO-(L-Ala)$_{10}$-$NHCH_3$ α-helices in the lowest-energy packing state (with a torsion angle of —154 °). The helix axes are indicated by arrows, with the head of the arrow pointing in the direction of the C-terminals of each helix (Chou *et al.*, 1983c).

IV. Some Calculations on Polypeptides and Proteins

Turning to natural polypeptides and proteins, we may cite our computations on gramicidin S, collagen-like poly(tripeptides), α-lactalbumin, melittin, and bovine pancreatic trypsin inhibitor (BPTI) as examples. In the case of gramidicin S, a cyclic decapeptide, the computed structure (Dygert *et al.*, 1975; Némethy and Scheraga, 1984) was found to agree (Rackovsky and Scheraga, 1980; Scheraga, 1984) with a subsequently-determined x-ray structure (Hull *et al.*, 1978) (see Fig. 2). Likewise, the computed triple-stranded helical structure of poly(Gly-Pro-Pro) (Miller and

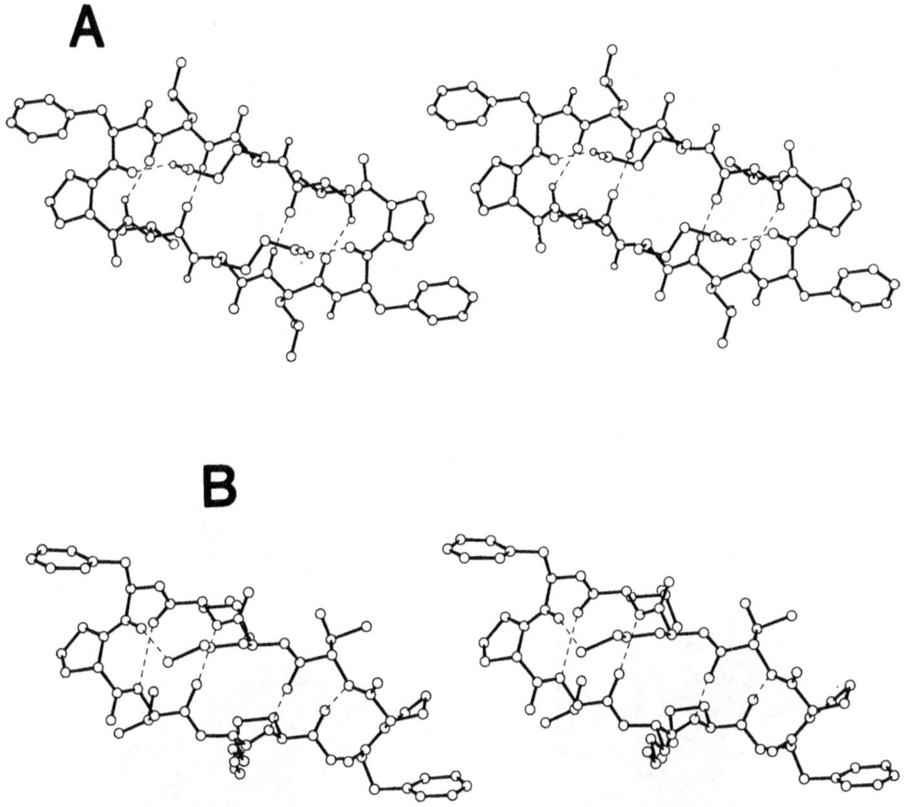

FIG. 2. (A) Computed (Dygert *et al.*, 1975; Némethy and Scheraga, 1984) and (B) X-ray (Hull *et al.*, 1978) structures of gramicidin S, showing (among other things) a hydrogen bond between the ornithine side chain and the phenylalanine backbone carbonyl group.

Scheraga, 1976) (Fig. 3) has cartesian coordinates that agree with those from a single-crystal x-ray diffraction study of (Pro-Pro-Gly)$_{10}$ (Okuyama *et al.*, 1976) within $\pm$ 0.3 Å.

A structure of α-lactalbumin was computed (Warme *et al.*, 1974) by taking advantage of its sequence homology to lysozyme and using the known structure of the latter protein as a starting point for energy minimization (Fig. 4). While the x-ray structure of α-lactalbumin has not yet been determined, the predicted structure is at least in agreement with various experimental results cited by Warme *et al.* (1974) and some (Berliner and Kaptein, 1981; Gerken, 1983) that were obtained subsequently.

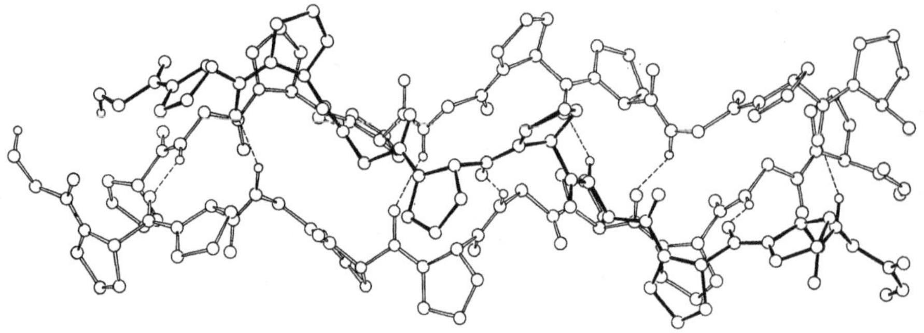

FIG. 3. Computed triple-stranded coiled-coil complex of poly (Gly-Pro-Pro) of lowest energy (Miller and Scheraga, 1976).

By use of a build-up procedure, in which small energy-minimized fragments were combined (with subsequent energy-minimization of the resulting large fragments), it was possible to compute the structure of the 20-residue membrane-bound portion of me-

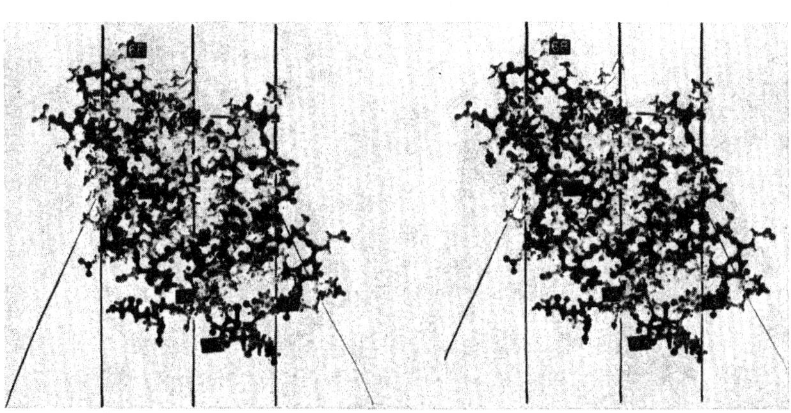

FIG. 4. Stereo view of computed structure of α-lactalbumin (Warme *et al.*, 1974).

littin (Pincus *et al.*, 1982). The computed structure (Fig. 5) is in satisfactory agreement with experiment (Brown *et al.*, 1982; Terwilliger *et al.*, 1982).

This methodology has also been used to carry out a potential energy-constrained real-space refinement of the structure of BPTI, using limited diffraction data (Fitzwater and Scheraga, 1982). Figure 6 illustrates how the energy-refined map on the right identifies more of the electron density of Arg-42 than the map on the left.

Finally, by using a series of constraints in an energy minimization, some progress has been made in computing the structure of BPTI (Wako and Scheraga, 1982). Thus far, a virtual-bond

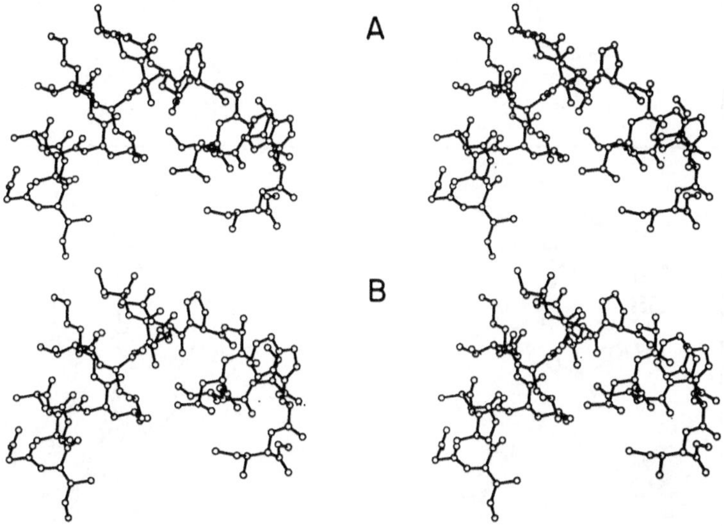

FIG. 5. Stereo view of the two lowest energy structures calculated for residues 1-20 of melittin (Pincus *et al.* 1982). (B) global minimum; (A) structure of slightly higher energy (1.5 kcal/mol).

FIG. 6. Comparison of sections of electron density maps around Arg-42 of BPTI, projected onto the x,y plane (Fitzwater and Scheraga, 1982). (Left) Map calculated by using the experimental 2.5-Å phases. (Right) Map calculated by using the phases from the structure factor calculation.

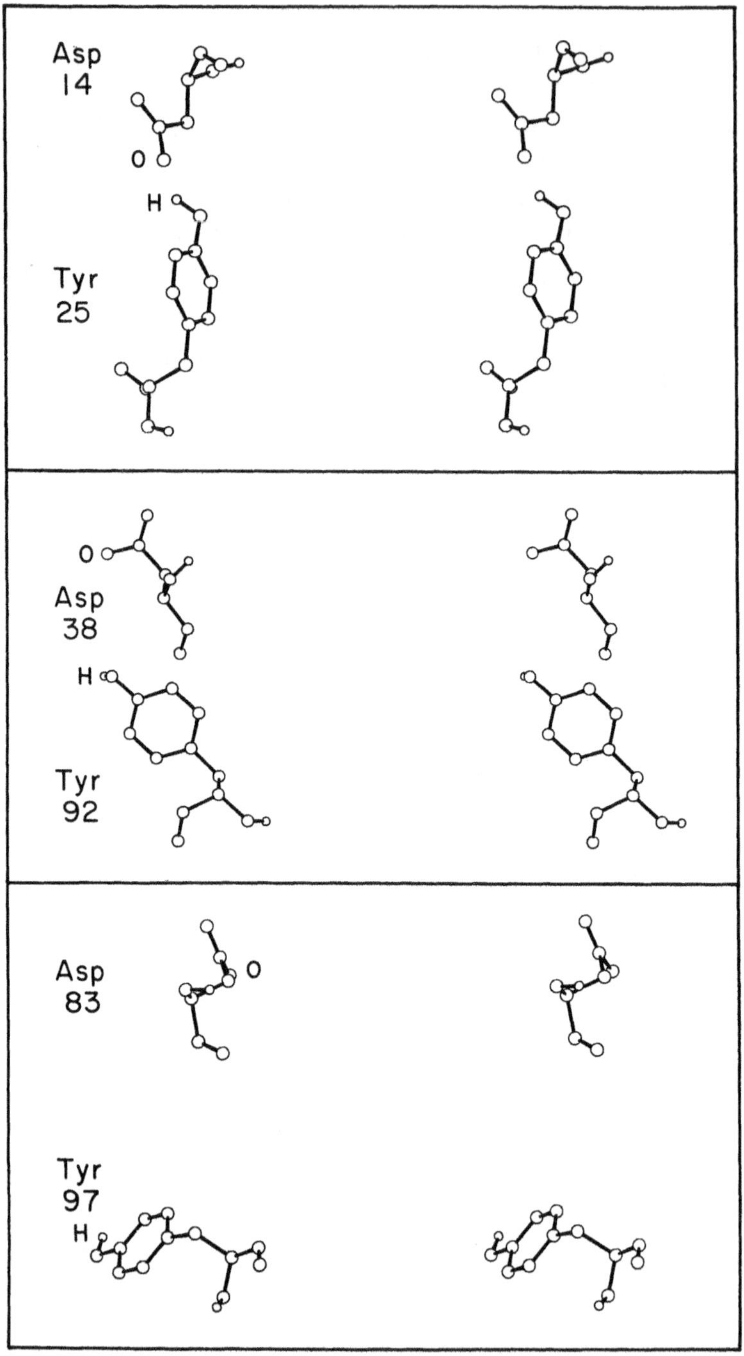

FIG. 7. Illustration of three Tyr. . .Asp interactions in bovine pancreatic ribonuclease A, deduced from various chemical and physical chemical experiments (Scheraga, 1967) before the X-ray structure was determined. The drawing was prepared from the X-ray coordinates of Wlodawer *et al.* (1982).

structure of this protein has an r.m.s. deviation of 2.2 Å from the experimental one, and work is in progress to try to improve the computational procedure, and hence the agreement between the computed and observed structures. Originally, distance constraints (used in these protein-folding algorithms) were obtained by a long series of chemical and physical chemical experiments, as illustrated for three tyrosyl... aspartyl interactions in ribonuclease A (Fig. 7) (Scheraga, 1967). More recently, nuclear Overhauser (Leach *et al.*, 1977; Braun *et al.*, 1981) and non-radiative energy transfer (McWherter *et al.*, 1983) measurements are being used to obtain similar information.

V. Chymotrypsin Plus Oligopeptides

With these introductory remarks, we may now consider the low-energy structures of enzyme-substrate complexes. In this section, we describe results for α-chymotrypsin and, in the next section, results for lysozyme. Further details may be found in a paper by Scheraga *et al.* (1982b).

The action of an enzyme such as chymotrypsin on an oligopeptide substrate may be described in terms of the notation of Schechter and Berger (1967), as shown in Figure 8. Here, P_1 represents the residues of the substrate and S_1 the subsites of the enzyme that interact with each residue; A and P stand for alanine and phenylalanine, respectively, in the three oligopeptide substrates illustrated. Computations on complexes of the enzyme with these three substrates also provide information about the binding of the pentapeptide AAPAA, wherein the flanking alanine residues influence the binding and, hence the rate of hydrolysis of the PA peptide bond in this substrate. Figure 9 shows the computed lowest-energy structure for PAA in the active site of the enzyme.

Such computations provide details of the interactions between the various parts of the enzyme and substrate and the relative affinities of different substrates. For example, the computed *relative* binding energies of the aromatic amino acids Trp, Tyr and Phe are quite similar to those determined experimentally

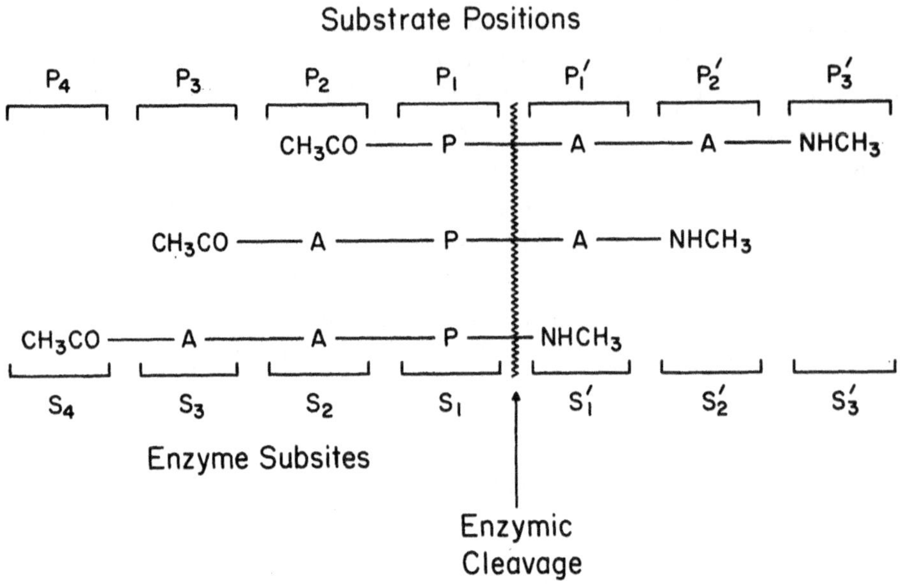

FIG. 8. Notation for the residues of the substrate, P_i, which interact with subsites S_i of the enzyme (α-chymotrypsin) (Schechter and Berger, 1967). In all cases, the peptide bond involving the C=O of phenylalanine is the one that is hydrolyzed, and phenylalanine (designated as P_1) occupies subsite S_1.

(Platzer *et al.*, 1972). In the case of the tryptophan peptide, the calculated coordinates of the indole ring in the active-site cleft agree closely with those determined for a similar substrate by X-ray crystallography (Steitz *et al.*, 1969).

During the course of catalysis, an acyl-enzyme intermediate is formed between Ser^{195} and Phe, as illustrated in Fig. 10. The computed lowest-energy conformation of this intermediate is shown in Fig. 11. These computations thus provide information about interactions in intermediates that are not readily accessible by experiment.

VI. LYSOZYME PLUS OLIGOSACCHARIDES

Lysozyme is a glycolytic enzyme which exhibits a strong specificity for β-1,4-linked N-acetylglucosamine (GlcNAc) and N-acetylmuramic acid (MurNAc) units (Imoto *et al.*, 1972) which

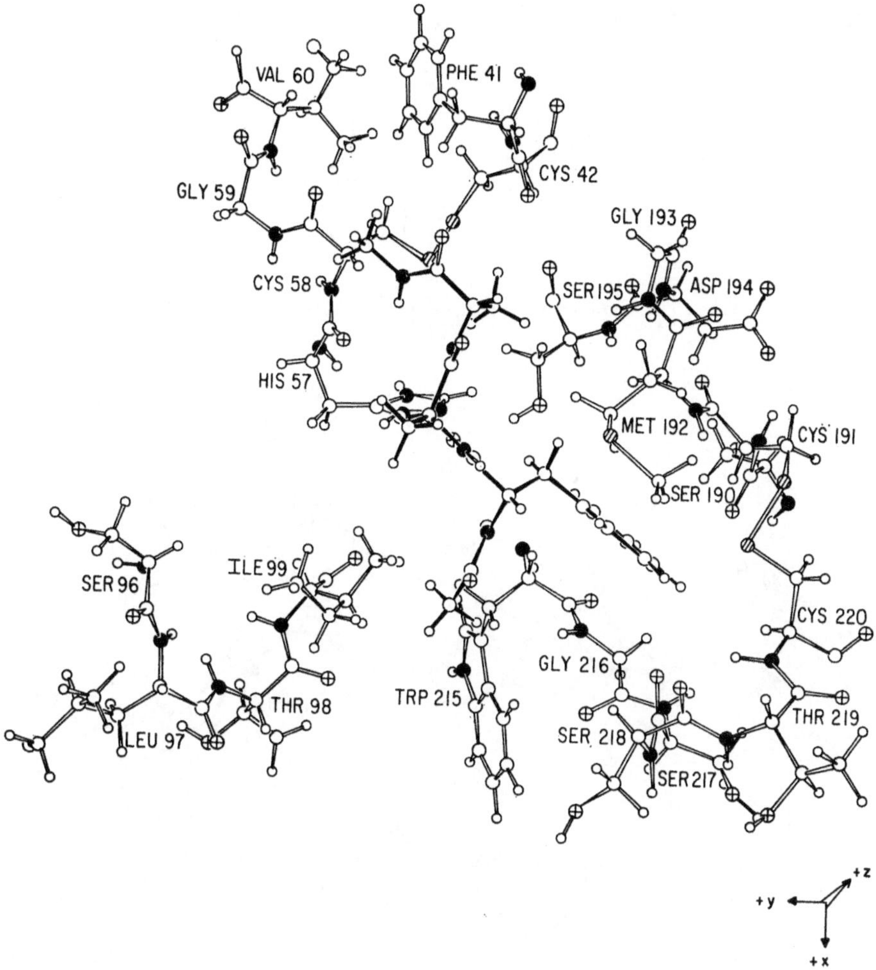

FIG. 9. Lowest-energy conformer for N-acetyl-Phe-Ala-Ala-NH_2 in the cleft binding region of α-chymotrypsin (Scheraga *et al.*, 1982b).

are shown in Fig. 12. Maximal rates of hydrolysis and highest substrate affinities are achieved when six saccharide units [as in $(GlcNAc)_6$ and $(GlcNAc{\cdot}MurNAc)_3$] are bound to the enzyme. The active site of this enzyme thus has six binding subsites, labelled A-F, the first three of which have been identified by x-ray crystallography on crystals of $(GlcNAc)_3$ bound to sites A-C (Imoto *et al.*, 1972). From solution studies, it is known that hydrolysis

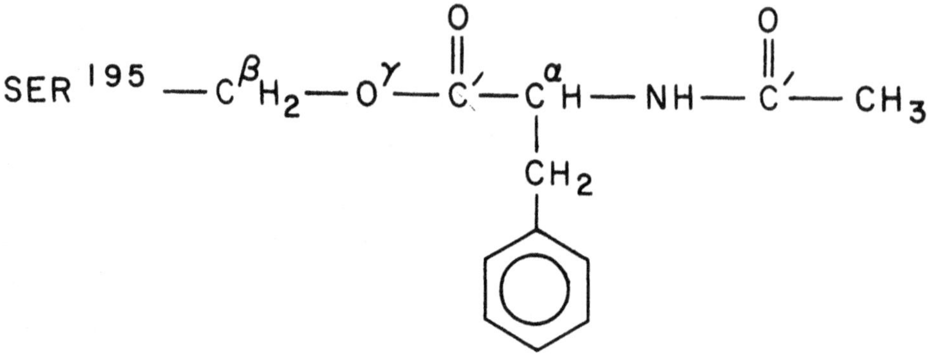

Fig. 10. Acyl-enzyme intermediate for N-acetyl-L-phenylalanine.

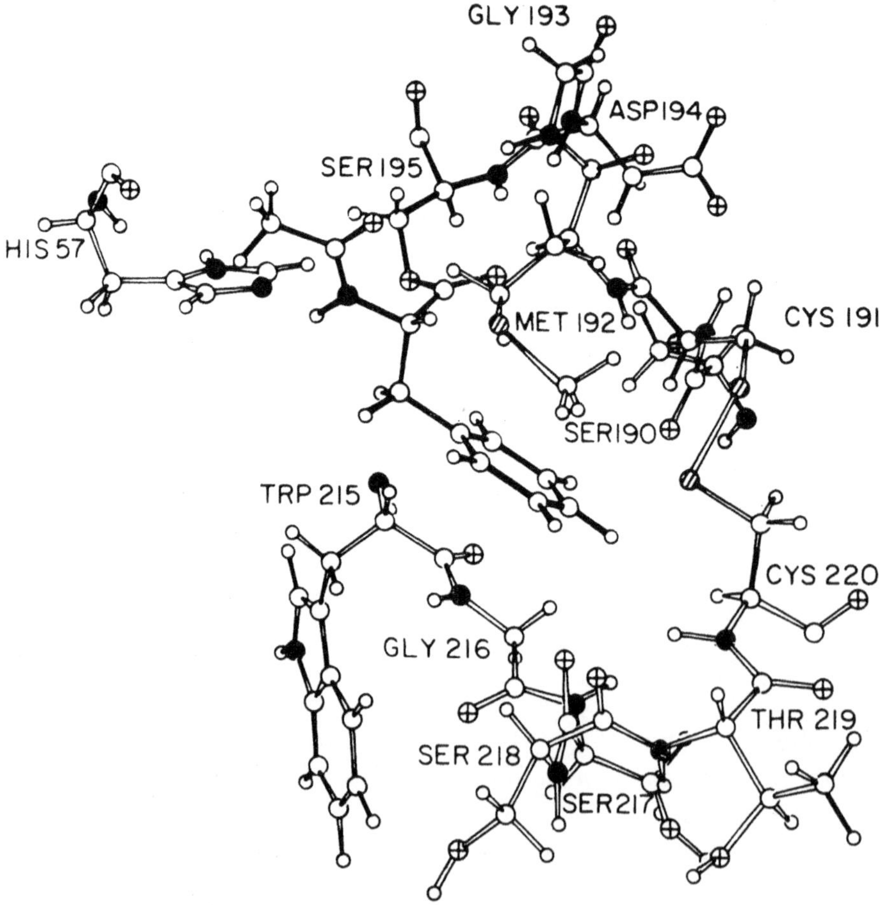

Fig. 11. Lowest-energy conformation for the acyl-enzyme intermediate of chymotrypsin with N-acetyl-L-phenylalanine (Scheraga *et al.*, 1982b).

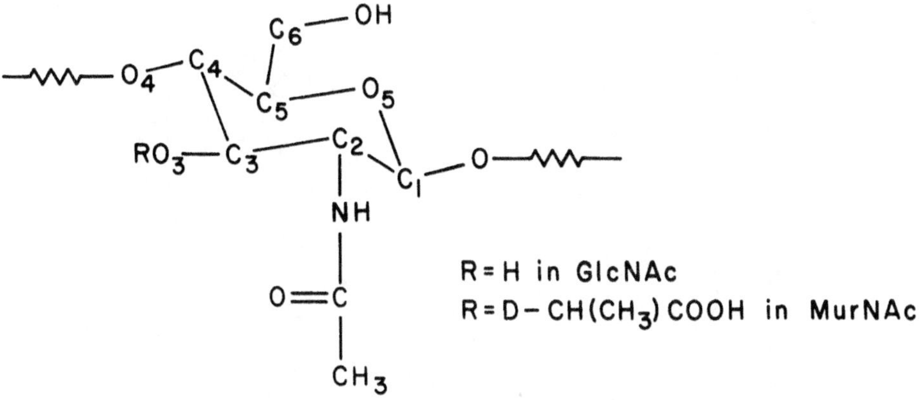

FIG. 12. Structures of GlcNAc and MurNAc.

occurs between sites D and E and is catalyzed by two acid residues Asp 52 and Glu 35. The proposed mechanism of hydrolysis involves the formation of a carbonium ion at C_1 of the D-site residue stabilized by the carboxylate of Asp 52 with Glu 35 serving as a general acid that protonates the oxygen of the departing saccharide (Imoto *et al.*, 1972).

In computations of lysozyme-oligosaccharide complexes, a total of six GlcNAc residues was found to bind to the active site, in agreement with experiment (Pincus *et al.*, 1977; Pincus and Scheraga, 1979). The structure of the calculated lowest-energy hexamer $(GlcNAc)_6$ is shown in Figure 13B. It may be noted that the last GlcNAc residue interacts with such residues as Arg 45, Asn 46 and Thr 47 on the " left side " of the cleft, whereas the position of the F site for this terminal residue has been postulated, on the basis of model-building (Imoto *et al.*, 1972), to occur on the " right side " of the cleft involving such residues as Arg 114 and Phe 34 as shown in Figure 13A. When the proposed model-built structure was subjected to energy minimization, the resulting structure was of significantly higher energy (Pincus and Scheraga, 1979). The two lowest-energy left- and right-sided $(GlcNAc)_6$ binding conformations are shown with space-filling models in Fig. 14.

In order to provide an experimental test of these calculations, i.e. to decide whether $(GlcNAc)_6$ binds preferentially to the left

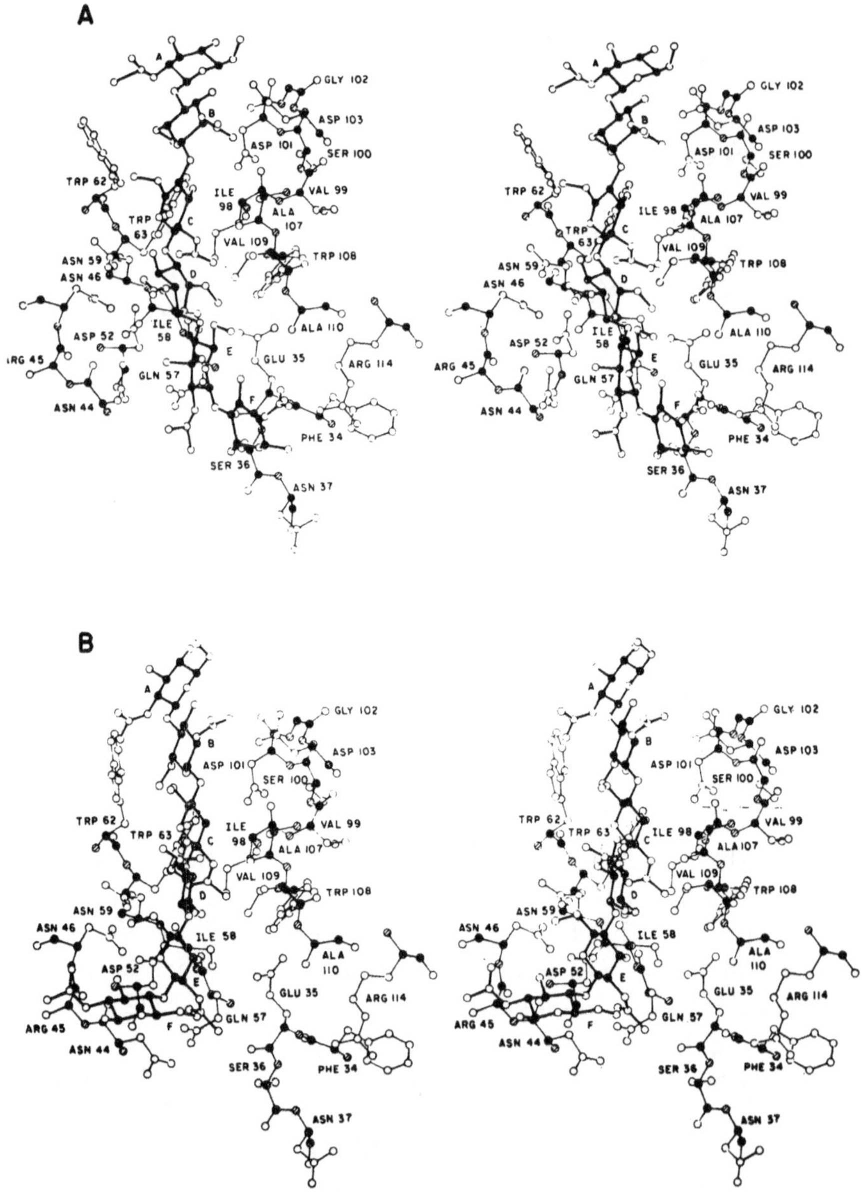

FIG. 13. Stereo views of: (A) Geometry- and energy-refined model-built structure binding to the right side of the active-site cleft. (B) Calculated lowest-energy structure of $(GlcNAc)_6$ bound to the left side of the active-site cleft (Pincus *et al.*, 1977; Pincus and Scheraga, 1979).

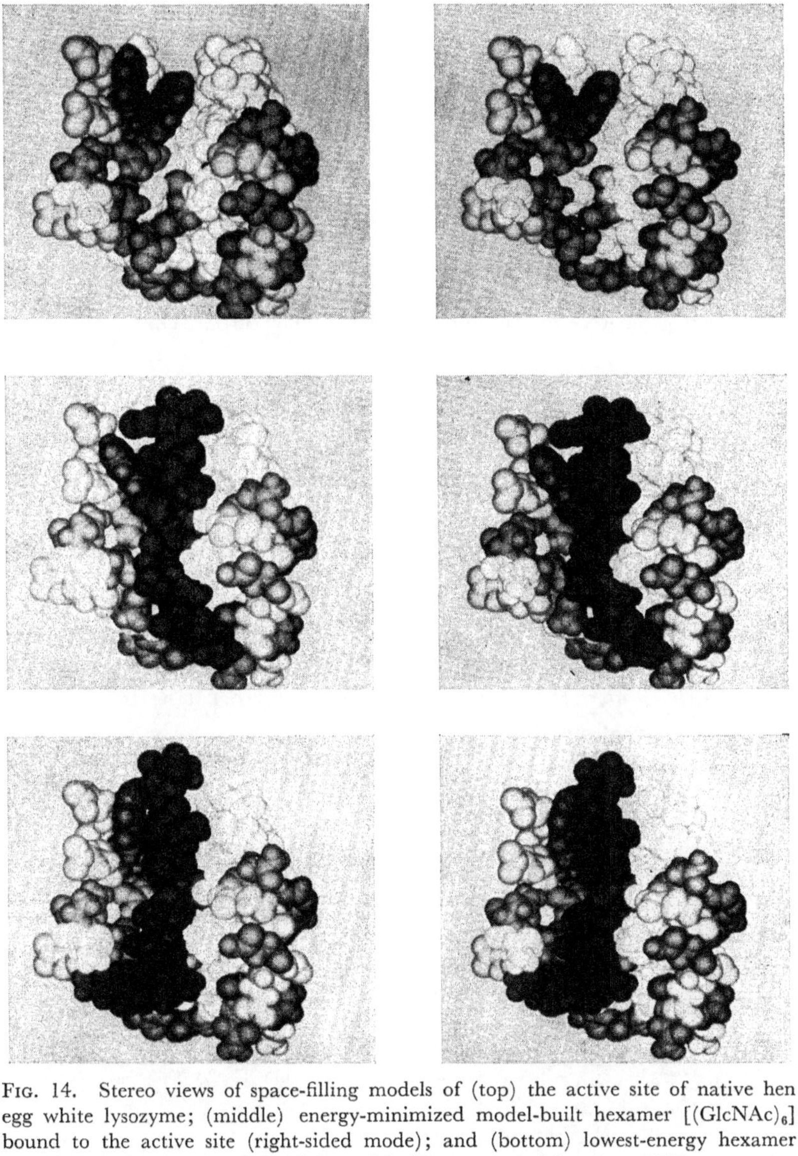

FIG. 14. Stereo views of space-filling models of (top) the active site of native hen egg white lysozyme; (middle) energy-minimized model-built hexamer [$(GlcNAc)_6$] bound to the active site (right-sided mode); and (bottom) lowest-energy hexamer bound to the active site (left-sided mode) (Pincus and Scheraga, 1979).

or right side, the following experiments were carried out (Smith-Gill *et al.*, 1984). These take cognizance of the fact that residues on the left and right side differ in different species of lysozymes, as indicated in Table 1 and illustrated in Fig. 15.

TABLE I. *Some Residues Involved in Binding Sites in Lysozymes*

Left Side	Right Side
Arg 45	Phe 34
Asn 46	Asn 113 (Lys)[2]
Thr 47	Arg 114 (His)[2]
Asp 48[1]	
Arg 68[1]	

[1] Additional residues included in epitope
[2] Substitutions in RNPL. All other residues are those in HEWL.

In ringed neck pheasant lysozyme (RNPL), Lys and His are substituted for Asn and Arg at positions 113 and 114, respectively. Therefore, the affinity of $(GlcNAc)_6$ for RNPL should differ from that for hen egg white lysozyme (HEWL) if binding occurs on the right side, but should not differ if binding occurs on the left side. From experiments on the kinetics of hydrolysis of this substrate by both enzymes, it appears that the values of the Michaelis-Menten parameter K_M are similar for both enzymes (Smith-Gill *et al.*, 1984). This suggests that the left-sided binding mode is preferred.

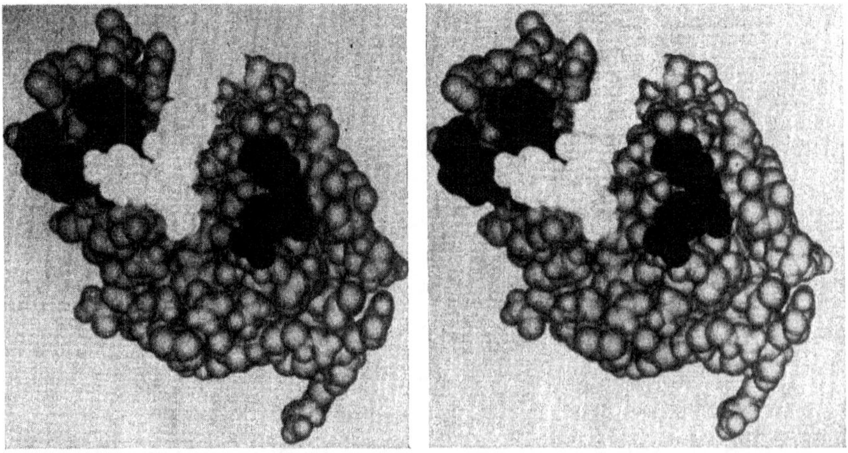

FIG. 15. Stereo view of the computed structure of the low-energy complex between $(GlcNAc)_6$ and the hen egg white lysozyme (same complex as in the bottom of Fig.14). The dark shaded regions contain the residues involved in left- and right-sided binding, respectively (Smith-Gill *et al.*, 1984).

In a second experiment, it was found that a monoclonal antibody that binds to an epitope on the left side can compete with $(GlcNAc)_6$ for the active site of HEWL (Smith-Gill *et al.*, 1984). This competition is not due to steric blockage of the left side because two different monoclonal antibodies can bind simultaneously to HEWL, at the two dark patches indicated in Fig. 15.

The results of both of these experiments suggest that $(GlcNAc)_6$ binds preferentially to the left side, as indicated by the computations. These results demonstrate the utility of conformational energy calculations in providing an understanding of the interactions involved in molecular recognition. They also indicate the importance of a *combined* theoretical/experimental approach to gain such an understanding. For example, the experiments with HEWL and RNPL would not have been carried out without the need to test the predictions of the calculations summarized in Fig. 14.

The calculations have also been extended to include the binding conformations of alternating copolymers of GlcNAc and MurNAc at the active site of the enzyme. The calculated lowest-energy structure for MurNAc · GlcNAc · MurNac bound to sites B, C, D is in excellent agreement with the x-ray crystal structure of this complex (Pincus and Scheraga, 1981a). If either the calculated or the X-ray structure is extended into the lower active site with GlcNAc residues, it is found that it binds with its F site residue on the left side of the cleft. The alternating hexamer copolymer $(GlcNAc \cdot MurNAc)_3$, however, binds in a preferred right-sided mode (similar to the homopolymer in Figure 13A) because of the absence of a favorable hydrogen bond in site F between the 3-OH of the substrate (the -OH of GlcNAc is now replaced with a lactic acid side chain in MurNAc) and the the backbone $>C{=}O$ of Arg 45; also, the lactic acid side chain makes unfavorable steric contacts with Arg 45. In addition, from an investigation of the allowed binding sites for MurNAc residues, it appears that this residue can be accomodated in sites B, D and F only, in good agreement with experiment (Pincus and Scheraga, 1981a).

Approximately one-third of the binding energy for hexasaccharide complexes is attributable to the interactions of the N-acetyl groups of the substrate with the enzyme (mainly with the three active site tryptophans 62, 63 and 108 in sites B, C, D, and through hydrogen bonding with Asn 59 and Ala 107 in site C and with Glu 35 in site E) in agreement with experimental results. Polymers of saccharides without the N-acetyl group (e.g. glucose) are thus predicted to bind with low affinities, again in agreement with experiment (Imoto *et al.*, 1972). The calculated hydrogen bonding scheme (hydrogen bonding accounts for about one-fourth of the total energy of stabilization) is identical in site C to that inferred from X-ray crystallography.

VII. Conclusions

It may be seen from the results described for α-chymotrypsin and lysozyme that conformational energy calculations can predict the three-dimensional structures of enzyme-substrate complexes. Information is obtained about the various pairwise interactions between enzyme and substrate that contribute to the ground-state binding energy. In fact, the binding energy results from a sum over *many* small pairwise interaction energies rather than a few large specific contributions; thus, the assumption of discrete sites (Figure 8) is an oversimplification. In addition, the calculations can predict the relative affinities of different substrates for a particular enzyme.

All of these calculations are based on the use of ECEPP potentials (Momany *et al.*, 1975; Némethy *et al.*, 1983) which have been obtained from crystal and gas-phase data. These potentials are ideally suited for computation of the structures of enzyme-substrate complexes because the interactions between enzymes and substrates are the same as those between the molecules of a crystal. They may therefore be employed (as we have done in the case of chymotrypsin and lysozyme) to identify the crucial interactions that lead to recognition. Once these interactions are known, they may be used to construct, from theoretical considerations alone, substrates and inhibitors that can bind with the highest affinities to the active site of the enzyme.

A related important question, discussed elsewhere (Pincus and Scheraga, 1981b), concerns the nature of the catalytic step, subsequent to the binding step. This question has thus far been approached by quantum mechanical calculations on model systems that simulate the binary enzyme-substrate complexes discussed here.

Figure 16 is a schematic representation of the kinds of problems that can be treated by the methodology described in section II.

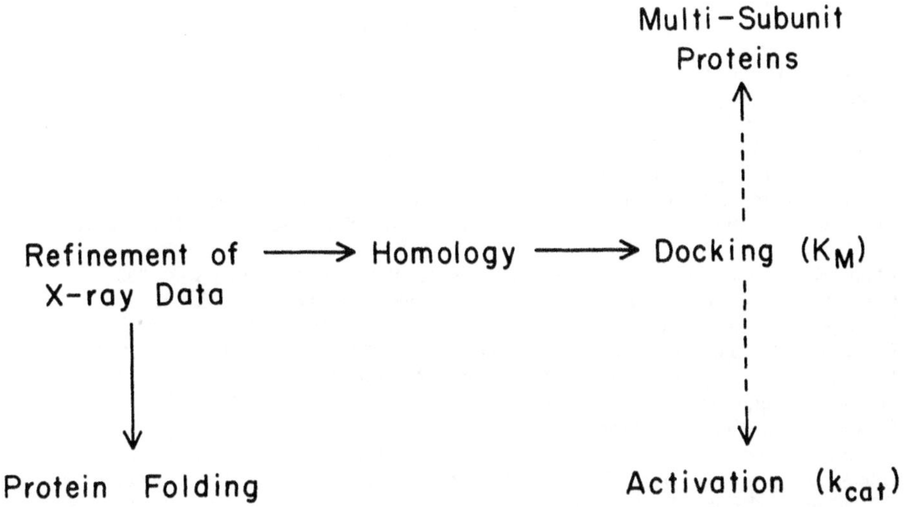

FIG. 16. Various problems treated by energy minimization. The solid arrows pertain to computations already carried out, while the dashed arrows pertain to potential new applications.

Direct application to x-ray data leads to low-energy structures (Warme and Scheraga, 1974; Swenson *et al.*, 1978; Fitzwater and Scheraga, 1982). Such refined structures provide starting points to compute the conformations of homologous proteins (Warme *et al.*, 1974; Swenson *et al.*, 1978), which in turn can be used (in a docking process) to treat the binding of enzymes to substrates (the process characterized by K_M), the binding of drugs and hormones to receptors, and the binding of proteins to nucleic acids. With sufficient experimental information about structure-activity relationships, this technique can be applied to map the active site of an unknown receptor. The same energy minimization algorithms can be used (together with constraints) in protein folding studies. Two potential extensions of this methodology are to intermolecular interactions in multi-subunit proteins such as hemoglobin, and to the structures and energetics of activation complexes (the process characterized by K_{cat}). In the latter case, the computations must allow for bond stretching and bond angle bending, in addition to the variation of dihedral angles that is allowed for in all of the other computations.

REFERENCES

Berliner, L. J. and Kaptein, R., 1981, Biochemistry *20*, 799.

Braun, W., Bösch, C., Brown, L. R., Gō, N. and Wüthrich, K., 1981, Biochim. Biophys. Acta *667*, 377.

Brown, L. R., Braun, W., Kumar, A. and Wüthrich, K., 1982, Biophys. J. *37*, 319.

Chothia, C., 1973, J. Mol. Biol. *75*, 295.

Chou, K. C. and Scheraga, H. A., 1982, Proc. Natl. Acad. Sci., U.S. *79*, 7047.

Chou, K. C., Pottle, M., Némethy, G., Ueda, Y. and Scheraga, H. A., 1982, J. Mol. Biol. *162*, 89.

Chou, K. C., Némethy, G. and Scheraga, H. A., 1983a, J. Mol. Biol. *168*, 398.

Chou, K. C., Némethy, G. and Scheraga, H. A., 1983b, Biochemistry, *22*, 6213.

Chou, K. C., Némethy, G. and Scheraga, H. A., 1983c, J. Phys. Chem. *87*, 2869.

Dunfield, L. G., Burgess, A. W. and Scheraga, H. A., 1978, J. Phys. Chem. *82*, 2609.

Dygert, M., Gō, N. and Scheraga, H. A., 1975, Macromolecules *8*, 750.

Fitzwater, S. and Scheraga, H. A., 1982, Proc. Natl. Acad. Sci. U.S. *79*, 2133.

Gerken, T. A., 1983, Fed. Proc. *42*, 2001.

Gibson, K. D. and Scheraga, H. A., 1969a, Physiol. Chem. and Phys. *1*, 109.

Gibson, K. D. and Scheraga, H. A. 1969b. Proc. Natl. Acad. Sci. U.S. *63*, 242.

Gō, N. and Scheraga, H. A., 1969, J. Chem. Phys. *51*, 4751.

Gō, N. and Scheraga, H. A., 1976, Macromolecules *9*, 535.

Gō, N., Lewis, P. N. and Scheraga, H. A., 1970, Macromolecules *3*, 628.

Hodes, Z. I., Némethy, G. and Scheraga, H. A., 1979, Biopolymers *18*, 1565.

Hull, S. E., Karlsson, P., Main, P., Woolfson, M. M. and Dodson, E. J., 1978, Nature *275*, 206.

Imoto, T., Johnson, L. N., North, A. C. T., Phillips, D. C. and Rupley, J. A., 1972, in "The Enzymes", Academic Press, New York and London, Vol. VII, p. 665.

Leach, S. J., Némethy, G. and Scheraga, H. A., 1977, Biochem. Biophys. Res. Commun. *75*, 207.

McWherter, C. A., Haas, E. and Scheraga, H. A., 1983, work in progress.

Meirovitch, H. and Scheraga, H. A., 1980, Macromolecules *13*, 1406.

Meirovitch, H. and Scheraga, H. A., 1981a, Macromolecules *14*, 340.

Meirovitch, H. and Scheraga, H. A., 1981b, Macromolecules *14*, 1250.

Meirovitch, H. and Scheraga, H. A., 1981c, Proc. Natl. Acad. Sci. U.S. *78*, 6584.

Miller, M. H. and Scheraga, H. A., 1976, J. Polymer Sci., Polymer Symp. *54*, 171.

Momany, F. A., McGuire, R. F., Burgess, A. W. and Scheraga, H. A., 1975, J. Phys. Chem. *79*, 2361.

Némethy, G. and Scheraga, H. A., 1984. Biochem. Biophys. Res. Commun., in press.

Némethy, G., Pottle, M. S. and Scheraga, H. A., 1983, J. Phys. Chem. *87*, 1883.

Okuyama, K., Tanaka, N., Ashida, T. and Kakudo, M., 1976, Bull. Chem. Soc. Japan *49*, 1805.

Ooi, T., Scott, R. A., Vanderkooi, G. and Scheraga. H. A., 1967, J. Chem. Phys. *46*, 4410.
Pincus, M. R., Burgess, A. W. and Scheraga, H. A. 1976, Biopolymers *15*, 2485.
Pincus, M. R., Zimmerman, S. S. and Scheraga, H. A., 1977, Proc. Natl. Acad. Sci. U.S. *74*, 2629.
Pincus, M. R., Klausner, R.D. and Scheraga, H. A., 1982, Proc. Natl. Acad. Sci. U.S. *79*, 5107.
Pincus, M. R. and Scheraga, H. A., 1979, Macromolecules *12*, 633.
Pincus, M. R. and Scheraga, H. A., 1981a, Biochemistry *20*, 3960.
Pincus, M. R. and Scheraga, H. A., 1981b, Accts. Chem. Res. *14*, 299.
Platzer, K. E. B., Momany, F. A., and Scheraga, H. A., 1972, Intntl. J. Peptide and Protein Res. *4*, 201.
Pottle, C., Pottle, M. S., Tuttle, R. W., Kinch, R. J. and Scheraga, H. A., 1980, J. Computational Chem. *1*, 46.
Rackovsky, S. and Scheraga, H. A., 1980, Proc. Natl. Acad. Sci. U.S. *77*, 6965.
Schechter, I. and Berger, A., 1967, Biochem. Biophys. Res. Commun. *27*, 157.
Scheraga, H. A., 1967, Fed. Proc. *26*, 1380.
Scheraga, H. A., 1981, in " Structural Aspects of Recognition and Assembly in Biological Macromolecules ", ed. M. Balaban, Elsevier, p. 87.
Scheraga, H. A., 1984, Annals. N.Y. Acad. Sci., in press.
Scheraga, H. A., Chou, K. C. and Némethy, G., 1982a, in " Conformation in Biology ", ed. R. Srinivasan and R. H. Sarma, Adenine Press, p. 1.
Scheraga, H. A., Pincus, M. R. and Burke, K. E., 1982b, in " Structure of Complexes Between Biopolymers and Low Molecular Weight Molecules ", eds. W. Bartmann and G. Snatzke, John Wiley, Chichester, p. 53.
Smith-Gill, S. J., Rupley, J. A., Pincus, M. R., Carty, R. P. and Scheraga, H. A., 1984, Biochemistry, in press.
Steitz, T. A., Henderson, R., and Blow, D. M., 1969, J. Mol. Biol. *46*, 337.
Swenson, M. K., Burgess, A. W. and Scheraga, H. A., 1978, in " Frontiers in Physicochemical Biology " ed. B. Pullman, Academic Press, p. 115.
Terwilliger, T. C., Weissman, L. and Eisenberg, D., 1982, Biophys. J. *37*, 353.
Wako, H. and Scheraga, H. A., 1982, J. Protein Chem. *1*, 5, 85.
Warme, P. K. and Scheraga, H. A., 1974, Biochemistry *13*, 757.
Warme, P. K., Momany, F. A., Rumball, S. V., Tuttle, R. W. and Scheraga, H. A., 1974, Biochemistry *13*, 768.
Wlodawer, A., Bott, R. and Sjolin, L., 1982, J. Biol. Chem. *257*, 1325.
Yan, J. F., Momany, F. A. and Scheraga, H. A., 1970, J. Am. Chem. Soc. *92*, 1109

MOLECULAR DYNAMICS OF THE DNA DOUBLE HELIX

MICHAEL LEVITT

Department of Chemical Physics, Weizmann Institute of Science,
Rehovot 76100, Israel

ABSTRACT

Molecular dynamics has been used to simulate the room temperature atomic motion of 12 and 24 base-pair DNA double helices for periods of up to 96 picoseconds. The hydrogen bonds between base-pairs are all found to be stable on this time-scale, and the motions of the torsion angles are found to be of small amplitude (< 10°). Both DNA fragments show cooperative overall bending motions of large amplitude that do not involve any major pertubation of the DNA torsion angles. This smooth bending differs from that expected of an isotropic elastic rod in that (a) it is asymmetric always acting to close the major groove of DNA and (b) it consists predominantly of the normal-mode that has a wavelength close to the helical repeat length. The stack of base-pairs is also seen to kink into the minor groove. The extent of this global motion is consistent with nuclear magnetic resonance measurements and explains the observed sensitivity of DNA conformation to local environment.

1. INTRODUCTION

Although its structure was determined almost 30 years ago (Watson and Crick, 1953), the DNA double helix is still full of surprises. In the past three years it has become clear that the number of base-pairs per turn is not the same in solution and in fibres (Arnott and Hukins, 1972: Wang, 1979; Rhodes and Klug, 1980), that the base and backbone atoms undergo angular motions of large amplitude ($> 25°$) on a time scale of nanoseconds (Hogan and Jardetsky, 1980; Early and Kearns, 1979; Bolton and James, 1980), and that there are cooperative conformational transitions mediated by changing environment or binding of drug molecules (Sobell *et al.*, 1977; Hogan *et al.*, 1979; Dattagupta and Crothers, 1981).

Here the nature of the dynamics of the DNA double helix is investigated using the technique of molecular dynamics that has been used on globular proteins (McCammon *et al.*, 1977; Levitt, 1981, 1983a). This calculation simulates the time-variation of the static X-ray structure and so provides information about the amplitudes and frequencies of vibrations, and the type, rate and pathway of conformational changes.

The results obtained here have implications for the way the DNA double helix may interact with repressors, polymerases and other cellular proteins (Anderson *et al.*, 1981; McKay and Steitz, 1981). The marked contrast between the overall flexibility and the stability of the hydrogen bonds between base-pairs, suggests that DNA may protect the integrity of the genetic message by absorbing thermal perturbations in bending motions rather than in base-pair opening motions.

2. Methods

a) *Molecular dynamics simulation*

Two different base-paired complexes of DNA are studied here: (a) the 12 base-pair fragment or dodecamer (CGCGAATT-CGCG)$_2$, whose static X-ray conformation was recently solved by single crystal diffraction (Drew *et al.*, 1981), and (b) the 24 base-pair fragment $(A)_{24}(T)_{24}$, whose static X-ray conformation is taken to be that found by fibre diffraction (Arnott and Hukins, 1972). All hydrogen atoms are added to the experimental heavy atom coordinates giving a total of 754 atoms for the dodecamer and 1530 atoms for $(A)_{24}(T)_{24}$ (the two 5′ terminal PO_4 groups are absent in both structures). The potential energy of these molecules is calculated as a function of the positions of the atoms using the same type of empirical force field used in previous energy calculations on nucleic acids (Levitt, 1978) and proteins (Levitt, 1981a and 1983a). This potential has the usual terms that allow for bond stretching, bond angle bending, hindered bond twisting, and van der Waals interactions. Hydrogen bonds are modelled by a specific short-range interaction that reproduces their expected strength and directionality (the well-depth is 4.5 kcal/mole). Electrostatic interactions are omitted here due to uncertainties about dielectric effects and atomic polarizability. The thousands of water molecules that would surround this molecule in solution are not included. This type of potential is the same first approximation to reality that has revealed a great deal about atomic motion in proteins (McCammon *et al.*, 1977; Levitt, 1981b and 1983a). Full details of the potential energy parameters and methods used have been given elsewhere (Levitt 1983a and b).

The forces on the atoms calculated from this potential energy function are used to iteratively solve the equations of motion for a time step of 0.002 picoseconds. Before starting the dynamic trajectory, the X-ray conformation is relaxed by between 1000 and 2000 cycles of conjugate gradient energy minimisation. The system is then heated until the average kinetic energy per atom is the room temperature thermal energy (300°K). This heating is accomplished by making small random velocity changes during the first few hundred time steps of the trajectories that are continued for up to 96ps (48,000 time steps). In spite of the relatively large time step (0.002ps) and the inclusion of all the rapidly moving hydrogen atoms, the sum of potential and kinetic energy is well conserved. The accurate solution of the equations of motions is confirmed by the persistence of low frequency motions that involve over 10,000 time steps per cycle (see below).

3. Results

a) *Stability of hydrogen bonds*

The thirty-two hydrogen bonds between the 12 base-pairs of the dodecamer are unexpectedly stable. The mean values of the O. .H hydrogen bond distances are between 1.77 Å and 1.74Å; the root mean square (r.m.s.) fluctuations in O. .H distance are between 0.07 Å and 0.1 Å. When the same hydrogen bond function was used in a molecular dynamics simulation on a small protein, pancreatic trypsin inhibitor (Levitt, 1981a), the interpeptide hydrogen bonds are much less stable than those found here for DNA. This indicates that the high stability of the DNA hydrogen bonds is due to the structure of the double helix rather than the nature of the potential function used.

The correlation coefficients of different hydrogen bond lengths were significant (greater than 0.2 in absolute value) only for the adjacent hydrogen bonds in a given base-pair (in C.. G pairs, the 1st hydrogen bond G,H1. .C,02 is *not* correlated with the 3rd hydrogen bond G,06. .C,H4). The values of the correlation coefficients are between + 0.2 and + 0.4, with hydrogen bonds in G. .C base-pairs generally showing higher correlations than those in A. .T base-pairs.

The variation of length with time for two typical hydrogen bonds is shown in Fig. 1. The low frequency power spectra, also in Fig. 1, show that the fluctuations of hydrogen bond length occur with a wide range of frequencies between 0.6 cm^{-1} and 33 cm^{-1}. There is no one characteristic frequency for this fluctuation. The forty-eight hydrogen bonds in $(A)_{24}(T)_{24}$ behave in the same way as those in the dodecamer, showing similar stabilities, correlations and fluctuation frequencies.

b) *Correlated torsion angle motion*

The single bond torsion angles are also unexpectedly stable and fluctuate about their mean values with r.m.s. amplitudes of 6° to 11°. During the trajectories, there are no major changes of torsion angles other than those associated with equilibration. The torsion angles show a clear pattern of correlation that is essentially the same for all nucleotides in the dodecamer or $(A)_{24}$ $(T)_{24}$. The significant correlations are all between the torsion angles of a single nucleotide as follows: (δ,χ) + 0.6, (δ,ζ) — 0.5, (α,γ) — 0.4, (ζ,χ) — 0.4, (ε,χ) — 0.4, (β,χ) + 0.3, (γ,χ) — 0.3, α to χ are as defined in Table 1, and the number after the correlated pair of angles is the correlation coefficient. The highest correlations involve the δ angle that is a measure of ribose pucker,

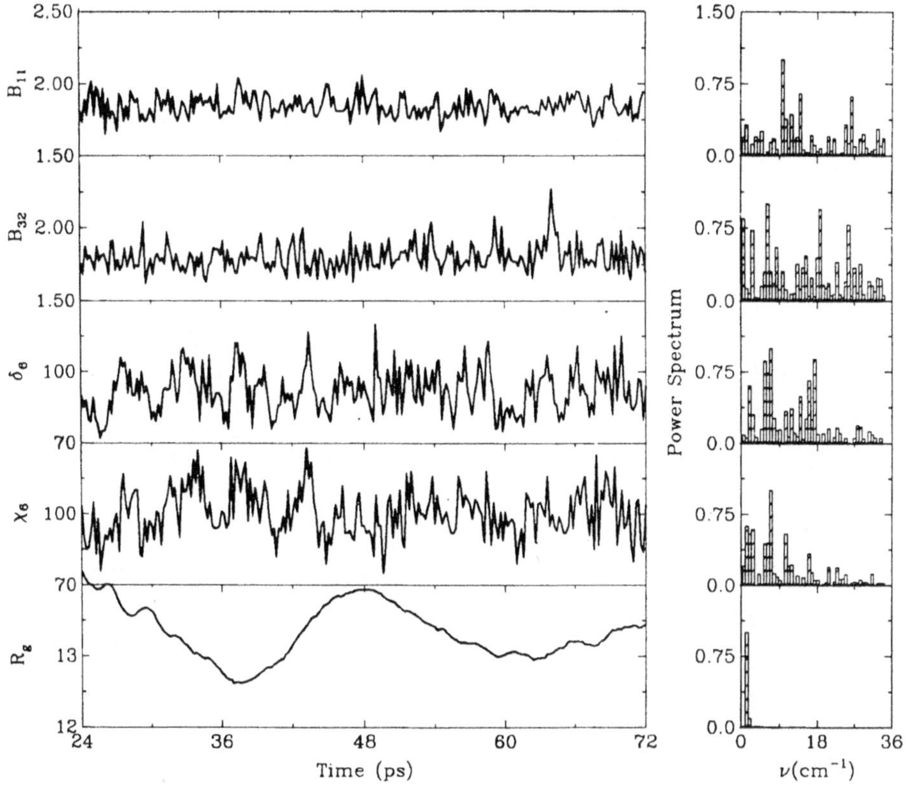

FIG. 1. Showing the time variation and spectral densities of two hydrogen bonds, two torsion angles and the radius of gyration calculated from the molecular dynamics simulation on the dodecamer $(CGCGAATTCGCG)_2$. Values are plotted at 0.2 picosecond intervals for the period 24 to 72ps of the trajectory. The power spectrum, defined as $g(\omega) = |\sum_t f(t)e^{i2\pi\omega t}|^2$ has been scaled so that the highest peak is 1.0. The values shown are (a) the hydrogen bond B_{11} between G_4, $H_5..C_{21}$, N_3, which is buried in the centre of the molecule; (b) the hydrogen bond B_{32} between G_{12}, $H_1..C_{13}$, 0_2, which is more exposed at one end of the molecule; (c) the torsion angle δ, A_6, which reflects changes of sugar puckers; (d) the torsion angle χ, A_6, which reflects base-ribose movement; and (e) the radius of gyration.

and the side chain χ angle that is correlated to all backbone angles except α.

The time variation and power spectra of two typical torsion angles, δ_6 and χ_6 are shown in Fig. 1. The frequencies of fluctuation of torsion angles are more characteristic than those of hydrogen bond lengths, with dominant frequencies of 1, 2, 7, 10 and 15 cm^{-1}.

TABLE I. *Mean values* of torsion angles and other properties*

	Torsion angle [1] (in degrees)							Base [2]	Helix [3]			
	α	β	γ	δ	ε	ξ	χ	τ_p	t_H	Θ_H	$\Sigma 1/R_{pp}$	R_9
Dodecamer												
X-ray	298	171	55	123	189	254	122	15	3.3	35.7	14.4	13.1
Ops	294	166	60	112	186	263	111	15	3.5	32.1	13.9	13.9
48ps	292	173	58	96	180	283	100	16	3.0	28.8	13.7	14.0
90ps	295	173	55	100	180	281	105	12	2.8	28.7	14.0	12.8
$(A)_{24}$ $(T)_{24}$												
X-ray	313	171	36	156	155	265	142	4	3.4	36.0	46.7	24.1
Ops	294	172	64	107	173	274	110	23	3.4	35.6	46.3	24.2
24ps	296	174	57	90	179	286	93	12	3.4	27.7	45.6	22.5
48ps	294	171	61	98	177	282	99	17	3.4	31.3	45.1	24.5
72ps	294	173	57	94	180	284	102	15	3.1	28.8	46.8	21.1
96ps	294	172	64	99	175	281	100	17	3.6	32.5	44.9	23.9

* The first and last base-pair are not included when calculating mean values.

[1] The torsion angles are defined to be zero in the *cis* conformation and increase with clock-wise rotation of the further bond. The atoms defining in each angle are α (02′-P-05′-C5′), β (P-05′-C5′-C4′), γ (05′-C5′-C4′-C3′), δ (C5′-C4′-C3′-03′), ε (C4′-C3′-03′-P), ξ (C3′-03′-P-05′), χ (C2′-C1′-N1-C6, for purimides and C2′-C1′-N9-C8, for purines).

[2] The base orientation is defined by τ_p, the propeller twist angle between the base normals of the two bases in a base-pair.

[3] The local helix defined by the main chain atoms of 4 base-pairs has a rise per base-pair of t_H(Å) and a rotation per base-pair of Θ_H (degrees).

c) *Smooth bending motion in the dodecamer*

The radius of gyration of the dodecamer (Fig. 1) shows a very regular slow oscillation (frequency of 1.3 cm^{-1}, period of 24ps) that has a large amplitude (4% of the mean value of 13.3 Å. This overall motion is seen in Fig. 2 as smooth bending in the plane that contains the dyad axis of the whole dodecamer (the dyad axis between the central base-pairs $A_6..T_{19}$ and $T_7..A_{18}$). The motion is not symmetrical about the straight conformation but tends to close the major groove relative to straight DNA.

The most significant deformation is bending and the angle of bending, defined as $\Theta_c = (180/\pi)L/R_c$ (where L = 40Å, the length of the dodecamer and R_c = the radius of curvature) is found to vary with time, t, as

$$\Theta_c(t) = 70^\circ - 47^\circ \cos(2\pi t/26)$$

The period of this motion (26ps), is the same as that observed for the radius of gyration and corresponds to a frequency of 33.4/26 = = 1.28 cm^{-1}. Although the time average value of Θ_c is not zero, the base-pairs do appear in Fig. 2 to tilt symmetrically about the long axis.

The bending motion persists for the entire trajectory. To eliminate all influence of the starting conditions, the conformation 24ps along the trajectory was cooled by energy minimisation and then used to start a parallel trajectory. The bending motion continues as before (Fig. 2) indicating it is not sensitive to initial conditions and that bending is the dominant slow motion mode of the dodecamer.

d) *Bending in $(A)_{24}(T)_{24}$ is complicated*

The global motions of the 24 base-pairs double-helix $(A)_{24}(T)_{24}$ are similar to those of the dodecamer but show greater diversity (Fig. 3). The motion seen in the y projection is like that of two dodecamers stacked on top of one another and has the same period (approximately 24ps). The bending is still into the major groove. It is as if there are pivots at the three points where the two phosphate chains intersect in the y projection. The motion in the x projection follows the same principle. There are pivots at the two intersection points and the motion is asymmetric with a period of about 48ps.

This pattern of bending can be analysed by considering a completely uniform elastic rod (Landau and Lifshitz, 1959). At

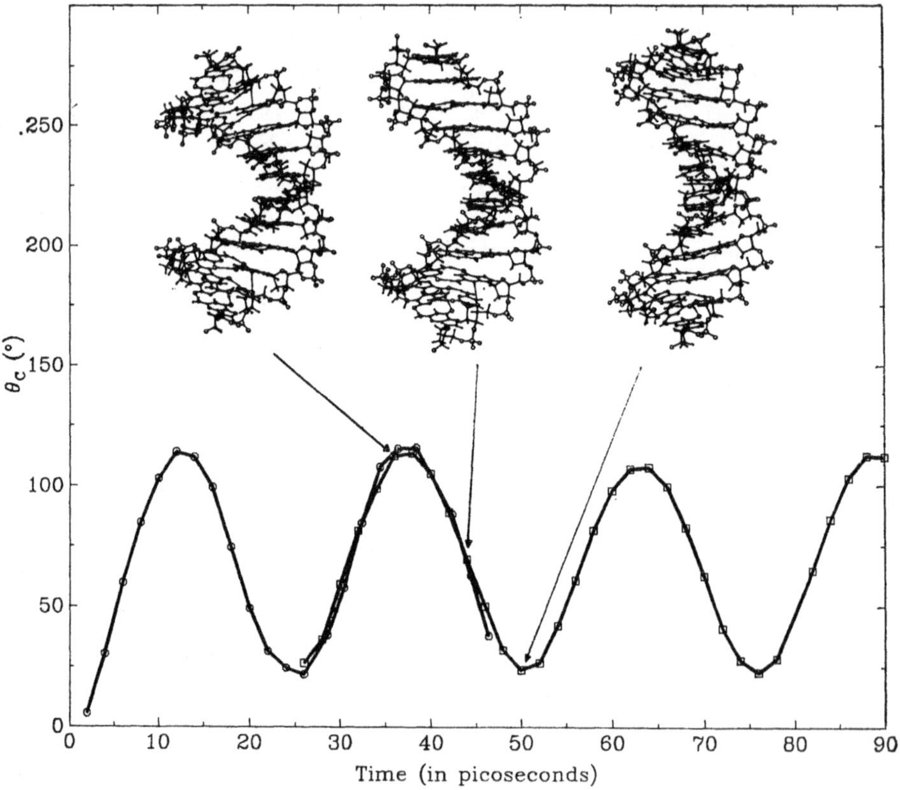

FIG. 2. Showing the periodic bending of dodecamer during the entire 90ps trajectory (45,000 steps). The three molecular drawings are viewed down the y axis and at (a) 37ps, (b) 44ps, (c) 49ps. The graph shows the time variation of the angle of bending Θ_c. The line -0-0 is for the first trajectory (0 to 46ps), while the line -□-□- is for the second trajectory (24 to 90ps The similarity of the two curves for the period 24 to 46ps shows that stopping the first trajectory at 24ps, cooling the conformation by energy minimisation and then starting the second trajectory has no effect on the bending motion. The bending angle was calculated by first generating a standard B-form double helix with the dodecamer sequence (Arnott and Hukins, 1972) and then bending, twisting, and stretching this standard helix by different amounts (16). Each of these deformed helices is then fitted to a particular dodecamer conformation as well as possible (Kabsch, 1976). The degree of bending that gives the best fit is plotted above. The best-fit r.m.s. deviation is typically about 2 Å; for A-form DNA the best-fit r.m.s. deviation is over 4 Å showing that the structure remains B-form.

any time, t, the axis of the rod lying initially along the z axis will have x and y coordinates for each normal mode of motion given by (Barkley and Zimm, 1979)

$$x(t,z,k) = \cos \omega_k t \{\cos K_k z + (-1)^{(K+1)/2} \cosh\sqrt{K_k L}\} \text{ for } k=1,3,5...$$

$$x(t,z,k) = \cos \omega_k t \{\sin K_k z + (-1)^{K/2} \sinh\sqrt{K_k L}\} \quad \text{ for } k=2,4,6...$$

where $K_k = (2k+1)\ \pi/2L$ and $\omega_k = K_k\sqrt{C/M}$ (the rod length is L, the bending force constant per unit length is C and the mass per unit length is M). The overall motion of the rod is a combination of all modes, with the amplitude of the mode being proportional to $1/K_k$.

In the present study of DNA bending does not follow this pattern in that a single mode dominates the x and y motion of both the dodecamer and $(A)_{24}(T)_{24}$. For the dodecamer the y-projection bending corresponds to k = 1 and has a period of 26ps. For the $(A)_{24}(T)_{24}$, the x-projection motion corresponds to k = 2 and has a period of 48ps, while the y-projection motion corresponds to k = 3 and has a period of 24ps. The wavelength of these modes is given by $\lambda_k = 2\pi/K_k$.

For an isotropic elastic rod, the wavelength of the dominant mode should increase with length, but here we find the same wavelength of about 50 Å for both DNA fragments. This occurs because the DNA double helix is not isotropic towards bending in a plane: it is much easier to bend the helix into the major groove than into the minor groove or in any other direction. The standing wave due to bending will have the highest amplitude when the weak points in the helix occur at the points of maximum curvature (the pivot points in Fig. 3). Because these weak points are separated by half a turn of helix, the wavelength of the dominant mode will be close to the helix repeat for DNA of any length.

The frequency of bending obtained from the molecular dynamics simulation can be used to estimate values for the bending force constant C. The frequency (in cm^{-1}) is given by $\nu_k = 108K_k^2\sqrt{C/M}$, where M is the mass per unit length (taken as 180 Daltons/Å). For the k = 1 mode of the dodecamer, $\nu =$

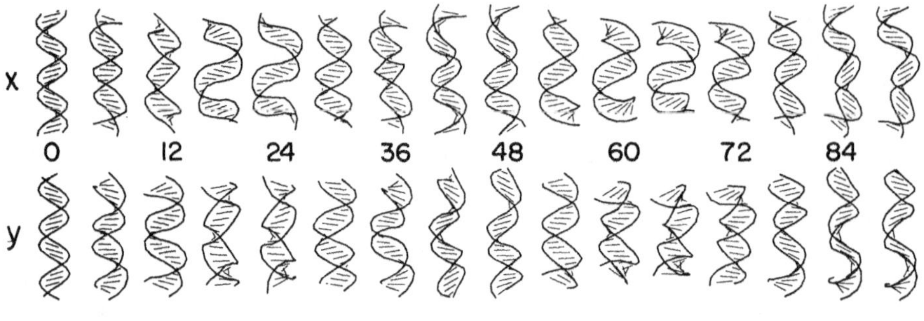

FIG. 3. Showing the overall motion of the $(A)_{24}$ $(T)_{24}$ double helix for the entire 96ps trajectory (48,000 time steps). The upper row shows the x projections of the backbones (drawn as thick P-P virtual bonds) and the base-pairs (drawn as thin C1′-C1′ virtual bonds) at 6ps intervals. The lower row shows the corresponding y projections. The motion seen in the x projection is an anti-symmetric bending motion (k = 2) with a period of about 48ps. The motion seen in the y projection is a symmetric bending motion (k = 3) with a period of about 24ps that is like the motion seen in the dodecamer (Fig. 2).

= 1.27 cm^{-1}, $K_1 = 3\pi/2L_{12}$ and L_{12} = 40Å giving C = 102 kcal $mole^{-1}$ $radian^{-2}$ Å. For the k = 2 mode of $(A)_{24}(T)_{24}$, ν_K = 0.69 cm^{-1}, $K_2 = 5\pi/2L_{24}$ and L_{24} = 80Å giving C = 78 kcal $mole^{-1}$ $radian^{-2}$ Å. For the k = 3 mode of $(A)_{24}(T)_{24}$, ν_K = 1.38 cm^{-1} and $K_3 = 7\pi/2L_{24}$ giving C = 81 kcal $mole^{-1}$ $radian^{-2}$ Å. This suggests that the dodecamer may be a little stiffer than $(A)_{24}(T)_{24}$.

e) *Bending does not perturb torsion angles*

The initial torsion angle values in $(A)_{24}(T)_{24}$ correspond to high energy partially eclipsed conformations as found by fibre diffraction (Arnott and Hukins, 1972); they are relaxed by energy minimisation and dynamics to almost perfectly staggered values. The initial torsion angle values in the X-ray conformation of dodecamer (Drew *et al.*, 1981) are influenced by the abnormal conformations of nucleotides G_4, G_{10} and G_{22} for which the mean values of δ, e, ζ and χ are 150°, 241°, 176° and 90°. Energy minimisation does not relax these values, but during the first 20ps of the trajectory these irregular torsion angles are dynamically annealed to become like those of other nucleotides.

After the initial equilibration, the mean torsion angles are essentially the same in both dodecamer and $(A)_{24}(T)_{24}$ at any time along the trajectory (Table 1). Any systematic variation in torsion angles due to sequence or overall bending is much smaller than the thermal fluctuation. The DNA double helix can undergo large scale conformational changes that are not reflected by any substantial changes (2°) of torsion angles, because there are 14 flexible torsion angles per base-pair that can change in a concerted fashion.

The propeller twist of the bases is close to 15° but due to the overall motion in $(A)_{24}(T)_{24}$ the mean angle between the base normal and long axis (z) is over 20° in agreement with experiment (Hogan *et al.*, 1978; Lee and Charney, 1982). The average rise per residue varies from 2.8 to 3.4 Å and the number of base-pairs per turn varies from 11 to 12.5 over 20ps. It is not clear why the turn angle is so much smaller than the experimental value of 36°; it may be due to the omission of solvent and long range electrostatic interactions.

f) *Kinking in* $(A)_{24}(T)_{24}$

Although much of the motion of the DNA double helix appears to be smooth bending, kinking occurs once during the 96ps $(A)_{24}(T)_{24}$ simulation. Inspection of the x projection in Fig. 3 for the period 60ps to 96ps shows that the kink is caused as the helix bends out of the major groove past the straight conformation. A more detailed drawing (Fig. 4) shows that the angle between the base-pairs at the kink is about 90°.

The kink formed here does not correspond in detail to either of the kinks proposed by model-building (Sobell *et al.*, 1977; Crick and Klug, 1975). The chain kinks by about 90° into the minor groove as proposed by Crick and Klug (1975), but the change in torsion angles found here is smaller ($< 25^{\circ}$) and not restricted to a single torsion angle. Kinking into the major groove as proposed by Sobell *et al.*, (1977) is not found here; bending into the major groove seems smooth and involves torsion angle changes of less than 2°.

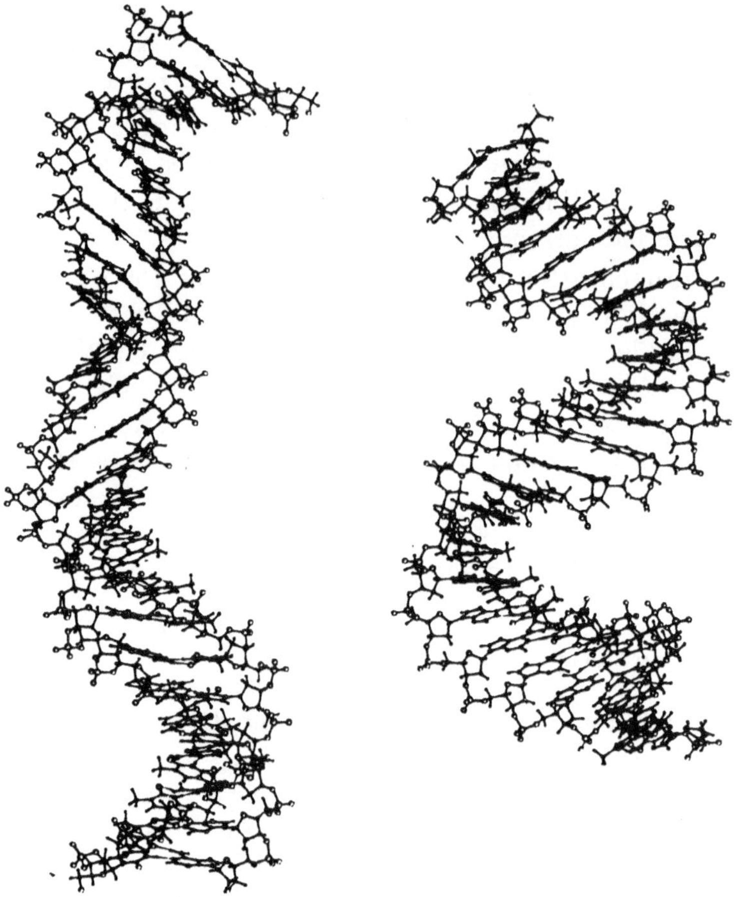

FIG. 4. All-atom drawings of some of the $(A)_{24}$ $(T)_{24}$ conformations that occur during the 96ps trajectory at 300° K. (A) After 24ps, showing one of the most compact conformations. (B) After 89.6ps, showing one of the most extended conformations that has a 90° kink between base-pairs $A_9..T_{40}$ and $A_{10}..T_{39}$. Both drawings are x-axis projections.

g) *Electrostatic interactions*

Electrostatic interactions between the negatively charged phosphate groups were not included in the energy function and these long-range interactions may be expected to affect the overall conformation significantly. The electrostatic energy can be approximated as $332q^2\ \Sigma 1/R_{pp}$ kcal/mole (see Table 1). The effective net charge on each phosphate group, q, will be less than 0.5 electrons due to tightly bound ions (Manning, 1978) and the effective dielectric constant, D, will be close to that of water (80, neglecting any Deybe-Huckle type shielding). For the straight and bent form of the dodecamer, this gives energies of 14.2 and 14.5 kcal/mole (at t = 48ps and t = 90ps, respectively, see Table 1), a difference of only 0.3 kcal/mole for 12 base-pairs. For the straight and bent form of $(A)_{24}(T)_{24}$, this gives energies of 46.6 and 48.5 kcal/mole (at t = 96ps and 72ps, see Table 1), a difference of only 1.9 kcal/mole for 24 base-pairs. Although these differences are small, they are significant in that the bent structures always have higher electrostatic energies.

4. Discussion

a) *The influence of solvent*

The present calculations relate to DNA in vacuum and it is important to consider the major effect of solvent, namely, the viscous damping of vibrations. The dominant bending mode can be approximated by considering a chain that consists of large spherical particles each corresponding to half a turn of helix. Assuming each sphere has a mass, m, of about 3000 Daltons and a radius, a, of 10 Å, gives rise to a Stokes frictional coefficient of $6 \pi a \eta$. The momentum of such a moving sphere will decay with a time constant $\tau = 6 \pi a \eta / m$ or less than 1ps (for water, the viscosity $\eta = 0.01$ poise), so that the 26ps bending motion observed here in vacuum will be damped. Instead there will be a diffusive motion with a relaxation time that has been calculated by Barkley and Zimm (1979) to be 110ps (assuming C = 100 and $\lambda = 50$ Å). The nature and amplitude of the motion will not change. In particular, conformations like those shown for $(A)_{24}$ $(T)_{24}$ in Fig. 3 may be important in water as they involve small changes of the long-range electrostatic energy and a small overall movements of the double helix.

b) *Agreement with experiment*

The present calculations show DNA motion to be cooperative with small correlated fluctuations of hydrogen bond lengths and torsion angles that result in large-scale overall bending and twisting.

The strongest correlations found here have also been observed in the dodecamer X-ray conformations (Drew *et al.*, 1981; Fratini *et al.*, 1983). The small effect of overall bending on the torsion angles has also been observed (Fratini *et al*, 1983).

The force constants calculated above from the frequencies of smooth bending can be compared with the stiffness of DNA in solution as monitored by the persistence length, P. For DNA that always bends along the dyad axis (Schellman, 1974 and 1980) as found here, P can be calculated from the bending force constant as 2C/RT (RT is the thermal energy, of 0.6 kcal/mole at T = 300 K), which gives calculated values of between 260 Å and 340 Å. The experimental value in high salt (Borochov *et al.*, 1981; Harrington, 1978) has been estimated at 260 Å.

The propeller twist of the bases is close to 15° but due to the overall motion in $(A)_{24}$ $(T)_{24}$ the mean angle between the base normal and long axis (z) is over 20° in agreement with experiment (Hogan *et al.*, 1979; Lee and Charney, 1982). It is of interest that the apparent angle between the base-pair and the helix axis as measured by electric dichroism decreases as the DNA is stretched by an electric field (Lee and Charney, 1982; Diekmann *et al.*, 1983) or aggregated into bundles (Mandelkern *et al.*, 1981). This led to the proposal that DNA might be supercoiled under normal conditions (Lee and Charney, 1982). Bending like that observed here for $(A)_{24}$ $(T)_{24}$ could also explain these findings as the amplitude of the bending waves would be decreased both by stretching or aggregation.

Nuclear magnetic resonance studies (Hogan and Jardetsky, 1980; Early and Kearns, 1979; Bolton and James, 1980) have shown that there are nanosecond motions of the base, ribose and phosphate groups of amplitude between 20° and 30°. For the central 14 base-pairs of $(A)_{24}$ $(T)_{24}$, the calculated r.m.s. fluctuation of the angle between the C8-H8 vector and the helix axis averages 17° for the period 20 to 96ps. The corresponding angle between the two protons on C2′ has an average r.m.s. fluctuation of 18°. Thus, the fluctuations obtained here on a time scale of 100ps are smaller but comparable to those observed on a time scale of 1 nanosecond.

The surprising sensitivity of DNA conformation to drug binding (Sobell *et al.*, 1977; Hogan *et al.*, 1979; Dattagupta *et al.*, 1981) can be understood on the basis of the cooperative motion between the very different conformations seen in Figs. 3 and 5. The energies of these conformations are the same to within a few times the room temperature thermal energy (0.6 kcal/mole) and low concentrations of a molecule that binds across a groove could have a major effect on the overall conformation.

REFERENCES

Anderson, W.F., Ohlendorf, D.H., Takeda, T. and Matthews, B.W., 1981, *Nature*, **290**, 745.

Arnott, S. and Hukins, D.W.L., 1972, *Biochem. Biophys. Res. Comm.* **47**, 1504.

Barkley, M.D. and Zimm, B.H., 1979, *J. Chem. Phys.* **70**, 2991.

Bolton. P.H. and James, T.L., 1980, *Biochemistry*, **19**, 1388.

Borochov, N., Eisenberg, H. and Kam, Z., 1981, *Biopolymers*, **20**, 231.

Crick, F.H.C. and Klug, A., 1975, *Nature*, **255**, 530.

Dattagupta, N. and Crothers, D.M., 1981, *Nucleic Acid Res.* **9**, 2971.

Diekmann, S., Hillen, W., Morgeneyer, B., Wells, R.D. & Pörschke, D., 1982, *Biophys. Chem.* **15**, 263.

Drew, H.R., Wing, R.M., Takano, T., Broka, C., Tanaka, S., Itakura, K. and Dickerson, R.E., 1981, *Proc. Natl, Acad. Sci. USA* **78**, 2179.

Early, T.A. and Kearns, D.R., 1979, *Proc. Natl. Acad. Sci. USA*, **76**, 4165.

Fratini, A.V., Kopka, M.L., Drew, H.R. and Dickerson, R.E., 1982, *Biochemistry*, **257**, 14686.

Harrington, R.E., 1978, *Biopolymers*, **17**, 919.

Hogan, M., Dattagupta, N. and Crothers, D.M., 1978, *Proc. Natl. Acad. Sci. USA*, **75**, 195.

Hogan, M., Dattagupta, N. and Crothers, D.M., 1979, *Nature*, **278**, 521.

Hogan, M.E. and Jardetsky, O., 1980, *Biochemistry*, **19**, 3460.

Kabsch, W., 1976, *Acta Crystallogr. sect. A*, **32**, 922.

Landau, L.D. and Lifshitz, E.M., 1959, *Theory of Elasticity* (Pergamon Press, London).

Lee, C.-H. and Charney, E., 1982, *J. Mol. Biol.* **161**, 289.

Levitt, M., 1978, *Proc. Natl. Acad. Sci. USA*, **75**, 640.

Levitt, M., 1981 a, *Nature*, **294**, 379.

Levitt, M., 1981 b, *J. Mol. Biol.* **145**, 251.

Levitt, M., 1983 a, *Cold Spring Harbor Symp. Quant. Biol.* **47**, 251.

Levitt, M., 1983 b, *J. Mol. Biol.* **168**, 595.

Mandelkern, M., Dattagupta, N. and Crothers, D.M., 1981. *Proc. Natl. Acad. Sci. USA*, **78**, 4294.

Manning, G.S., 1978, *Q. Rev. Biophys.* 11,179.

McCammon, J.A., Gelin, B.R. and Karplus, M., 1977, *Nature*, **267**, 585.

McKay, D.R. and Steitz, T.A., 1981, *Nature*, **290**, 744.

Rhodes, D. and Klug, A., 1980, *Nature*, **286**, 573.

Schellman, J.A., 1974, *Biopolymers*, **13**, 217.

Schellman, J.A., 1980, *Biophys. Chem.* **11**, 321.

Sobell, H.M., Tsai, C.-C., Jain, S.C. and Gilbert. S.G., 1977, *J. Mol. Biol.* **114**, 333.

Wang, J.C., 1979, *Proc. Natl. Acad. Sci. USA*, **76**, 200.

Watson, J.D. and Crick, F.H.C., 1953, *Nature*, **171**, 737.

SIMULATING THE ENERGETICS AND DYNAMICS OF ENZYMATIC REACTIONS

ARIEH WARSHEL

Department of Chemistry
University of Southern California

ABSTRACT

Two methods for computer simulation of enzymatic reactions are described. The first method allows one to estimate activation-free energies of enzymatic reactions by using available X-ray structures of enzyme-substrate complexes. This method is based on the empirical valence bond (EVB) approach which correlates reaction rates with the difference in electrostatic stabilization of the relevant resonance structures in the enzyme and in solution. The second method is based on a molecular dynamics extension of the EVB approach. This method allows one to study fundamental aspects of enzymatic reactions, including analysis of the role of structural fluctuations and estimates of entropic effects. The dynamical method is used here for examination of the relationship between reaction free energies (ΔG_o) and activation free energies ($\Delta g^{\neq}$) in enzyme catalysis. The preliminary results presented here indicate that the rates of proton transfer reactions in enzymes might be correlated in a simple way with the corresponding pK_a differences.

I. INTRODUCTION

Understanding the molecular origin of enzyme catalysis is one of the basic problems in biochemistry. Although this complicated question must be related to the exact three-dimensional structure of the enzyme substrate complex, it was not resolved even when X-ray studies provided detailed information about the active sites of several enzymes. Basically, the rate of an enzyme reaction is determined by the interaction energy between the active site and the substrate, but available experimental information does not discriminate between the various components of these interactions and cannot determine their relative importance. Thus it is of great importance to develop theoretical approaches that allow one to correlate available X-ray structures with the energetics of the corresponding enzymatic reactions.

In addition to the interest in a direct structure function correlation, there has been recently a significant interest in the dynamical aspects of enzyme catalysis. Part of this interest is is associated with the possibility that structural fluctuations play an important role in catalysis (Careri *et al.*, 1979). However, despite extensive advances in studies of structural fluctuations (Fraunfelder *et al.*, 1979; Karplus and McCammon, 1981) it is questionable whether or not they play an important role in catalysis. Obviously, enzymatic reactions, as well as other reactions, involve fluctuations of the reacting system. However, it has not been demonstrated that the reactive fluctuations are fundamentally different in enzyme and solution reactions (catalysis is determined by the difference between a given reaction in solution and in enzyme active sites).

In addition to the interest in the dynamics of enzyme catalysis there is also a great interest in the role of entropic factors in enzyme action. Such effects have long been considered to be major factors in enzyme catalysis (see, for example, Page and Jencks, 1971). However, while entropic factors are probably very important in substrate binding, there is no clear demonstration of their contributions to the catalytic rate constant (k_{cat}) in bond breaking reactions (Warshel, 1981).

In view of the complexity of the factors involved in enzymatic reactions, it might be very difficult to design an experiment that will determine conclusively which dynamical effects are important. It is possible, however, that computer simulation approaches will eventually provide an important link between experimental observation and the detailed dynamics of enzyme-substrate complexes.

This paper will describe computer simulation approaches for both correlating X-ray structures to catalytic energy and for simulating the dynamics of actual bond breaking enzymatic reactions. The first part of the paper will concentrate on statical approaches.

II. Potential Surfaces and Energetics of Enzymatic Reactions

Early attemps to study bond breaking enzymatic reactions used different quantum mechanical approaches to evaluate the potential surface of the isolated reaction region (Loew and Thomas, 1972; Umeyama *et al.*, 1973; Scheiner *et al.*, 1975; Scheiner and Lipscomb, 1976). These pioneering calculations were concerned mainly with the properties of the reacting system and did not include most of the enzyme.

The importance of including the microenvironment provided by the enzyme and the surrounding water molecules became clear in a study of the catalytic reaction of lysozyme (Warshel and Levitt, 1976). This study indicated that no properly calibrated quantum mechanical method could simulate a hetrolytic bond cleavage process in solution or protein. In proteins and solutions bond breaking processes lead in most cases to ionic fragments. If the huge effect of " solvation " of these fragments by the environment is not included, one obtains covalent fragments which are less stable by 20 to 50 kcal/mol than the solvated ionic fragments. This environmental effect was introduced in the quantum mechanical calculations by a microscopic dieletric model (Warshel and Levitt, 1976) that includes the potential from the environment (the potential from the induced and permanent dipoles of the enzyme and the surrounding water molecules was included in the effective ionization potential of the reacting atoms). This treatment gave reasonable results for the relevant bond breaking energies.

Subsequent quantum mechanical studies by other workers made attempts to simulate part of the environmental effects. Van Duijnen *et al.* (1979) studied the formation of the $Im^{+} \ldots S^{-}$ ion pair in papain including the permanent dipoles of the protein α-helix. Bolis *et al.* (1978) studied the reaction of papain including two protein residues that form hydrogen bonds with the reacting system. Hayes and Kollman (1980) studied the catalytic system of several serine proteases including the permanent dipoles of the corresponding enzymes. Allen and co-workers also considered the effect of the protein permanent dipoles on several enzymatic reactions (Allen, 1981). Other relevant studies are reviewed by Naray-Szabo and Bleha (1982). Including the protein permanent dipoles is clearly a step in the right direction. However, such a treatment still misses two key factors: the " solvation " by the protein induced dipoles and by the surrounding water molecules. The error associated with such a partial treatment could amount to $\sim$ 30 kcal/mol (see Warshel, 1981).

In principle it is possible to extend the approach of Warshel and Levitt (1976) to *ab initio* quantum mechanical calculations (Tapia, 1981). This, however, does not solve another problem; It appears that even calculations of the isolated reacting system in the gas phase are associated with very significant errors. Reliable results may be obtained by using a double zeta plus polarization basis set which is extremely expensive for even small reacting systems. Attempts to correct the gas phase error led to introduction of an empirical " basis set correction " factor (Scheiner *et al.*, 1975; Kollman and Hayes, 1981).

The difficulties associated with the reliability of current quantum mechanical approaches (even when the environment is properly treated) point strongly toward the need for simpler and more reliable approaches. This led recently to a change in focus and to the development of an alternative approach which uses reactions in solutions as a reference. This approach, which is called the Empirical Valence Bond (EVB) approach (Warshel and Weiss, 1980; Warshel, 1981a), will be used here to describe the physics of electrostatic effects in enzyme catalysis. As a simple example, we will consider a proton transfer reaction.

$$AH + B \rightarrow A^{-} + BH^{+} \qquad (1)$$

The basic quantum mechanics of this system can be described by considering three alternative bonding forms (resonance structures):

$$\begin{aligned} \Psi_1 &= (A\text{——}H \quad\quad B) \\ \Psi_2 &= (A^{-} \quad H\text{——} \; B^{+}) \\ \Psi_3 &= (A^{-} \quad H^{+} \quad\quad B) \end{aligned} \qquad (2)$$

Since the energy associated with Ψ_3 is much higher than that of the other forms we will neglect it in the presentg simplified treatment. The different resonance structures represent different limiting cases for the bonding arrangement, i.e., Ψ_2 represents a fully ionic state and Ψ_1 represents a pure covalent state. Thus we can describe each structure using experimental information about the pure related systems. That is, in the gas phase we can write the energy of the two states as:

$$
\begin{aligned}
E_1^g &= M\,(r_{A-H}) + V_{nb}\,(r_{H\cdot\cdot B}) \\
E_2^g &= H_{PT}^g - 332/r_{A\cdot\cdot B} + M\,(r_{H-B}) + \\
&\quad + V_{nb}\,(r_{H\cdot\cdot A}) + (D_{B^+-H} - D_{A-H})
\end{aligned}
\qquad (3)
$$

where M is a Morse type potential for the corresponding bond, V_{nb} is a repulsive non-bond potential, ΔH_{PT}^g is the energy of the corresponding gas phase proton transfer reaction (forming $A^- BH^+$ from AH B at infinite separation in the gas phase) and D is the dissociation energy of the corresponding bond. The potential surface of the reaction in the gas phase, $\bar{E}^g$, is obtained by solving the secular equation:

$$
\begin{pmatrix} E_1^g - \bar{E}^g & H_{12}^g \\ H_{12}^g & E_2^g - \bar{E}^g \end{pmatrix} = 0 \qquad (4)
$$

where the mixing term, H_{12}, can be evaluated using experimental information (see Warshel and Weiss, 1980). To obtain the potential surface for the reaction in solution, $\bar{E}^s$, one has to solve the secular equation for the system in solution. The first order effect of the solution is to leave the non-polar structure Ψ_1 unchanged while changing drastically the energy of the ionic state Ψ_2. This gives:

$$
\begin{aligned}
E_1^s &\simeq E_1^g \\
E_2^s &\simeq E_2^g + \Delta G_{sol}^{(2)}
\end{aligned}
\qquad (5)
$$

where $\Delta G_{sol}^{(2)}$ is the solvation energy of the ionic structure. Approximating H_{12}^s by H_{12}^g, one obtains the potential surface for the reaction in solution in a very convenient way. The relation between the gas phase and solution potential surfaces is described in Fig. 1.

The key advantage of the EVB approach is in allowing one to compare reactions in solutions to reactions in enzymes. That is, instead of using gas phase experimental results (which are sometimes hard to obtain) it is more convenient to calibrate the energies of the reactions in solution using the corresponding experimental energies. For example, in treating proton transfer

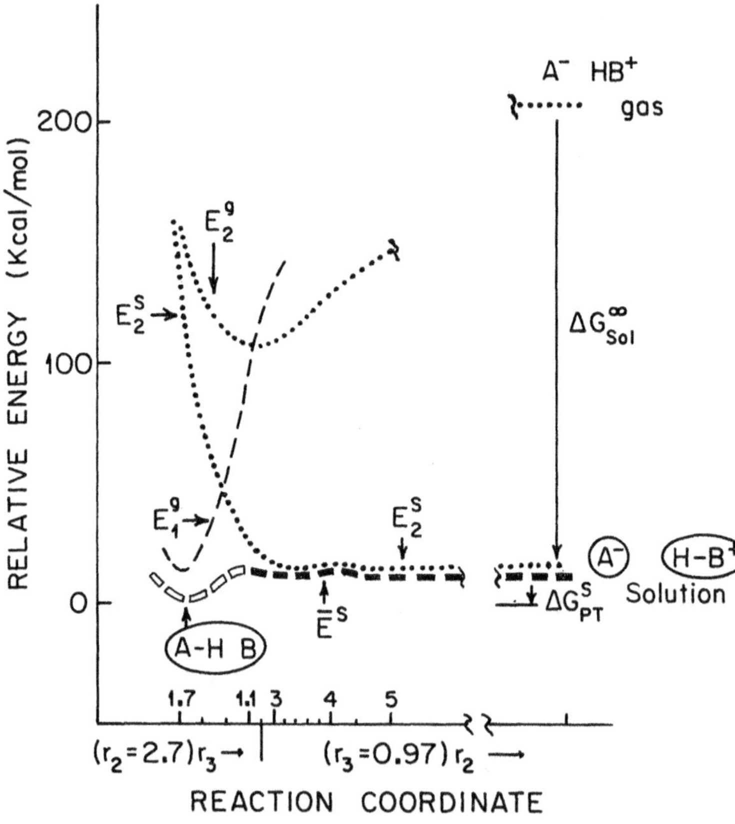

FIG. 1. The energetics of proton transfer reactions in the gas phase (g) and in solution (s). E_1 and E_2 are the energies of the covalent and ionic states (A-H B) and (A^- H-B^+), respectively. E^s is the ground state potential surface obtained from the mixing of E_1^s and E_2^s. The reaction coordinate is given in terms of the change of the H. .B (r_3) and A. .B (r_2) distances. The energies in the figure correspond to calculations of a proton transfer between CH_3OH and H_2O in water.

reactions one can force the energy of A^- BH^+ at infinite separation to reproduce the corresponding experimental energy which is given by:

$$\Delta G_{PT}^{s} = 2 \cdot 3RT\,(pK_a(AH) - pK_a(BH^+)) \tag{6}$$

With this calibration, which can be accomplished by changing E_2^s by a constant, one obtains:

$$E_2^p = E_2^s + (\Delta G_{sol}^p - \Delta G_{sol}^s) \quad (7)$$

Since E_2^s is calibrated by experimental information, the only task left for actual calculations is the evaluation of $\Delta G_{sol}^p - \Delta G_{sol}^s$. This point, which is demonstrated in Fig. 2, indicates that the key problem in simulating enzymatic reactions, is not the quantum mechanical description of the reacting bonds but rather the evaluation of the relevant difference in electrostatic energies. In fact, using the EVB approach one finds that the quantum mechanical aspects of the problem nearly cancel out. Thus the problem of enzyme catalysis is reduced to an understanding of the origin of the difference in interactions between the reacting bonds and their environment in the enzyme active site and in the corresponding reference reaction in solution.

In view of the above discussion the reliability of studies of enzymatic reactions depends on the reliability of the calculated solvation energy. A reasonable way to estimate solvation energies in solution is provided by the Surface Constrained Soft Sphere Dipoles (SCSSD) model described elsewhere (Warshel, 1979). The error range associated with this model is ± 3 kcal/mol. Experimental estimates (Warshel, 1981a) can bring the error range to the 2 kcal/mol range. The solvation of charged fragments by the protein can be estimated by simulating the protein environment using the microscopic dielectric model of Warshel and Levitt (1976) and representing the surrounding water molecules by a combination of all atom type models for some bound water molecules and a Langevin dipole model for the rest (Warshel, 1981a). The error associated with this model is $\pm$ 5 kcal/mol. This error can be reduced if the water molecules are represented by a Langevin model with a dense grid spacing (Russell and Warshel, to be published).

Actual EVB calculations of enzymatic reactions are reported elsewhere (Warshel, 1981a; Warshel, 1981b; Warshel, *et al.*, 1982). These calculations seem to indicate that the difference between the activation free energy of a given enzymatic reaction

$\Delta G^{\neq}_{p}$, and the corresponding reference reactions in a solvent cage, $\Delta G^{\neq}_{s}$ (see Warshel, 1981a), is correlated with the free energy difference ($\Delta G^{p}_{sol} - \Delta G^{s}_{sol}$). For example, as seen from Fig. 2, the differences between $\Delta G^{\neq}_{s}$ and $\Delta G^{\neq}_{p}$ are very similar to the differences between E^{s}_{2} and E^{p}_{2}. If such a " free energy relationship " can be established as a general feature of enzymatic reactions, it might

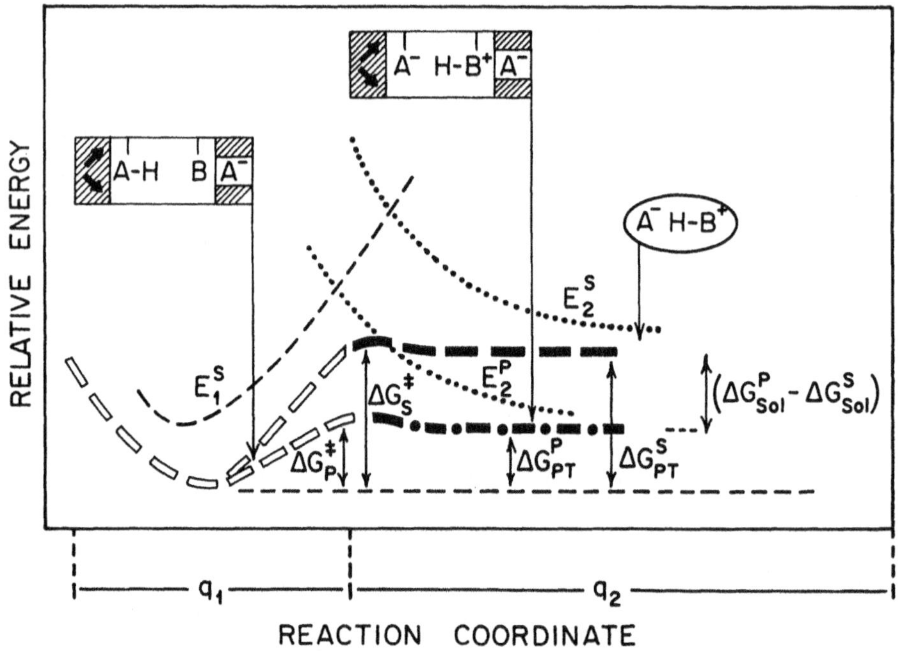

FIG. 2. Comparing the potential surface for proton transfer reactions in solution and in an enzyme. The figure considers the energy of the covalent and ionic states in the protein (p) and, in solution (s). The energy of the covalent state, E_1, is similar in solution and in proteins while the energy of the ionic state, E_2, depends strongly on its interaction with the corresponding environment. The figure indicates that the difference between the free energy for proton transfer, ΔG_{PT}, in the protein and in solution is determined by the corresponding change in solvation free energy. The reaction coordinate is given in terms of q_1 and q_2 that correspond to the change of r_3 and r_2 in Fig. 1.

provide an important concept for structure-function correlation in bioenergetics. However, the above static approach cannot be used to prove the existence of such a relationship, since it does no give the reaction rate from a first principle. Moreover, our static approach introduces free energies in the Hamiltonian, such a treatment is justified for the asymptotic states but clearly not based on any first principle. To obtain a more rigorous dynamical treatment (at the price of much more extensive calculations) one has to be able to simulate the dynamics of bond breakingbond making reactions. Here again the EVB approach provides a very powerful and relatively simple model that will be described next.

III. Simulating the Dynamics of Enzymatic Reactions

The calculations presented in the previous section are based on a static approach, estimating the activation free energy of a given reaction by finding the least energy path for the reaction. In order to deal more rigorously with the fluctuations of the enzyme-substrate complex one should simulate the molecular dynamics of bond breaking reactions. This can still be done by the EVB approach. The main difference is that now the interaction of the various resonance forms with the surrounding environment is included as a conformational dependent interaction potential rather than as an effective solvation free energy. The term ΔG^{p}_{sol} in eq. (7) is replaced by an interaction potential, U, and E^{p}_{i} is given by:

$$E^{p}_{i}(\underset{\sim}{R}, r) = E^{g}_{i}(\underset{\sim}{R}) + U_{i}(\underset{\sim}{R}, r) \qquad (8)$$

where R are the coordinates of the reacting bonds and r are the coordinates of the protein and the surrounding water molecules. The interaction potential, U_i, includes the interaction between the ith resonance form of the reacting bonds and their environment (protein and water) plus the interaction between the environment atoms and themselves. In this approach the free energy is obtained as a result of averaging the relevant potential (see below).

The effect of the fluctuations of the protein is introduced in the calculations through the dependence of U(R, r) on the protein fluctuations. For example, for the case of a proton transfer reaction considered in Fig. 3, the energy of the ionic resonance structure (A^{-} HB^{+}) changes upon change of the orientation of the protein dipole while the energy of the covalent structures stay almost constant.

In simulating the dynamics of the system one has to consider the fluctuations of both the protein and the surrounding water molecules. Since including many water molecules in molecular dynamic calculations is very expensive, we introduce in thc present

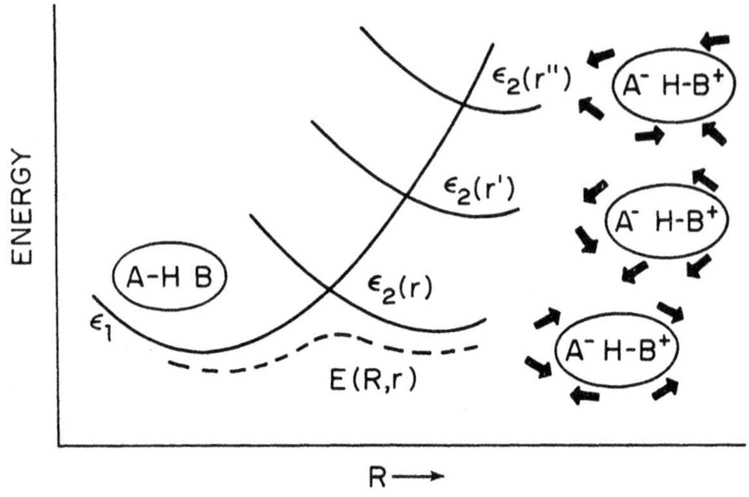

FIG. 3. Energetics of a proton transfer between an acid (A) and a base (B).The reaction is described in terms of a covalent resonance structure (A-H B) and an ionic resonance structure (A^- H-B^+). The energies of the valence bond structures are given in the text and depend on the coordinates of the reacting atoms, R, and the coordinates of the rest of the system, r. Only the ionic resonance structure interacts strongly with the dipoles and charges of the surrounding solvent or protein cage. Thus, as shown schematically in the figure, the energy of the ionic structure, E_2, changes strongly upon fluctuations of the surrounding dipoles. The figure also shows the reaction potential surface, E(R, r), obtained by mixing of the relevant resonance forms.

simulation the following simplification. (i) The induced dipoles of the protein were treated using the approach of Warshel and Levitt (1976). (ii) Few bound water molecules were included in the dynamics. (iii) The electrostatic effect of the rest of the surrounding water molecules was introduced by evaluating its average value (using the Langevin dipole model (Warshel and Levitt, 1976)) for protein configurations generated by the trajectory calculations. A more consistent approach that involves molecular dynamics of the Langevin water model is in progress. This, however, is not a crucial point in the present work which is mainly concerned with examining free energy relationships rather than determining the absolute value of the rate constant (see below). Thus our current practical simplification replaces E_i^s of eq. (8) by:

$$E_i^p(R, r) = E_i^g(R) + U_i^p(R, r) + \Delta_i^L \qquad (9)$$

where U_i^p is the contribution to U_i from the protein and the bound water molecules, while Δ_i^L is the Boltzmann average of the solvation energy, of the protein and the substrate, by the surrounding Langevin dipoles.

1. *Calculating Rate Constants From Molecular Dynamics*

The availability of potential surfaces for an enzyme-substrate complex allows one to simulate the molecular dynamics of the corresponding reaction. This can be done by solving the classical equations of motion for all the atoms in the system, starting with a given set of initial conditions and keeping the average kinetic energy at a value that corresponds to a given temperature. The dynamical simulation is similar to that used to explore the thermal fluctuations of proteins (Karplus and McCammon, 1981; Levitt, 1981) and the dynamics of the chromophore of the visual pigment (Warshel, 1976). However, in the present study, we simulate the protein fluctuations that lead to real bond breaking processes. As a test case, we simulate the transfer of a proton from glutamic acid 35 (Glu 35) to the oxygen atom of the $C_{(1)}-O_{(4)}$ bond that connects two sugar residues in the active site of lysozyme. This reaction is the first step in the catalytic reaction of lysozyme (see, for example, Warshel, 1981a).

In exploring the dynamics of a given enzymatic reaction, we start with initial conditions that correspond to the bound enzyme-substrate complex. This initial state is referred to here as the reactant state. Following the dynamics of the thermally equilibrated system we monitor the formation of configurations which correspond to the reaction product (e.g., protonate $O_{(4)}$ in our proton transfer reaction). These configurations are referred to as the product state. The rate constant of the reaction is given by $k_{a\to b} = 1/\tau_{a\to b}$ where $\tau_{a\to b}$ is the average time it takes for trajectories which start from the reactant state (with different initial conditions) to reach the product state. A typical example of such a direct simulation is shown in Fig. 4. The figure describes a trajectory that leads to a proton transfer. The trajectory fluctuates in the reactant state and then a productive fluctuation of the system transfers the proton to the donor ($O_{(4)}$). After some fluctuations in the product state the proton returns to the reactant state. One can classify the coordinates of the system in terms of a reaction coordinate, X, which describes the degree of proton transfer. The reaction coordinate can be divided into three regions: a reactant region, $X < 0.5$, a product region, $X > 0.5$ and a transition state region $X = X^{\neq} = 0.5$. (A more unique definition of a reaction coordinate will be given below.) Note that many different protein configurations give the same value of X.

A direct evaluation of the rate constant in terms of the average time $\tau_{a \to b}$ is practical only for reactions with small activation barriers between the reactant and product states. Here, however, we are interested in more general cases where the activation barrier might be significantly higher than the average thermal energy. In such cases the relevant productive fluctuations are very rare and their simulation requires extremely long computer time. To evaluate the rate constant of rare productive fluctuations we exploit recent advances in studies of reaction dynamics (Keck, 1966; Anderson, 1973; Tully, 1981) and in statistical mechanics of many-particle systems (Valleau and Torrie, 1977). These studies established that the rate constant can be expressed as:

$$k_{a \to b} = \bar{k}_{a \to b} F \tag{10}$$

where $\bar{k}_{a \to b}$ is the equilibrium rate constant of the well-known transition rate theory (Glasstone *et al.*, 1941) and F is a dynamical correction term that relates the number of trajectories that reach the transition state from the reactant state to those that end up at the product region (see, for example, Tully, 1981). The equilibrium rate constant, $\bar{k}_{a \to b}$, can be expressed as a product of the probability $P(X^{\neq})\, dX$ that a trajectory of the system will fall in a region between $X^{\neq}$ and $X^{\neq} + \Delta X$ and the time $(\Delta X / v^{\neq})$ it takes to pass this region (where $v^{\neq}$ is the average velocity along

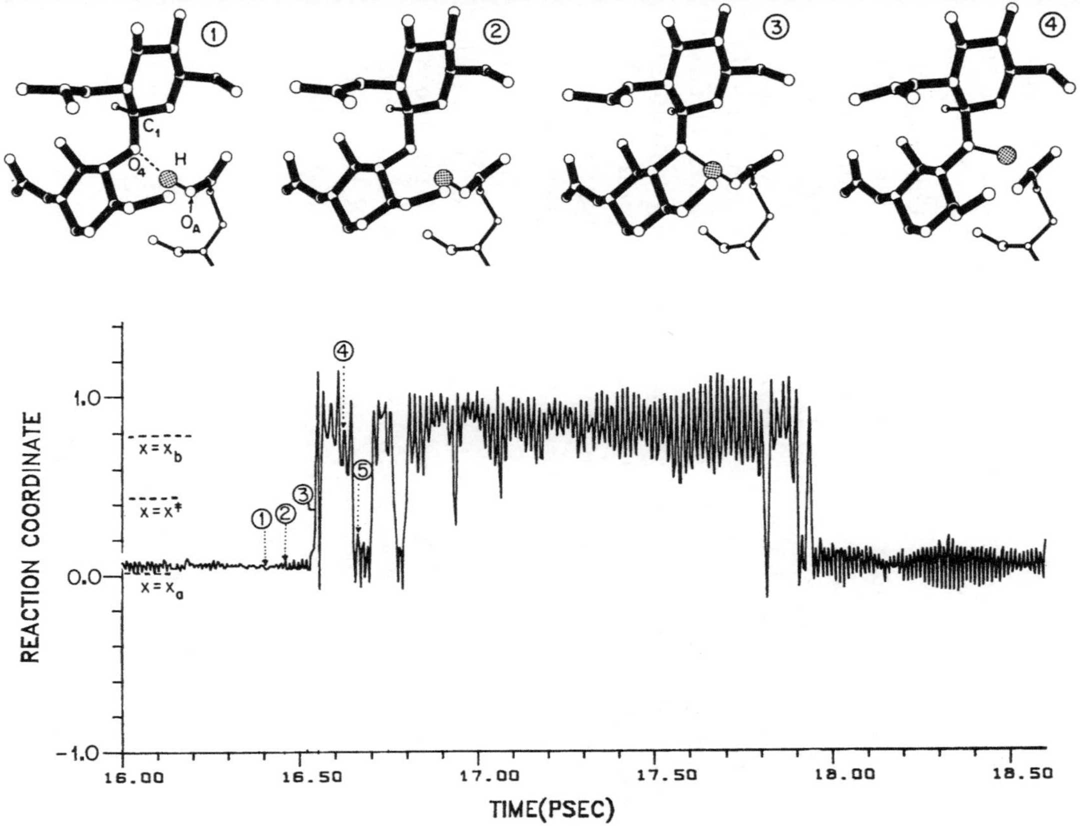

FIG. 4. Simulation of the dynamics of proton transfer between the Oá of Glu 35 and the 0_4 of the $C_{(1)}$-$0_{(4)}$ bond of the substrate of lysozyme. The reaction coordinate, X, is chosen here as the degree of proton transfer $(X = \Delta R_1/(\Delta R_1 + \Delta R_2))$ where R_1 and R_2 are the Oá-H and $O_{(4)}$-H bonds and ΔR is the difference between R and its equilibrium value. X_a and X_b are the average values of X in the reactant and product regions. The transition state coordinate $X{\neq}$ is chosen here, for simplicity, as the point where $X = 0.5$ (a much more rigorous definition can be obtained using the points where $E_1 - E_2 = 0$). The figure describes the fluctuations of the proton in a productive trajectory that starts at X_a, crosses the transition state to X_b and returns to X_a.

the reaction coordinate at the transition state.) Thus we have:

$$k_{a \to b} = v^{\neq} P(X^{\neq}) F \tag{11}$$

The factors $v^{\neq}$ and F (whose evaluation will be considered in the next section) are not expected to change drastically in related reactions in enzyme and in solution. The main difference between the rate constants of related reactions is expected to be in the probability factor $P(X^{\neq})$.

a) *Evaluating The Probability Factor Using Umbrella Sampling*

In order to determine the probability of being at $X^{\neq}$ or any other value of X (e.g., $X = X_l$), we use the so-called " umbrella sampling " approach (Valleau and Torrie, 1977). This approach is in some respects a dynamical version of the " adiabatic mapping " method used in exploring barriers for conformational changes in cyclopentane (Lifson and Warshel, 1968) and in the substrate of lysozyme (Warshel and Levitt, 1976). The implementation of the umbrella sampling approach to chemical reactions in solutions is described in Warshel (1982) which considers electron transfer reactions in solutions. Here we basically extend the same treatment to proton transfer reactions in proteins. We start by dividing the total partition function of the system into the partial partition functions that correspond to the given X's.

$$\bar{q}_0(X_l) = \int \exp\{- E_0(X_l)\beta\} dS \tag{12}$$

where S is the subset of coordinates that span the configuration space for a fixed X, $\beta = 1/k_b T$ and k_b is the Boltzmann constant and E_0 is the reaction potential surface. It is convenient to take the partition function of the most probable value of X at the reactant region ($X = X_a$) as a reference. In this way $P(X_l)$ can be expressed in terms of the relative partition function ($q_0(X_l) = \bar{q}_0(X_l)/\bar{q}_0(X_a)$) by:

$$P(X_l) = \int_{X_l}^{X_l + \Delta X} q_0(X) dx / Q_a \tag{13}$$

$$X^{\neq}$$

$$Q \!-\!\!= X$$

$$({}_a q x_0) \quad \int \quad d\infty$$

In principle one can evaluate eq. (13) by running trajectories and finding the ratio between the time spent at the region ($X_1 > X \geq X_1 + \Delta X$) and the total trajectory time. This direct approach is, however, not practical in cases of significant activation barrier between the reactants and products. In such cases it might take an enormous amount of time for a trajectory that started at the reactant region to reach the product region. Thus, it is impossible in many cases to use molecular dynamics on the reaction potential surface E_0 to accumulate information about configurations near the transition state. The umbrella sampling approach offers a solution to this problem, where instead of running trajectories on the reaction surface, E_0, one runs trajectories on a sampling potential, E_1, that forces the system to spend most of its time near a chosen as one of the coordinates of the system (e.g., the O-H bond length) and the sampling potential can be chosen as ($E_1 = E_0(S, X) + V_1(X)$). However, in cases of chemical reactions in solutions and proteins (e.g., electron transfer and proton transfer reactions) the proper reaction coordinate involves concerted change of many coordinates. In such cases it is much more convenient computationally and conceptually to use sampling potentials of the form:

$$E_1 = (1 - \Theta_1)\, E_1^p + \Theta_1 E_2^p \qquad (14)$$

where E_1 and E_2 are the energies of the resonance forms that give the largest contributions to the reactant and product regions respectively (e.g., for proton transfer reactions E_1 and E_2, are, respectively, the energies of the covalent and ionic resonance structures). The parameter, Θ, changes from 0 to 1 upon motion from the reactant to the product region. Instead of taking the reaction coordinate, X, as a function of the length of the reacting bonds it is more convenient to divide the space of the system into classes, S, that satisfy the relation $E_2(S^1) - E_1(S^1) = C^1$, where C^1 is a constant. The locus of point $\overline{X}_1$ that minimizes $E_1(S^1)$ for positive C^1 and $E_2(S^1)$ for negative C^1 is taken as the reaction coordinate (Warshel, 1982). Since this type of reaction coordinate is not directly related to intuitive chemical concepts, we will discuss the results in terms of the mapping parameter Θ.

The ratio between the partition function that corresponds to E_1 and the one that corresponds to E_0 is given by:

$$Q_1/Q_0 = \iint \exp\{-E_1\beta\}\, dSd\bar{X}/Q_0$$
$$Q_0 = \iint \exp\{-E_0\beta\}\, dSd\bar{X} \qquad (15)$$

Multiplying and dividing the numerator by $\exp\{-E_0\beta\}$ gives:

$$Q_1/Q_0 = \iint \exp\{-(E_1 - E_0)\beta\} \exp\{-E_0\}\beta$$
$$/Q_0 dSd\bar{X} = <\exp\{-(E_1 - E_0)\beta\}>_{E_0} \qquad (16)$$

where $<>_{E_0}$ indicates an average over trajectories that are propagated on E_0. This useful identity (Valleau and Torrie, 1977) is based on the fact that averaging over trajectories of E_0 corresponds to integration over the distribution function $(\exp\{-E_0\beta\}/Q_0)$. Now the relative probability factor $q_0(\bar{x}_1)$ is given by:

$$q_o^l = q_o^l/\bar{q}_o^a = (q_l^l/Q_1)(Q_o/q_o^a)(Q_1/Q_0)(q_o^l/q_l^l)$$
$$= (q_l^l/Q_1)(Q_0/q_o^a) < \exp\{-(E_1 - E_a)\beta\} >_{E_0} \qquad (17)$$
$$\exp\{(E_1(\bar{X}_1) - E_0(\bar{X}_1))\beta\}$$

where we use the abbreviation $q_m^l = q_m(\bar{X}_1) \cdot E_1(\bar{X}_1)$ is the value of the sampling potential at $\bar{X}_1$. The ratio q_l^l/Q_l is evaluated very easily numerically from the ratio between the time a trajectory on E_1 is spent between $\bar{X}_1 + \Delta\bar{X}$ and the total trajectory time. Q_0/q_o^a is evaluated in the same way from a trajectory on E_0. Note that the term $(\exp\{(E_1(\bar{X}_1) - E_0(\bar{X}_1))\beta\}$ should be replaced by the average $\langle \exp\{(E_1(\bar{X}_1) - E_0(\bar{X}_1))\beta\} \rangle_{E_1}$ if $\bar{X}$ is taken as a given bond length rather than the present selection of $\bar{X}$ (which corresponds to a constant value of $(E_1 - E_0)$). Note also that if E_1 is taken as $E_0 + V_1(\bar{X})$, then the average $\langle\,\rangle_E$ becomes simply $\exp\{V_1(\bar{X}_1)\beta\}$.

An important practical point about the evaluation of eq. (17) is that (Q_1/Q_0) *should not* be evaluated by a direct use of eq. (16). Such a direct evaluation converges very slowly for large values of $(E_1 - E_0)$. The real advantage of the umbrella sampling approach can be exploited when Q_1/Q_{1-1} is evaluated for closely spaced

$\bar{x}_l$ and $\bar{x}_{l-1}$ and then Q_l/Q_0 is evaluated by the relation $Q_l/Q_0 = (Q_l/Q_{l-1})(Q_{l-1}/Q_{l-2}) \ldots (Q_1/Q_0)$. Such an approach converges relatively fast since trajectories on E_{l-1} reach points which contribute significantly to Q_l quite frequently. The relative probability Q_l/Q_{l-1} can be expressed in terms of the corresponding free energy difference:

$$\Delta G = G_l - G_{l-1} = -k_b T \ \ln(Q_l/Q_{l-1}) \qquad (18)$$

ΔG is the reversible work associated with changing E_{l-1} to E_l. This gets a clear physical meaning in our choice of E_l where the sampling potential is associated with changing the charge distribution of the reacting bonds from that of (O H-O) to that of ($^+$O-H O$^-$). In this representation ΔG is related to the reversible work involved in charging the reacting atoms.

The reaction probability is determined by the relative probability $q_0(\bar{X}_l)$ which can also be expressed in terms of the corresponding free energy $\Delta g_0(\bar{X}_l) = g_0(\bar{X}_l) - g_0(\bar{X}_a) = k_b T \ \ln(q_0(\bar{X}_l))$. The function $\Delta g_0(\bar{X}_l)$ can then be considered as the potential of mean force for the reaction.

The selection of the exact position of $\bar{X}^{\neq}$ is quite arbitrary in the present treatment, and the value of $k_{a \to b}$ is invariant for different values of $\bar{X}^{\neq}$ (Anderson, 1973). However, if $\bar{X}^{\neq}$ is chosen as the point where $\Delta g_0(\bar{X})$ reaches its maximum, the corresponding value of the dynamical factor F is close to unity. Thus we chose $\bar{X}^{\neq}$ by searching for a maximum in $\Delta g_0(\bar{X})$ and define the value of Δg at $\bar{X}^{\neq}$ as the activation free energy of the reaction. Once $\bar{X}^{\neq}$ is defined, it is possible to use $\Delta g(\bar{X})$ in a convenient way to evaluate the free energy difference ΔG_0 between the product and reactant states. This can be done by the relation:

$$\exp\{-\Delta G_0 \beta\} = Q_b/Q_a = \int_{\bar{X}^{\neq}}^{\infty} q_0(\bar{X})d\bar{X} \Big/ \int_{-\infty}^{\bar{X}^{\neq}} q_0(\bar{X})d\bar{X} \qquad (19)$$

where a and b indicate reactant and product states respectively. This equation can be written as:

$$\exp\{-\Delta G_0\beta\} = \int_{X^{\neq}}^{\infty} \exp\{-\Delta g_0(\bar{X})\beta\} d\bar{X} / \int_{-\infty}^{X^{\neq}} \exp\{-\Delta g_0(\bar{X})\beta\} d\bar{X} \quad (20)$$

A preliminary study of the nature of $\Delta g^{\neq}$ for proton transfer reactions in proteins is summarized in Fig. 5. In this study we took the test case of proton transfer in lysozyme and examined the dependence of $\Delta g^{\neq}$ on the free energy difference between the reactants and the product states (ΔG_0). This was done by using modified energies E^p_1 and E^p_2 given by:

$$\begin{aligned} \bar{E}^p_2(\bar{X}, S) &= E^p_2(\bar{X}, S) + (\Delta - E^p_2(\bar{X}^o_2, S)) \\ \bar{E}^p_1(\bar{X}, S) &= E^p_1(\bar{X}, S) \end{aligned} \quad (21)$$

This relation forces the calculated difference between the minimum of E^p_2 at the product state ($\bar{X}^o_2$) and the minimum of E^p_1 at the reactant state ($\bar{X}^o_1$) to correspond to an assumed energy difference Δ. Changing Δ as a parameter for a set of calculations can be used to evaluate the dependance of $\Delta g^{\neq}$ on ΔG_0. Fig. 5 compares calculations with three different values of Δ. Despite the premilinary nature of the calculations, they seem to point toward a simple relationship between the activation free energies, $\Delta g^{\neq}$, and the corresponding free energy differences, ΔG_0. For $\Delta G_0 \geq 0$ the *change* in $\Delta g^{\neq}$ is linearly correlated with the change in ΔG_0 while for ($\Delta G_0 < 0$)$\Delta g^{\neq}$ approaches zero. A more quantitative relation can be obtained by considering the behavior of the solution of eq. (4). Since ΔG_0 is given approximately by the pK_a difference between the proton donor and the acceptor (Warshel, 1981a) we can write (for $\Delta G_0 > 0$) the approximate relation:

$$\Delta\Delta g^{\neq} \simeq \Delta\Delta G_0 \simeq (2.3RT\Delta\Delta pK_a) \quad (22)$$

A verification of this relationship requires further studies.

2. *Simulating the Dynamics of rare reactive Fluctuations.*

In order to evaluate the average velocity $v^{\neq}$ and the dynamical factors F in eq. (10) one has to be able to simulate the dynamics of productive fluctuations at the transition state region. In cases of high activation barriers it is impossible to simulate rare produc-

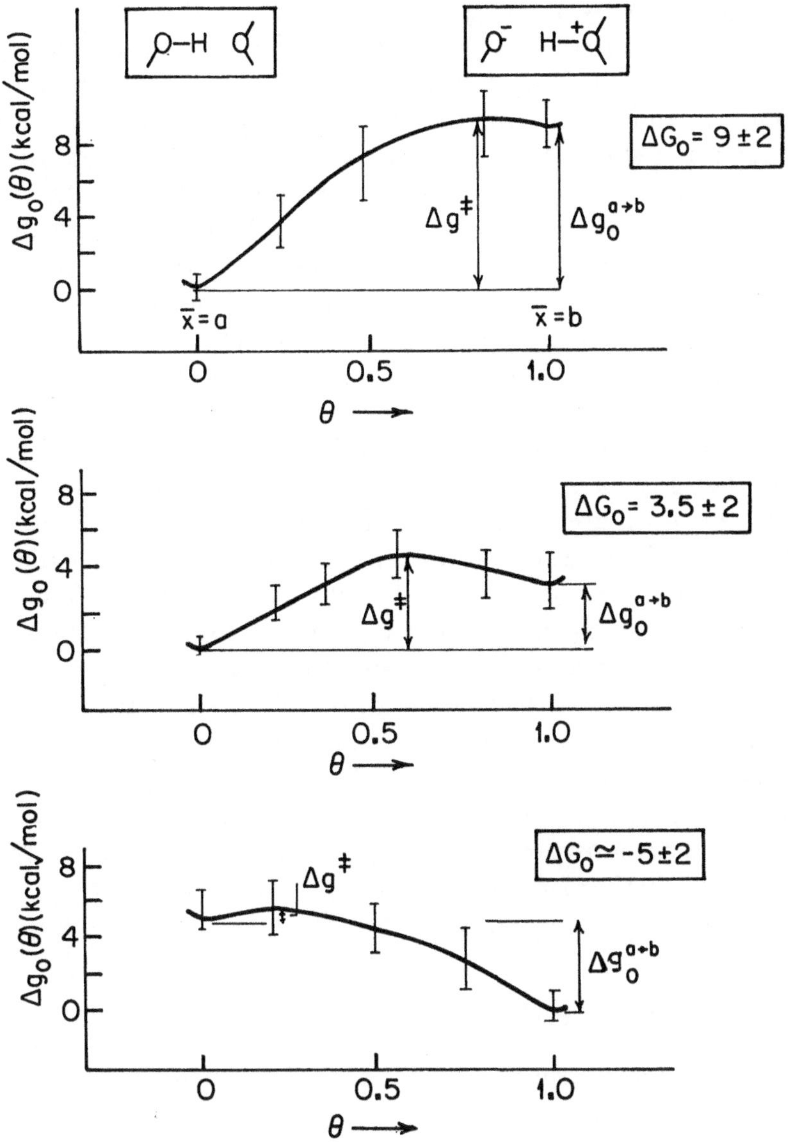

FIG. 5. Evaluation of the activation free energy, $\Delta g^{\neq}$, for proton transfer reactions in proteins. The figure presents the results of the umbrella sampling approach for several assumed values of the free energy differences, ΔG_0, between the products and reactant states. The error bars correspond to an estimate of the convergence of the calculations. Calculations of the type presented here can be used to correlate $\Delta g^{\neq}$ of proton transfer reactions with the corresponding pK_a's differences.

tive trajectories by starting the trajectories at the reactant region. Yet, one can still simulate such trajectories in a rather convenient way. That is, the trajectories can be run on the constraint potential, E_l, that forces the trajectories to pass frequently through the transition state, $X^{\neq}$, at a kinetic energy that corresponds to a given temperature. When a trajectory passes through the transition state one can switch the constraint potential, E_l, to the real surface, E_0. This produces trajectories that proceed downhill to the reactant or product regions. Those trajectories that move from $X^{\neq}$ to the reactant state are the same rare reactive trajectories that lead from the reactant to the transition state, except that the time is reversed. Such a time reversal approach can be used for a direct simulation of the $v^{\neq}$ and F of eq. (10). A preliminary simulation gave a value of about 10^{13}Å sec^{-1}. This value is similar to the estimate of $v^{\neq} = (k_b T/2\pi m)^{1/2} = 6.3 \cdot 10^{12}$Å sec^{-1} (at 300°K), obtained using transition state theory. The value of F for our proton transfer reaction is not much smaller than 1 but many more trajectories are needed for exact determination of $v^{\neq}$ and F. Regardless of the exact value of F and $v^{\neq}$ it seems unlikely that they will change drastically between reactions in enzyme to the corresponding reaction in solutions.

The use of the above approach for simulating the rare fluctuation involved in the C_1-O_4 bond cleavage in the reaction in lysozyme is demonstrated in Fig. 6. Such a simulation can be used to explore whether or not the proton transfer step in the reaction of lysozyme is concerted with the C-O bond breaking process.

IV. Concluding Remarks

This paper considers static and dynamical approaches for computer simulation of enzymatic reactions. The first part of the paper reviews static approaches for correlating structure and function of proteins. It is demonstrated that the calculations can be formulated in a way that expresses rate accelerations in terms of electrostatic energy differences, thus overcoming the difficulties associated with quantum mechanical calculations. The static approach seems to suggest (Warshel, 1981b) that complicated en-

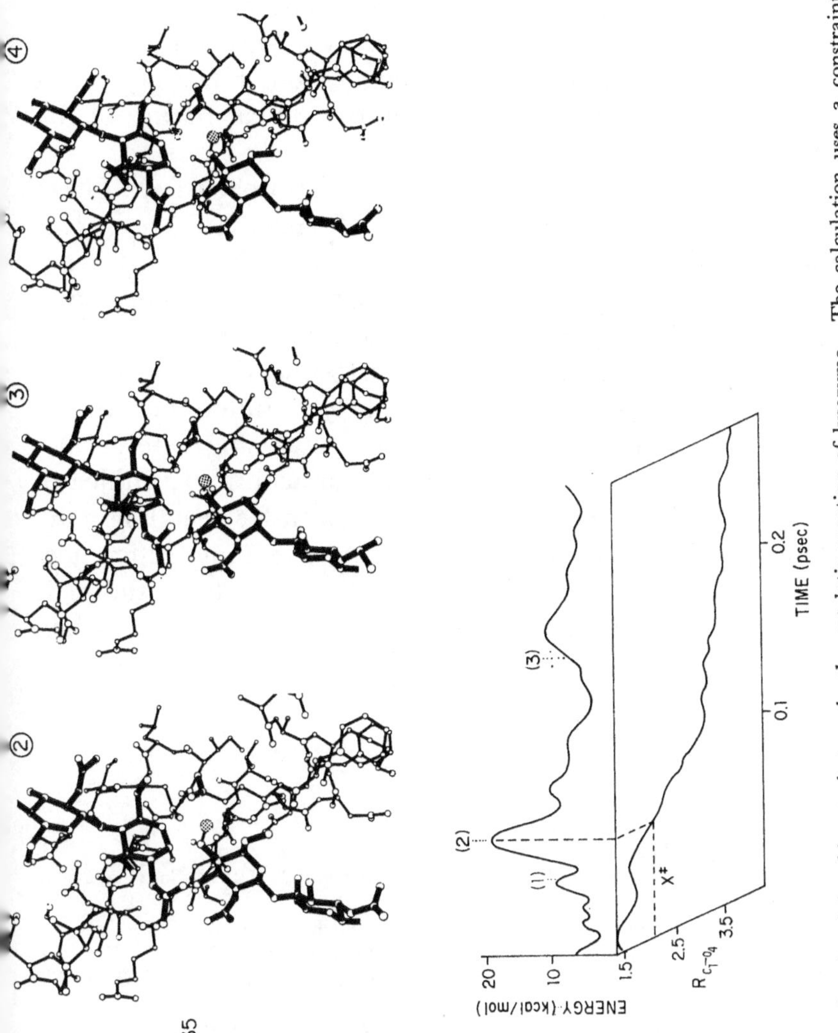

FIG. 6. Simulation of a rare $C_{(1)}$-$O_{(4)}$ bond breaking trajectory in the catalytic reaction of lysozyme. The calculation uses a constraint potential, E_1, to prepare a thermally equilibriated trajectory that reaches the transition state. Switching the potential surface back to the unconstrained E_o yields trajectories that run downhill from the transition state to the product's and reactant's regions. Connecting two such trajectories and inverting the time for the trajectory from the transition state to the reactants gives the desired productive bond breaking trajectory. The figure describes the reaction potential surface along the productive trajectory. Since the fluctuations in the potential energy of the protein (V) are very large, we draw only the relevant part of the potential surface (E-V). The figure also describes the change in the $C_{(1)}$-$O_{(4)}$ distance along the trajectory and displays the configuration of the enzyme substrate complex at representative points along the trajectory.

zymatic reactions can be studied by simply considering the *change* in stabilization of the ionic resonance forms between the enzyme active site and a solvent cage. This suggestion, however, is not based on a rigorous consideration of the dynamical nature of the corresponding reactions.

The second part of the paper presents an approach for a direct simulation of the molecular dynamics of enzymatic reactions. This approach can be used for detailed studies of the role of structural fluctuations and for estimating entropic factors in enzymatic reactions. The work described here concentrates on two aspects of the dynamical calculations: (i) demonstrating the feasibility of such calculations (ii) using the calculations to examine the validity of using linear free energy relationships in analyzing rate acceleration by enzymes. Our preliminary dynamical study indicates that the activation-free energies of proton transfer reactions may be related in a simple way to the corresponding differences in pK_a's. If this relation is confirmed by further studies it might provide a powerful concept in bioenergetics (Warshel, 1981b).

REFERENCES

Allen, L.C. (1981) Ann. N.Y. Acad. Sci., *367*, 383-392.
Anderson, J.B. (1973) J. Chem. Phys., *58*, 4684-4692.
Bolis, G., Ragazzi, M., Salvaderi, D., Ferro, D.R. and Clementi, E. (1978) Gazzetta Chem. Ital., *108*, 425-443.
Careri, G., Fasella, P. and Gratton, E. (1979) Ann. Rev. Biophys. Bioeng., *8*, 69-97.
Fraunfelder, H., Petsko, G. and Tsernoglu, A. (1979) Nature, *280*, 558-563.
Glasstone, S., Laidler, K.J. and Eyring, H. (1941) *The Theory of Rate Processes*, McGraw-Hill, New York, 153-201.
Karplus, M. and McCammon, A.J. (1981) CRC Crit. Rev. Biochem., *9*, 293-349.
Keck, J.C. (1966) Adv. Chem. Phys., *13*, 85-121.
Kollman, P.A. and Hayes, D.M. (1981) J. Am. Chem. Soc., *103*, 2953-2961.
Levitt, M. (1981) Nature, *294*, 379-380.
Lifson, S. and Warshel, A. (1968) J. Chem. Phys., *49*, 5116-5124.
Loew, G. and Thomas, D.D. (1972) J. Theor. Biol., *36*, 98-104.
Naray-Szabo, G., Bleha, T. (1982) Progress in Theoretical Organic Chemistry, ed. Csizmadia, I.G., Vol. 3, 267-336.
Page, M.I. and Jencks, W.P. (1971) Proc. Nat. Acad. Sci. USA, *68*, 1678-1683.
Scheiner, S., Kleier, D.A. and Lipscomb, W.N. (1975) Proc. Natl. Acad. Sci. USA, *72*, 2606-2610.
Scheiner, S. and Lipscomb, W.N. (1976) Proc. Natl. Acad. Sci. USA, *73*, 432-436.
Tapia, O., Johanin, G. (1981) J. Chem. Phys., *75*, 3624-3634.

Tully, J.C. (1981) Computers and Chemistry, *4*, 159-156.
Umeyama, H., Imamura, A., Nagata, C. and Hanano, M. (1973) J. Theor. Biol., *41*, 485-502.
Valleau, J.P. and Torrie, G.M. (1977) in Modern Theoretical Chemistry, ed. B. Berne, Vol. 5, Plenum Press, New York, 137-193.
Van Duijnen, P.T., Thole, B.Th. and Hol, W.G.J. (1979) Biophys. Chem., *9*, 273-280.
Warshel, A. (1976) Nature, *260*, 679-683.
Warshel, A. and Levitt, M. (1976) J. Mol. Biol., *103*, 227-249.
Warshel, A. (1979) J. Phys. Chem., *83*, 1640-1652.
Warshel, A. and Weiss, R.M. (1980) J. Am. Chem. Soc., *102*, 6218-6226.
Warshel, A. (1981a) Biochemistry, *20*, 3167-3177.
Warshel, A. (1981b) Acc. Chem. Res., *14*, 284-290.
Warshel, A., Russell, S. and Weiss, R.M. (1982) in *Chemical Approaches to Understanding Enzyme Catalysis*, ed. Green, B.S., Ashani, Y. and Chipman, D., Studies in Organic Chemistry, Vol. 10, Elsevier Scientific Publishing Co., Amsterdam, 267-279.
Warshel, A. (1982) J. Phys. Chem., *86*, 2218-2224.

STRUCTURAL STUDIES OF DNA-PROTEIN INTERACTIONS

B. W. MATTHEWS

Institute of Molecular Biology and Department of Physics University of Oregon
Eugene, Oregon 97403

ABSTRACT

The structures of three proteins that regulate gene expression have been determined recently and suggest how these proteins may bind to their specific recognition sites on the DNA. One protein (Cro) is a repressor of gene expression, the second (CAP) usually stimulates gene expression and the third (λ repressor) can act as either a repressor or activator. The three proteins contain a substructure consisting of two consecutive α-helices that is virtually identical in each case. Structural and amino acid sequence comparisons suggest that this bihelical fold occurs in a number of proteins that regulate gene expression. DNA-protein recognition appears to be based primarily on a network of hydrogen bonds between side-chains of the protein and the parts of the base-pairs exposed within the major groove of the DNA.

1. INTRODUCTION

The control of the genetic information encoded in DNA is of critical importance in all living systems. It has been known for some time that, at least in simple organisms, this control is achieved by proteins (" switch molecules ") that recognize and bind to specific sites on the DNA. Until recently, little was known of the structures of these molecules or the way in which they work. However, within the past few years, the three-dimensional structures of three DNA-binding switch molecules (" Cro", " CAP " and " λ repressor ") have been determined and have suggested how these proteins bind to their specific recognition sites on the DNA.

In this paper we will briefly review the structures of these three proteins and discuss the modes of DNA-protein interaction that are suggested. As will become apparent, there are similarities between the three proteins, but there are also striking differences as well. For additional background and more detailed information, reference can be made to the reviews of Ohlendorf and Matthews (1983), Takeda *et al.*, (1983) and Pabo and Sauer (1984).

2. Three DNA-binding proteins

Cro, λ-repressor and CAP are all dimeric DNA-binding proteins, but have substantial differences in their overall structures. For both CAP and λ repressor, the respective polypeptides form two domains. In λ repressor, the amino-terminal domain binds to the DNA whereas in CAP it is the carboxyl-terminal part of the molecule that has this function.

Both Cro and λ repressor bind to six similar but non-identical 17-base-pair sites on the chromosome of bacteriophage λ. The site at which Cro binds most tightly is the weakest binding site for λ repressor, and *vice versa.* The two proteins are involved in the adoption by the phage of either the lytic or the lysogenic mode of development. Although this decision-making process is complicated and not yet understood in its entirety, the respective roles of Cro and λ repressor have been analyzed in detail (Johnson *et al.*, 1981). Cro is a straight-forward repressor of gene expression. It binds preferentially to the DNA at its highest affinity site (0_R3) and, when bound, prevents transcription from the adjacent repressor maintenance promoter. λ repressor is more versatile. In common with Cro it can act as a repressor but it can also stimulate the expression of its own gene. CAP (" catabolite gene activator protein," also known as " cyclic AMP receptor protein ") participates in the regulation of a number of genes in *Escherichia coli.* In the presence of cyclic AMP, CAP promotes transcription of these genes, and, in some circumstances, can act as a negative regulator as well. The DNA sequence that CAP recognizes is approximately 15 nucleotides long and more than ten such sites are known in *E. coli* (de Crombrugghe *et al.*, 1984).

3. Structures of DNA-binding proteins

A sketch of the structures of Cro (Anderson *et al.*, 1981), the amino terminal domain of λ repressor (Pabo & Lewis, 1982) and the carboxyl terminal domain of CAP (McKay & Steitz, 1981) as determined from the respective crystal structures is shown in Fig. 1.

The structure of Cro is quite simple, consisting of three α-helices (α_1, α_2, α_3) and a three-stranded antiparallel β-sheet. In the crystal, four polypeptides associate as a tetramer with approximate 222 symmetry. Ultracentrifugation indicates that Cro is dimeric in solution. This dimer is presumed to be the one shown in Fig. 1. Residues 55-61 of each monomer extend and lie against

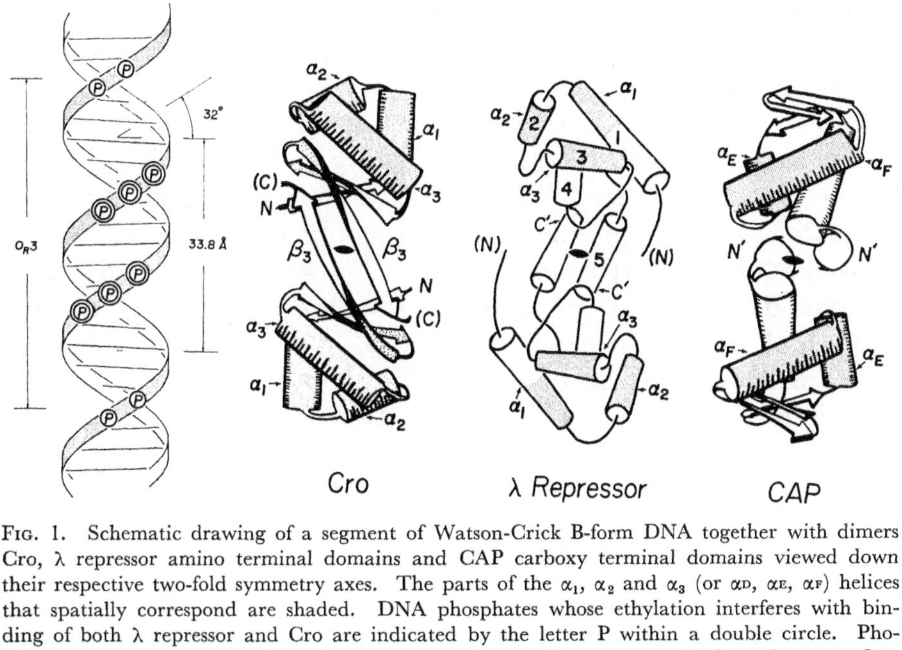

Fig. 1. Schematic drawing of a segment of Watson-Crick B-form DNA together with dimers Cro, λ repressor amino terminal domains and CAP carboxy terminal domains viewed down their respective two-fold symmetry axes. The parts of the α_1, α_2 and α_3 (or α_D, α_E, α_F) helices that spatially correspond are shaded. DNA phosphates whose ethylation interferes with binding of both λ repressor and Cro are indicated by the letter P within a double circle. Phosphates whose ethylation effects λ repressor (and also P22 repressor) binding, but not Cro, are indicated by a P in a single circle (after Ohlendorf & Matthews, 1983; Takeda *et al.*, 1983).

the surface of the other monomer. Phe 58, in particular, makes hydrophobic contact with its partner subunit. The carboxy-terminal residues 63-66 are disordered in the crystals and, presumably, in solution as well (Anderson *et al.*, 1981; Matthews *et al.*, 1983).

The DNA-binding form of intact λ repressor is predominantly a dimer. Proteolytic cleavage separates the protein into an amino-terminal domain of 92 residues and a carboxyl terminal domain consisting of residues 132-236. The amino-terminal domain is responsible for DNA recognition and can act as both a positive and negative regulator of transcription. Although monomeric in solution, the amino terminal domain dimerizes as it binds to DNA and protects exactly the same bases against chemical modifications as intact λ repressor. It is this amino terminal domain whose structure has been determined (Pabo & Lewis, 1982; Lewis *et al.*, 1983) and is shown in Fig. 1.

The structure contains five α-helices with no β-sheet. In the crystal, the two amino-terminal domains make contact via the fifth α-helix in each subunit (Fig. 1). Studies of mutants of λ-repressor suggest that a similar helix-helix contact may occur in intact λ repressor.

The complex of CAP with cyclic-AMP, i.e. the DNA-binding form of CAP, was shown by McKay and Steitz (1981) to be a two-domain structure. Fig. 1 includes only the carboxyl-terminal domains, i.e. the presumed DNA-binding region. Not shown is the larger amino-terminal domain to which the cyclic AMP is bound. The structure in the crystals is dimeric with the amino-terminal domains related by a local two-fold axis. However, the carboxyl-terminal domains adopt somewhat different conformations and are not exactly two-fold-related (McKay *et al.*, 1982; Steitz *et al.*, 1983b).

4. Models for DNA-binding

To date, no structure has been determined of a sequence-specific DNA-binding protein complexed with its target DNA sequence. Structural models for DNA-protein recognition therefore rely primarily on inspection of the uncomplexed proteins as seen in the respective crystal structures (Fig. 1).

For Cro protein, the 34Å spacing between the two-fold-related α_3 helices, together with their angle of tilt (Fig. 1), strongly suggests that these α-helices bind within successive major grooves of right-handed Watson-Crick DNA as illustrated in Figure 2. It is presumed that the flexible carboxy-terminal residues of Cro participate in DNA binding by lying along the minor groove.

This model for Cro binding is not only attractive stereochemically, it also is consistent with chemical protection and modification studies of the DNA as well as the protein, and is supported by recent NMR studies (Arndt *et al.*, 1983). A characteristic feature of the model is the match between the two-fold symmetry of the protein and the (approximate) two-fold sequence and spatial symmetry of the DNA binding site (Anderson *et al.*, 1981; Matthews *et al.*, 1983; Ohlendorf *et al.*, 1982).

A very similar mode of DNA binding has been proposed for λ repressor. Here also there is a pair of two-fold-related α_3-helices (Fig. 1) that are presumed to bind within successive major grooves of Watson-Crick DNA. Furthermore, the amino-terminal residues of λ repressor form two long " arms " with flexible ends that " wrap around " the DNA when the protein binds. In this case the " arms " contact the major groove of the DNA. The presumed mode of DNA binding is consistent with DNA protection and modification studies (Pabo & Lewis, 1982; Lewis *et al.*, 1983). Also the model is strongly supported by the observed locations of mutations in λ repressor that alter DNA binding (Fig. 3) (Nelson *et al.*, 1983; Hecht *et al.*, 1983).

The mode of interaction of CAP with DNA is not obvious. As can be seen in Fig. 1, the α_F helices are 34Å apart, but their tilt (c.f. Cro) is not complementary to the grooves of right-handed B-form DNA. Faced with this dilemma, Steitz and coworkers initially proposed that the DNA changes from right-handed to left-handed when CAP binds to its sequence-specific sites (McKay & Steitz, 1981). However, experiments designed to test this hypothesis indicate that there is no unwinding of the DNA when CAP binds to such sites (Kolb & Buc, 1982). Current models

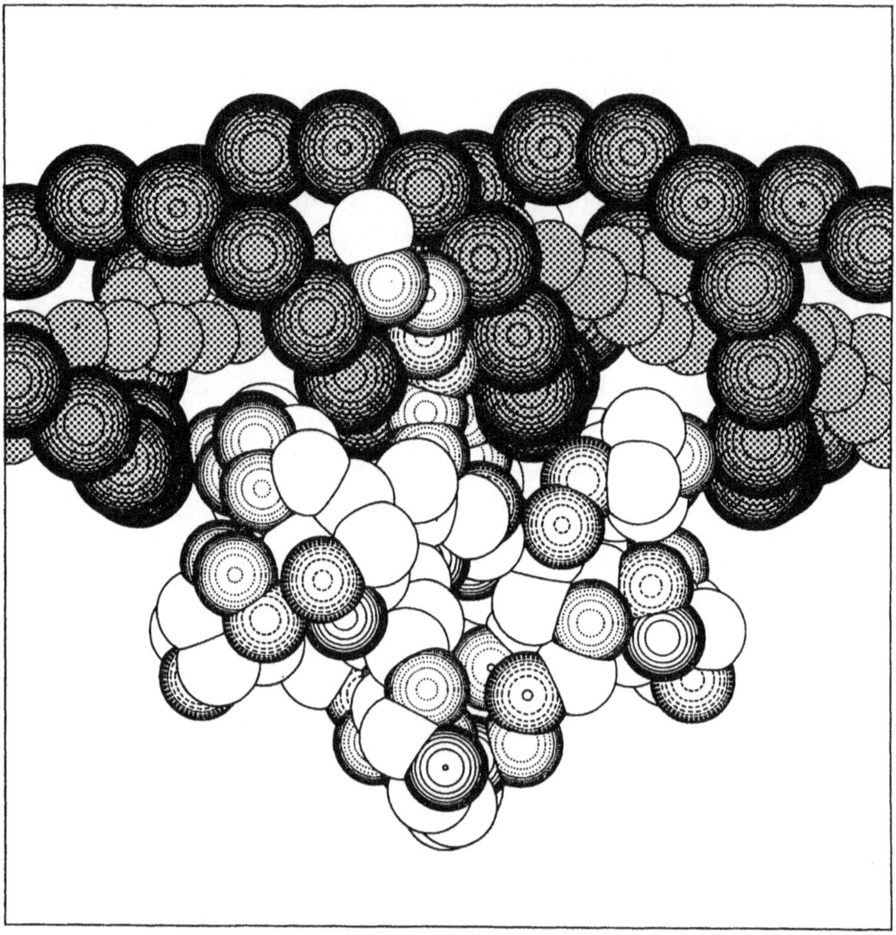

FIG. 2. Stylized drawing showing the complementarity between the structure of Cro repressor protein and DNA. In the presumed sequence-specific complex, the protein is assumed to move closer to the DNA, with the α_3 helices penetrating further into the major grooves of the DNA than is shown in the figure. The carboxyl terminal residues of the protein are presumed to bind in the vicinity of the minor groove of the DNA. In order to maximize the contacts between Cro and DNA, the protein may undergo a hinge-bending motion and/or the DNA may bend (as shown), although these are not essential features of the model. The DNA is represented stylistically by large dotted sphere centered at the phosphate positions and small dotted spheres that follow the bottom of the major groove. In the protein, each residue is represented by a single sphere. Acidic residues have solid concentric circle shading, basic residues have broken circle shading, uncharged hydrophilic residues have dotted circle shading and hydrophobic residues have no shading (after Anderson *et al.*, 1981; Ohlendorf *et al.*, 1982).

suggest that CAP may indeed bind to right-handed DNA with the N-terminal parts of the α_F helices placed within successive major grooves (Pabo & Lewis, 1982; Steitz *et al.*, 1983a). Recent studies of mutants of CAP that recognize altered DNA sequences support the right-handed DNA model (personal communication from R. Ebright).

5. Sequence similarities in DNA-binding proteins

Following the structure determinations of Cro, CAP and λ repressor, it has become apparent that these three proteins have features in common which, presumably, extend to a number of other DNA binding proteins.

The suggestion that several DNA-binding proteins might have structural similarities came first from comparisons of their amino acid sequences. In some cases, such as Cro and λ repressor, the sequence homology is poor, and was not apparent on first inspection. However, with additional sequences available, the overall homology becomes obvious (Fig. 3). The sequence homology includes not only repressors and activator proteins from different phages, but also other DNA binding proteins such as the *lac* repressor from *E. coli* (Anderson *et al.*, 1982; Matthews *et al.*, 1982; Sauer *et al.*, 1982; Ohlendorf *et al.*, 1983b; Weber *et al.*, 1982).

The region of best sequence homology occurs within the parts of the sequences that align with the α_2 and α_3 helices of Cro and of λ repressor, i.e. within the part of the respective proteins that are assumed to interact with the DNA. Thus, it is reasonable to infer that the homologous proteins contain an α-helical DNA-binding supersecondary structure similar to the α_2-α_3 fold seen in Cro and λ repressor. The locations of known mutants of *lac* repressor are consistent with such a hypothesis (Fig. 3) and additional support in this case has subsequently come from NMR studies (Arndt *et al.*, 1982; Zuiderweg *et al.*, 1983).

6. Structural similarities in DNA-binding proteins

Independent of the above sequence relationships, it was found that Cro and CAP have a striking structural correspondence in their presumed DNA binding regions (Steitz *et al.*, 1982). The

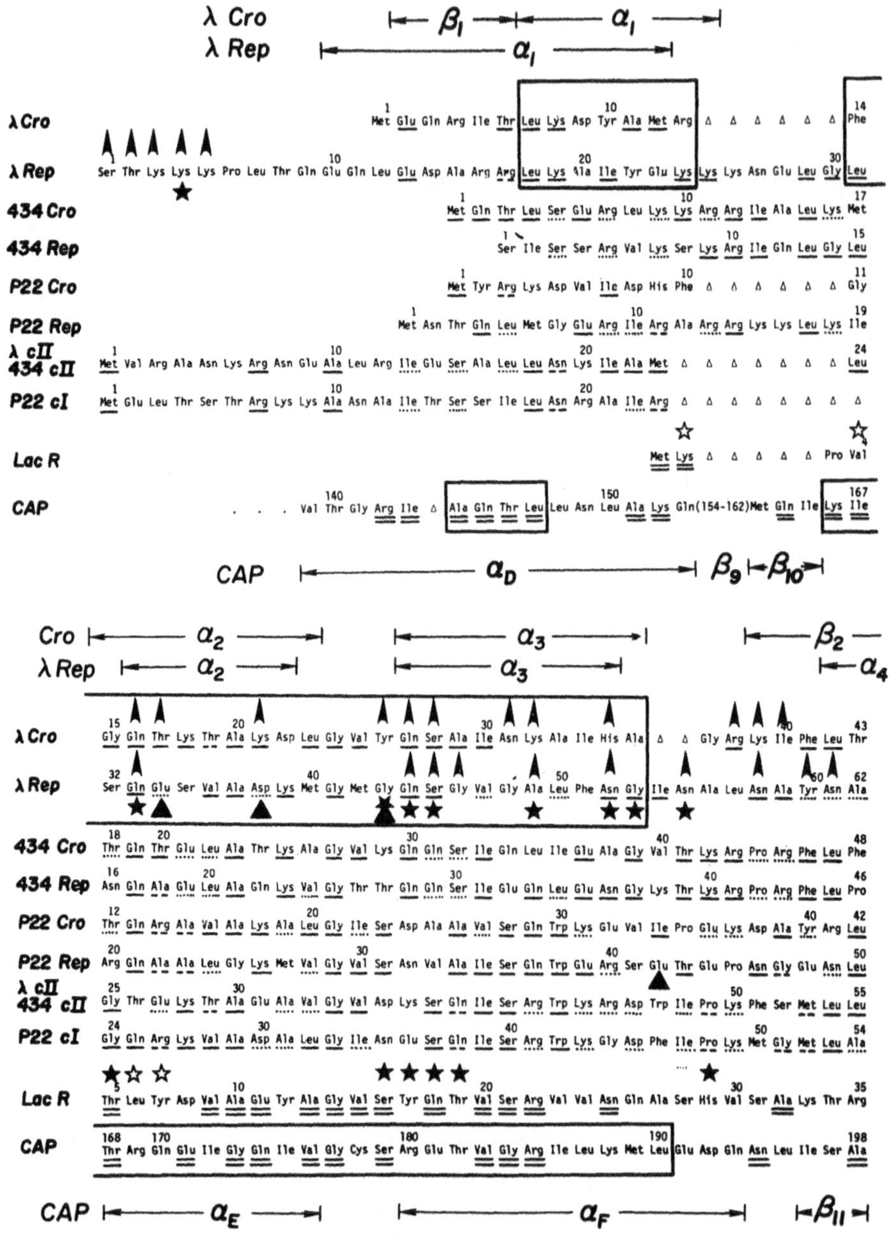

FIG. 3. Alignment of a series of DNA-binding proteins based on sequence homology and correspondence of the three-dimensional structures of Cro, CAP and λ repressor. The first eight sequences are DN[A-] binding proteins from bacteriophages. Within this set, residues identical in two or more sequences a[re] underlined. Residues in the final two sequences (*lac* repressor and CAP) that agree with the phage protei[ns] have a double underline. Residues that have similar three-dimensional structures in Cro and λ repress[or] or in Cro and CAP are boxed. Residues in Cro and λ repressor that are presumed to contact D[NA] are capped by arrow heads. Additional presumed contacts occur for residues near the carboxyl termin[us] of Cro. The solid stars underneath the λ repressor sequence show locations where surface amino ac[id] substitutions (excluding proline) interfere with the function of the protein. Similarly, the locations of re[si]dues of *lac* repressor for which all known mutants dramatically reduce DNA-binding but do not simp[ly] destabilize the protein are indicated by solid stars. The open stars indicate locations where amino ac[id] substitutions may reduce DNA-binding, or where the reduction in DNA-binding affinity is weak. Locatic[ons] in λ repressor and P22 repressor at which amino acid substitutions result in non-stimulation of transcripti[on] are indicated by triangles. Approximate locations of α-helices and β-sheets in the three known structur[es] are shown (after Takeda *et al.*, 1983).

three α-helices (α_D, α_E, α_F) in the carboxyl-terminal domain of CAP can be approximately superimposed on the α_1, α_2 and α_3 helices of Cro (Figs. 1, 3). For the α_E-α_F and α_2-α_3 helical units the superposition is striking. There are 24 α-carbons in the respective units that superimpose with an average discrepancy of 1.1Å. An exhaustive search through all protein structures in the Brookhaven Data Bank failed to find a similar two-helical unit. Following the determination of the gene sequence and the inferred amino acid sequence of CAP it has been shown that there is an amino acid sequence homology between CAP and other DNA-binding proteins corresponding with the observed structural homology between CAP and Cro (Weber *et al.*, 1982; Ohlendorf *et al.*, 1983b).

It has also been shown for Cro and λ repressor that their α_2 and α_3 helices, and parts of their α_1 helices as well, spatially superimpose (Ohlendorf *et al.*, 1983a) (Figs. 1, 3). Again, as with Cro and CAP, it is the α_2-α_3 helical units of the two proteins that have virtually identical conformations.

The alignment of the sequences of the DNA binding proteins included in Fig. 3 is based in part on sequence homology and also on structural homology between Cro, CAP and λ repressor. The residues that are buried tend to be most strongly conserved. Recent amino acid sequence comparisons made in the light of the known structures of Cro, CAP and λ repressor, suggest that the two-helical unit may occur in additional DNA-binding proteins not included in Fig. 3, that are involved in gene regulation at the level of transcription (Ohlendorf *et al.*, 1983b; Pabo & Sauer, 1983).

7. Similarities in DNA binding

The amino acid sequence comparisons and the structural comparisons both point to a special role for the two-helical "α_2-α_3" unit in DNA recognition and binding.

The mode of interaction of this unit with DNA, as inferred from the structure of Cro, is sketched in Fig. 4. The presumed mode of binding of the α_2-α_3 helical unit in λ repressor is very similar al-

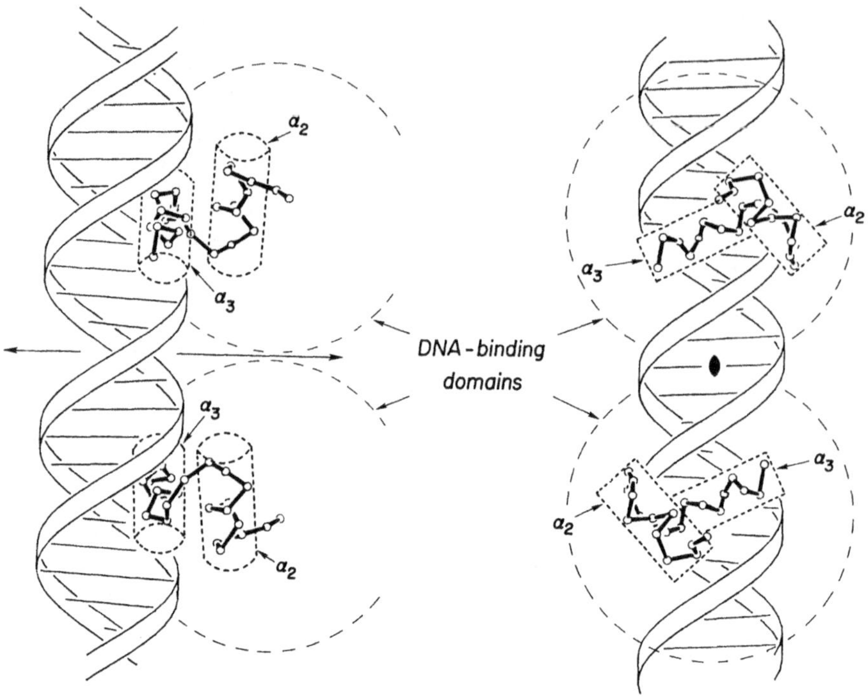

FIG. 4. The figure illustrates the general nature of the interaction presumed to occur in many DNA-regulatory proteins between a common α_2-α_3 helical unit and right-handed B-form DNA. At left is a "side view" with the two-fold axis of symmetry (arrowed) extending from left to right. On the right the view is "face on" (after Ohlendorf *et al.*, 1983b).

though not identical. The α_3 helix occupies the major groove of the DNA with its amino acid side chains positioned so as to make sequence-specific interactions with the exposed parts of the DNA base pairs. Side chains of the α_2 helix are also presumed to contact the DNA, these interactions being primarily to the phosphate backbone.

It is reasonable to anticipate that similar modes of DNA binding will be found for a number of other gene regulatory proteins whose sequences have been shown to be homologous with those included in Fig. 3. DNase foot-printing and chemical protection experiments have estimated the sizes of the recognition sites for such proteins to be about two turns of the DNA helix (usually 15 to 20 nucleotides). This is the DNA length which would be expected to be covered by the DNA recognition super-secondary structure.

Although Cro and λ repressor bind to the same sites on the DNA (albeit with different relative affinities), the amino acid sequences of their α_3 " recognition helices " are substantially different (Fig. 3). In contrast, the comparison of the three-dimensional structures of the two proteins show that the backbone conformations of their respective α_2-α_3 units are practically identical. However, the structural comparison also shows that there are stereochemical restrictions that prevent these units binding to the DNA in exactly the same way. Cro and λ repressor apparently use a non-identical set of amino acid side chains organized in a somewhat different spatial arrangement to recognize the same DNA sequence (Ohlendorf *et al.*, 1983a). This suggests that there is not a simple one-on-one recognition code between amino acids and bases.

8. Electrostatic binding

The phosphate backbone of DNA is highly charged and it is to be anticipated that the interaction between proteins and DNA will have a significant ionic component.

For CAP and Cro, the known three-dimensional structures have been used to calculate the electrostatic potential surface surrounding the respective proteins. The calculated potential for Cro is shown in Fig. 5 (Ohlendorf *et al.*, 1983). There is a dominant, elongated positive region that coincides remarkably well, both in position and orientation, with the presumed DNA binding site. For CAP there is also strong calculated positive potential in the presumed DNA binding region. The distribution seems to be more consistent with CAP binding to right-handed rather than left-handed DNA (Steitz *et al.*, 1983b).

9. Recognition of specific sites

The above models for DNA-protein interaction are based in large part on an overall structural complementarity between the protein and the DNA. An underlying assumption of such models is that the DNA does not change its conformation very much when the protein is bound. This has been experimentally verified for *lac* repressor, λ repressor and CAP. The complementarity between the shape of the protein and the shape of the DNA is

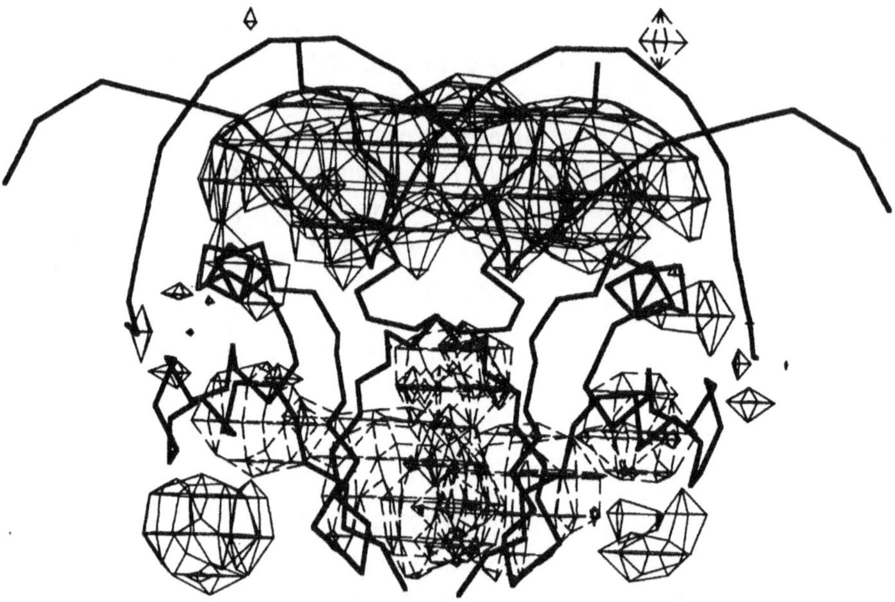

FIG. 5. Calculated electrostatic potential energy surface for a dimmer of Cro. Contours are drawn at intervals of 1.5 kT with positive potential energy shown as solid lines and negative as broken lines. The backbones of Cro (below) and DNA (above) are superimposed as solid thick lines (after Ohlendorf *et al.*, 1983c).

also consistent with the observation that these proteins bind to non-specific sites on the DNA ($K_D \sim 10^{-5}$ -10^{-8}M) in addition to their specific sites ($K_D \sim 10^{-10}$ -10^{-13}M).

In order to obtain an insight into sequence-specific recognition, model building and refinement were used to develop a detailed model for the complexes between Cro and DNA (Ohlendorf *et al.*, 1982). This model is consistent with the know affinities of Cro for its six binding sites on λ DNA and for mutant sites as well. The model suggests that the recognition of a specific base sequence on the DNA is due, in large part, to a complementary network of hydrogen bonds between amino acid side chains of the protein and DNA base-pair atoms exposd within the grooves of the DNA.

The hydrogen bond network that is presumed to exist between Cro and its tightest known binding site, 0_R3, is shown schematically in Fig. 6. In the figure, the successive base pairs are imagined to be seen edge-on, with all the possible hydrogen bond acceptor and donor atoms indicated. Atoms that do not hydrogen bond to the protein are presumed to remain hydrogen bonded to solvent. One striking feature of the model is the multiple hydrogen bonding of the amino acid side chains, for example, Arg 38, Lys 32, Gln 27 and even Ser 28 (Fig. 6). Such bi- and multi-dentate interactions provide a clear rationale for enhancing the specificity of DNA-protein recognition. Also, as has been shown for *lac* repressor, hydrophobic interactions can play an important part in recognition.

While protein-DNA interactions of the sort shown in Figure 6 are presumed to be responsible for recognition of a specific base sequence, it is understood that the overall energy of interaction of the complex comes primarily from interactions with the DNA that do not depend on the base sequence, including the ionic interactions suggested by Fig. 5.

10. Principal features of DNA recognition

Taken together, the structural studies of Cro, CAP and λ repressor suggest the following principal features for DNA-protein recognition (Matthews *et al.*, 1983; Ohlendorf *et al.*, 1982; Steitz *et al.*, 1983a; Pabo & Sauer, 1984):

a) The (approximate) two-fold symmetry of the DNA binding sites is matched by two-fold symmetry of the protein.

b) A pair of two-fold-related α-helices bind within successive major grooves of the DNA.

c) A number of proteins that recognize specific sites on DNA contain a similar two-helical DNA-binding structure.

d) Recognition of specific binding sites on DNA is accomplished, in large part, by a network of multiple hydrogen bonds

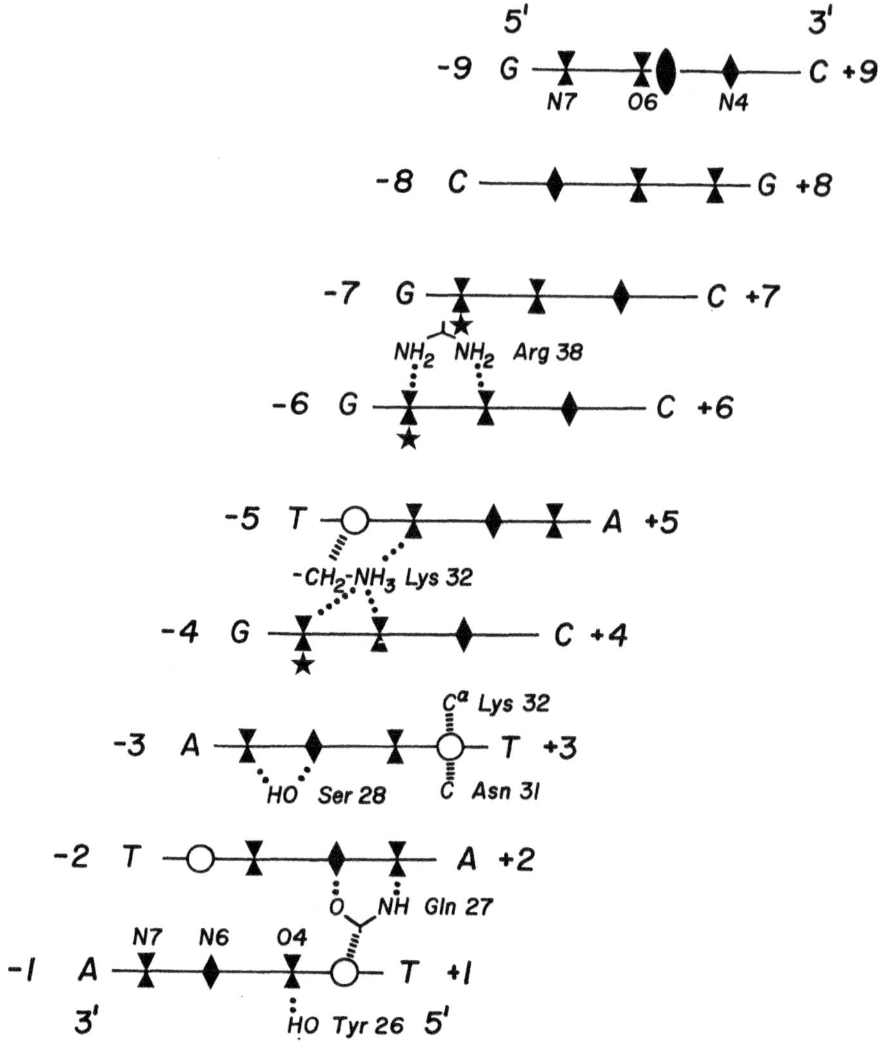

FIG. 6. Schematic representation of the presumed sequence-specific interactions between Cro and the parts of the base-pairs exposed within the major groove of the DNA. The direction of view is imagined to be directly into the major groove of the DNA with the base-pairs seen edge-on. The dyad symbol within the topmost base pair indicates the center of the overall 17-base-pair binding region. The symbols are as follows: ⧗, hydrogen bond acceptor; ◆, hydrogen bond donor; ○, methyl group of thymine; ★, guanine N7 which is protected from methylation when Cro is bound. Presumed hydrogen bonds between Cro side chains and the bases are indicated (lllll). Apparent der Waals contacts between Cro and the thymine methyl groups are not shown (after Ohlendorf *et al.*, 1982).

between side chains of the protein and parts of the DNA base pairs exposed within the grooves of the DNA.

e) A specific site on DNA does not need to change its conformation in order to be recognized.

f) Flexibility of DNA-binding proteins may be important in finding the correct site on the DNA and binding to it.

11. Evidence in support of the structural models

Considerable evidence, some very recent, provides support for the general features of DNA-protein recognition as summarized in the previous section.

a) DNA protection and modification experiments of Johnson *et al.* (1981) and Pabo *et al.* (1982).

b) Lysine protection and carboxypeptidase digestion experiments for Cro (personal communication from Y. Takeda).

c) NMR studies of Cro (Arndt *et al.*, 1983) and *lac* repressor (Arndt et al., 1982; Zuiderweg *et al.*, 1983).

d) Mutants of λ repressor, Cro and CAP that lack the ability to bind to DNA (Hecht *et al.*, 1983; Nelson *et al.*, 1983; Pabo & Sauer, 1984; R.T. Sauer, personal communication; R. Ebright, personal communication).

Presumably the best way to confirm the detailed basis of DNA-protein recognition will be to determine the structures of specific DNA-protein complexes. Suitable crystals of such complexes have been obtained in several laboratories and detailed structural analyses are underway (Anderson *et al.*, 1983; personal communications from J. Rosenberg, M. Ptashne and S. Harrison).

Acknowledgements

This work was supported in part by grants from the NIH (GM20066), the M.J. Murdock Charitable Trust and the NSF (PCM-8014311). The structural studies of Cro protein included here were in collaboration with Dr. W. F. Anderson, D. H. Ohlendorf and Y. Takeda.

REFERENCES

Anderson, W.F., Cygler, M., Vandoselaar, M., Ohlendorf, D.H., Matthews, B.W., Kim, J. and Takeda, Y., 1983, J. Mol. Biol. *168*, 903.

Anderson, W.F., Ohlendorf, D.H., Takeda, Y. and Matthews, B.W., 1981, Nature *290*, 654.

Anderson, W.F., Takeda, Y., Ohlendorf, D.H. and Matthews, B.W., 1982, J. Mol. Biol. *159*, 745.

Arndt, K.T., Boschelli, F., Cook, J., Takeda, Y., Tecza, E. and Lu, P., 1983, J. Biol. Chem. *258*, 4177.

Arndt, K., Nick, H., Boschelli, F., Lu, P. and Sadler, J., 1982, J. Mol. Biol. *161*, 439

de Crombrugghe, B., Busby, S. and Buc, H., 1984, In "Biological Regulation and Development," Vol. III-B, K. Yamamota, ed., in press.

Hecht, M.H., Nelson, H.C.M. and Sauer, R.T., 1983, Proc. Natl. Acad. Sci. USA *80*, 2676.

Johnson, A.D., Poteete, A.R., Lauer, G., Sauer, R.T., Ackers, G.K. and Ptashne, M., 1981, Nature *294*, 217.

Kolb. A. and Buc, H., 1982, Nucl. Acids Res. *10*, 473.

Lewis, M., Jeffrey, A., Ladner, R., Ptashne, M., Wang, J. and Pabo, C.O., 1983, Cold Spring Harbor Symp. Quant. Biol. *47*, 435.

Matthews, B.W., Ohlendorf, D.H., Anderson, W.F. and Takeda, Y., 1982, Proc. Natl. Acad. Sci. USA *79*, 1428.

Matthews, B.W., Ohlendorf, D.H., Anderson, W.F., Fisher, R.G. and Takeda, Y., 1983, Cold Spring Harbor Symp. Quant. Biol. *47*, 427.

McKay, D.B. and Steitz, T.A., 1981, Nature *290*, 744.

McKay, D.B., Weber, I.T. and Steitz, T.A., 1982, J. Biol. Chem. *257*, 9518.

Nelson, H.C.M., Hecht, M.H. and Sauer, R.T., 1983, Cold Spring Harbor Symp. Quant. Biol. *47*, 441.

Ohlendorf, D.H., Anderson, W.F., Fisher, R.G., Takeda, Y. and Matthews, B.W., 1982, Nature *298*, 718.

Ohlendorf, D.H., Anderson, W.F., Lewis, M., Pabo, C.O. and Matthews, B.W., 1983a, J. Mol. Biol. *169*, 757.

Ohlendorf, D.H., Anderson, W.F. and Matthews, B.W., 1983b, J. Molec. Evol. *19*, 109.

Ohlendorf, D.H., Anderson, W.F., Takeda, Y. and Matthews, B.W., 1983c, J. Biomolecular Structure and Dynamics I, 553.

Ohlendorf, D.H. and Matthews, B.W., 1983, Ann. Rev. Biophy. Bioen. *12*, 259.

Pabo, C.O., Krovatin, W., Jeffrey, A. and Sauer, R.T., 1982, Nature *298*, 441.

Pabo, C.O. and Lewis, M., 1982, Nature *298*, 443.

Pabo, C.O. and Sauer, R.T., 1984, Ann. Rev. Biochemistry, in press.

Sauer, R.T., Yocum, R.R., Doolittle, R.F., Lewis, M. and Pabo, C.O., 1982, Nature *298*, 447.

Steitz, T.A., Ohlendorf, D.H., McKay, D.B., Anderson, W.F. and Matthews, B.W., 1982, Proc. Natl. Acad. Sci. USA *79*, 3097.

Steitz, T.A., Weber, I.T., Ollis, D. and Brick, P., 1983a, J. Biomolecular Structure and Dynamics, in press.

Steitz, T.A., Weber, I.T. and Matthew, J.B., 1983b, Cold Spring Harbor Symp. Quant. Biol. *47*, 419.

Takeda, Y., Ohlendorf, D.H., Anderson, W.F. and Matthews, B.W., 1983, Science *221*, 1020.

Weber, I.T., McKay, D.B. and Steitz, T.A., 1982, Nucl. Acids. Res. *10*, 5085.

Zuiderweg, E.R.P., Kaptein, R. and Wuthrich, K., 1983, Proc. Natl. Acad. Sci. USA *80*, 5837.

PROTEIN-NUCLEIC ACID INTERACTIONS MODELS AND REALITY *

WOLFRAM SAENGER

Institut für Kristallographie, Freie Universität Berlin
Takustr. 6, 1000 Berlin 33, FRG

Abstract

The simple schemes which can be drawn on a theoretical basis for protein-nucleic acid interactions are only rarely observed in reality. Complexes between individual amino acids and nucleotides or their constituents are hard to crystallize and even then, interactions are often not as expected. If nucleotides or single stranded polynucleotides bind to proteins, they unfold with torsion angle $\gamma(O_{5'}\text{-}C_{5'}\text{-}C_{4'}\text{-}C_{3'})$ in *-sc* or *ap* range instead of the usually preferred *+sc*. In this form, phosphate-sugar-base can interact better with protein active sites. On the other hand, double stranded DNA maintains its helical structure and proteins associate by inserting α-helices in major grooves as suggested by model building studies with repressor molecules.

Introduction

In biology, the most important molecules are nucleic acids and proteins. Whereas the latter are composed of 20 different units, occur in a large variety of shapes and sizes and take part in essentially all processes associated with life, the nucleic acids are less versatile. They consist of only 4 units, play a role either as single strands or as double helices, and are found in two varieties, RNA and DNA. Nevertheless, they have a crucial function because DNA as site of genetic information and RNA as messenger and as transfer RNA are essential for the expression of genetic information. This expression comprises DNA replication, DNA tran-

* To be presented at the Pontifical Academy of Sciences, Meeting of the Study Group on Specificity in Biological Interactions, City of Vatican, November 9-11th, 1983.

scription into complementary messenger RNA and translation of messenger RNA nucleotide sequence into protein sequence, this step being mediated by transfer RNA.

These events are so complex that they are only partly understood. In any case, they depend on protein-nucleic acid recognition which is also essential in other biological contexts where nucleic acids or their constituents interact with proteins or other molecules and is therefore of general importance.

During the past two decades, many studies have been reported which are of relevance for the understanding of protein-nucleic acid interactions. These include model systems employing amino acids and nucleotides or only bases, more realistic cases where polypeptides are investigated, " real " systems including protein-nucleotide and protein-nucleic acid complexes, and model building studies where DNA was fitted to known protein structures. It is the aim of this paper to give an assessment of the present state of the art.

1. Model considerations

At the outset, let us investigate what parts of amino acids and of nucleotides and what kinds of interactions are relevant if protein-nucleic acid interactions are considered.

Nucleotides are composed of the three moieties phosphate-sugar-base which can interact with a protein. In general, contacts to phosphate and sugar are said to be unspecific because these groups belong to the " inert " backbone whereas contacts to bases are said to be specific. This notion, however, is questionable because in most instances a protein interacts with base, sugar and phosphate simultaneously and proper alignment is required in order to produce a biologically active complex. On the other hand, amino acid side chains are taken to be specific sites in protein structure. Looking at the protein as a whole, however, we know that its function is intimately correlated with the three-dimensional folding (Schulz & Schirmer, 1979) and therefore the " inert " backbone conveys specificity which can be even more important than individual side chains. In essence, if considering specific protein-nucleic acid interactions, one should not only look at amino acid side chains and nucleotide bases, but be aware of the intricate folding of the " inert " backbone in proteins and in nucleic acids which can as well express specificity. As we shall see later, backbone interactions can even be more specific than those involving side chains.

On a theoretical basis, many interactions between proteins and nucleic acids can be formulated. They comprise:

1) salt bridges between charged groups, e.g. phosphate and lysine, arginine, histidine;

2) hydrogen bonds between functional groups, i.e. amino-, imino-, keto-, hydroxyl groups. These can also involve charged phosphate, ammonium or guanidinium groups;

3) hydrophobic interactions between non-polar groups;

4) stacking of aromatic amino acid side chains with bases.

Some of these interactions are presented in Figure 1. In all cases " cyclic " hydrogen bonds or salt bridges are drawn because they appear to be more preferred than single, individual interactions as has been demonstrated for base-base association (Kyogoku et al., 1976). Of course, many more examples can be thought of, especially if dipeptides or longer oligopeptides are folded properly to allow for multiple contacts with bases or if besides bases, sugar and phosphate moieties are included as well. Obviously, there are many possibilities and theoretical predictions are bound to fail because the system is too complex.

2. Larger model systems studied theoretically

The situation is different if well-defined secondary or tertiary structures of nucleic acids and of proteins serve as matrix to which the other partner is fitted. If we consider a Watson-Crick double helix, specific recognition of base sequence is only possible in major and minor groove sites if sugar-phosphate backbone is

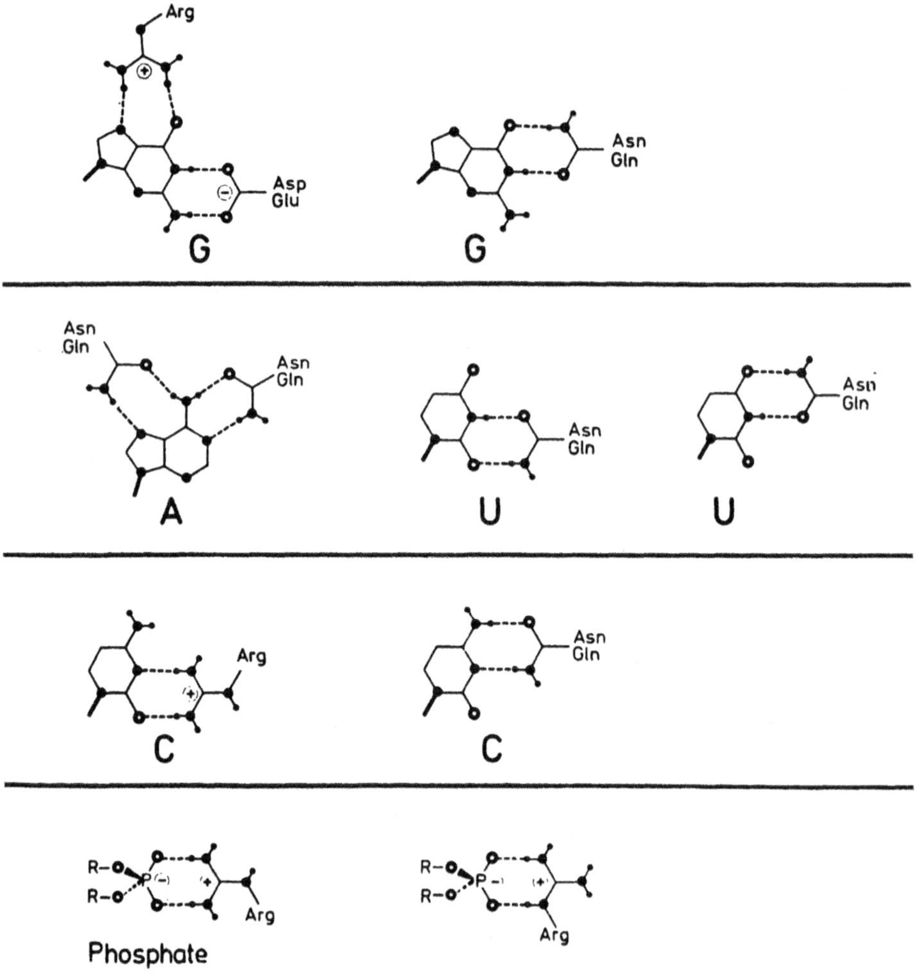

FIG. 1. Schematic representation of interactions between amino acid side chains and nucleic acid bases, and phosphate groups. Taken from (Saenger, 1983).

taken as regular, " inert " part of the helix [1]. In a theoretical study, it was proposed that the major groove should be the preferred target site because here distinction between A/T and G/C base-pairs is far better than in the minor groove, (Seeman *et al.*, 1976).

[1] This is probably incorrect, because we know from the published oligonucleotide crystal structures that base sequence is to some extent correlated with sugar pucker (R.E. Dickerson, 1983).

We can go further and try to model-fit nucleic acid double helices and protein β-pleated sheet and α-helix structures. In protein structures, antiparallel β-sheets generally display a right-handed twist which corresponds closely to the curvature of grooves in A-RNA or in A-DNA (Chothia, 1973). It was indeed possible to construct a model complex where a parallel or antiparallel β-sheet is inserted in the minor groove, with peptide N-H acting as hydrogen bond donors to A-DNA (or A-RNA) 03′-atoms. Even B-DNA is able to accommodate an antiparallel β-sheet if certain steric requirements are observed (Church *et al.*, 1977), and we shall see later that in *cro* repressor such a scheme appears to be verified. An important point is, that the dyad symmetry inherent in the DNA (or RNA) double helix is also present in antiparallel β-sheet, so that the overall symmetry is maintained after complex formation. This is not the case for the parallel β-sheet, but two such β-sheets arranged in antiparallel orientation in consecutive minor grooves could satisfy two-fold symmetry requirements. In this context, it is of interest that most of the proteins interacting with the DNA double helix are dimeric so that the dyad symmetry of the overall complex can be preserved (Sobell, 1976).

3. Experimental model systems for nucleic acids and polypeptides or their constituents

There have been numerous attempts to study protein-nucleic acid interactions by looking at individual nucleotides and amino acids using spectroscopic and crystallographic methods. Because the latter provide for a clearly defined picture of the mutual interactions between the two partners, results obtained crystallographically are discussed here. In detail, it was observed:

A) Interactions between bases and carboxylic acids occur if, for instance, N_3 in cytosine is protonated and the negatively charged carboxylate accepts two H-bonds from N_4-H and N_3^+-H (Ohki *et al.*, 1980). This scheme is not invariant, however, because in 5-bromocytosine, the bases are dimerized cyclically via two N_4-H···N_3 bonds. Therefore, N_3 cannot be protonated;

the partner glutamic acid does not dissociate and hydrogen bonds (carboxyl) O-H···O_2 (cytosine) (Ohki *et al.*, 1977a). A comparable case is also found in the self-complex formed by thymidine-5'-carboxylic acid (Suck *et al.*, 1974).

B) In solution, tryptophan forms strong charge-transfer complexes with bases, as indicated by spectroscopic methods (Morita 1975). Surprisingly, however, if cocrystallized with 9-methyladenine, no stacking is observed and the indole N-H forms only hydrogen bonds with adenine N_3 and N_7 (Ohki *et al.*, 1977b).

C) Stacking can, however, be induced if bases are suitably modified as in case of 1,9-dimethyladenine, nicotinamide and isoalloxazine which were cocrystallized with tyramine, an analog of tyrosine (Ishida *et al.*, 1980). Stacking between naturally occurring constituents was only observed in one example, uracil and phenylalanine. The two partners are, however, covalently linked to form a nucleoside peptide 5-[N-(L-phenylalanyl)-amino]uridine with the amino acid attached at the 5-position of uracil, and therefore the situation is again artificial (Berman *et. al.*, 1973).

D) In the crystal structure, the above-mentioned molecule also displays base-peptide hydrogen bonds. Similar interaction are observed as well as in another " hybrid " molecule, 3-(adenin-9-yl) propionamide with propionamide bound to N_3 of adenine, (Takimoto *et al.*, 1981). It should be noted, however, that both cases are no cocrystallizations and that the partners, if not linked covalently, might have crystallized differently, if not separately.

E) In addition, there are several instances of " hybrid " protein-nucleic acid constituents which crystallized without any interaction between the partners and display only contacts protein-protein and nucleic acid-nucleic acid Ohki *et al.*, 1977c; Narayanan and Berman, 1977). This means that there is no tendency to organize in a pattern where protein-nucleic acid contacts are possible, although the two different moieties coexist in the dame crystal.

In conclusion, all these examples suggest that complexes between constituents of proteins and nucleic acids do not form readily, that patterns of interactions cannot be predicted and that, even if they are successfully cocrystallized, there might be no inter-partner contacts. Needless to say that there have been numerous frustrating crystallization experiments in many laboratories (including our own) which are not reported in literature.

The situation is more clear if interactions between charged groups like phosphate and primary ammonium (= lysine) and guanidinium groups (= arginine) are studied. Phosphate or diethylphosphate were cocrystallized with arginine, propylguanidine, putrescine, and complexes with structures sketched in Figure 2 were obtained (Furberg and Sølbakk, 1974). They suggest how the geometry of complexes between DNA/RNA and charged amino acid side chains can look like and they were used as a basis in the investigations of complexes between double helical nucleic acids and synthetically available polylysine and polyarginine which may be considered as models for protamines.

Polylysine and polyarginine complex with DNA in a cooperative and irreversible mode (von Hippel and McGhee, 1972). The positively charged polypeptide side chains neutralize negative charges on phosphates and, at appropriate concentrations, DNA is condensed into rod-like structures and precipitates. In all cases, the melting temperatures T_m of the DNA or RNA double helices are raised after complex formation because negative charges are now shielded and repulsion is diminished. The two polypeptides exert a different influence on DNA structure, polyarginine preserving B-DNA under all humidity (or salt) conditions whereas polylysine induces some conformational change attributed to A, C and distorted B-forms (Suau and Subirana, 1977; Prescott *et al.*, 1976).

Aromatic amino acids are able to intercalate between base pairs, a process called " bookmarking " (Gabbay *et al.*, 1973; Hélène and Maurizot, 1981). It occurs easily if the aromatic residue X is flanked on one or on both sides by basic amino acids, e.g. Lys-X-Lys. NMR studies revealed that binding takes place in two steps, one fast (on the periphery) and the other slow (intercalation). There is even modest selectivity for intercalation in DNA, Trp > Phe > Tyr > Ileu.

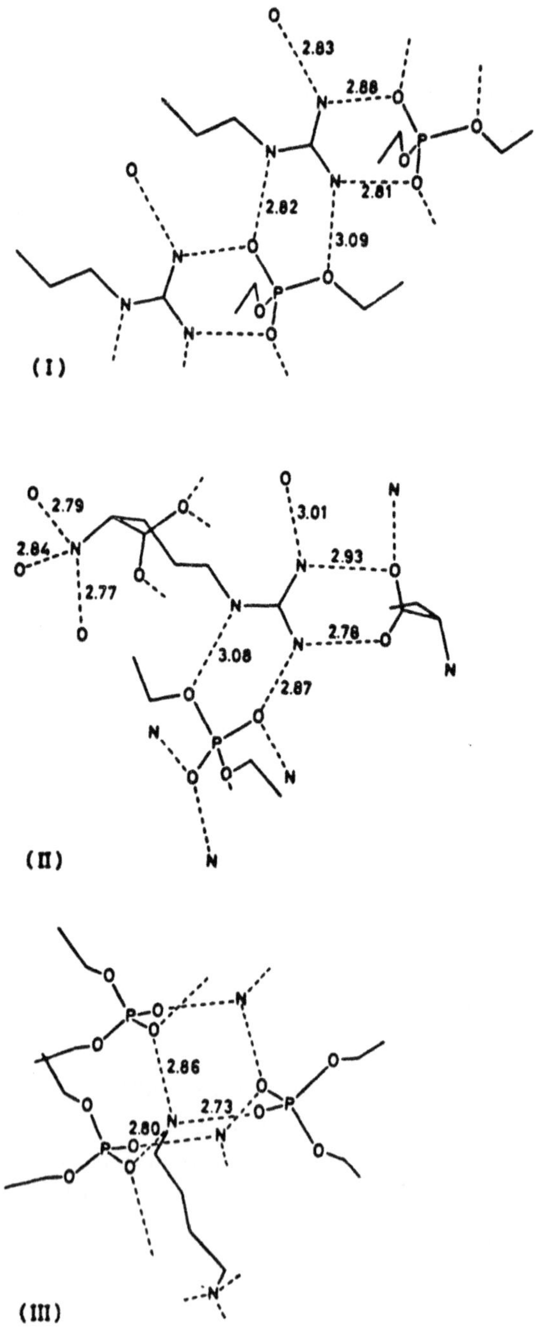

FIG. 2. Interactions between diethyl phosphate and propylguanidine (I), arginine (II) and putrescine (III). Taken from (Furberg and Sølbakk, 1974).

Besides these complexes formed between DNA, RNA and oligo- or polypeptides, interaction between protein α-helix and transfer RNA (tRNA) was observed when crystals of phenylalanyl-specific $tRNA^{Phe}$ were soaked with protamine AI (Warrant and Kim, 1978). This is a polypeptide mixture of 4200 daltons on average and composed primarily of Lys and Arg. Protamine adopts a helical structure, i.e. its positively charged groups are arranged like bristles of a brush and interact with phosphate groups in minor (wide and shallow) grooves of tRNA. This information was extended to DNA and models were put forward where a protein α-helix binds to the major groove of B-DNA although binding to minor groove sites could not be ruled out. Recalling that the major groove is better suited for specific sequence recognition than the minor groove (as outlined above), major groove binding of α-helix to B-DNA should be preferred and is probably crucial in repressor-DNA binding as outlined further below.

4. Examples for specific protein-nucleic acid interactions

We shall now discuss systems where the topology of both partners or at least of one of the (macromolecular) partners is known from detailed X-ray crystal studies so that the mutual recognition can be inferred with some safety. These systems are grouped as follows:

A) Protein interacting with nucleotides and coenzymes;

B) Complex between RNA single strand and protein in tobacco mosaic virus (TMV);

C) Model building of DNA single strand to crystallographically derived structure of phage fd gene 5 protein;

D) Fitting of B-DNA to crystal structure models of netropsin and of *cro* and λ repressors and catabolite activator protein (CAP).

Since we are now dealing in more detail with structures of nucleotides and of single and double stranded nucleic acids, let us first consider what the most preferred conformational parameters of a nucleotide are (Saenger, 1983.) On the basis of spectroscopic, theoretical and crystallographic data it was concluded that in a 5′-nucleotide which constitutes the basic building block of ADP, ATP, coenzymes and nucleic acids,

(i) The base favours the *anti* orientation, i.e. the bulk of the base (O_2 in pyrimidine and C_4 in purine) is directed away from the sugar, with torsion angle χ ($O_{4'}$-$C_{1'}$-N_1-C_2 or $O_{4'}$-$C_{1'}$-N_9-C_4) at $-90^o \leq \chi \leq -160^o$. The *syn* form with $0^o \leq \chi \leq 100^o$ is sometimes observed for purines, very rarely for pyrimidines.

(ii) The sugar puckering mode is $C_{2'}$-*endo* or $C_{3'}$-*endo*;

(iii) Torsion angle γ ($O_{5'}$-$C_{5'}$-$C_{4'}$-$C_{3'}$) describing orientation about $C_{4'}$-$C_{5'}$ bond is $+sc$ ($\sim 60^o$)[2];

(iv) C-O torsion angles β (P-$O_{5'}$-$C_{5'}$-$C_{4'}$) and ε($C_{4'}$-$C_{3'}$-$O_{3'}$P) are both *ap* ($\sim 180^o$);

(v) In helical polynucleotides with right-handed screw sense, P-O torsion angles α and ζ are in *-sc* range ($\sim 300^o$).

These structural features are maintained for a nucleotide in solution, in the crystalline state or if incorporated in DNA and RNA double helices, i.e. the nucleotide can be treated as a " rigid " unit (Sundaralingam and Westhof, 1979). It is stabilized by a (base)C-H...$O_{5'}$ interaction (Taylor and Kennard, 1982) which pulls $O_{5'}$ " over " the sugar and gives the nucleotide a compact form. If a nucleotide as such or as constituent of a coenzyme or single stranded RNA or DNA interacts with a protein, a conformational change occurs in most cases which rotates $O_{5'}$ away from the sugar (γ in *-sc* or *ap*) and thus produces a nucleotide in extended conformation (Figure 3). Now all parts of the nucleotide, i.e. phosphate-sugar-base, are exposed and can interact much better with the protein surface than the nucleotide in its " rigid " form. Let us discuss in more detail how these interactions look like.

Proteins Interacting with Nucleotides and Coenzymes

In Table 1, conformations of nucleotides bound to several different proteins are gathered. There are only two examples, the binding of GDP to elongation factor Tu and NAD^+ bound to lobster glyceraldehydephosphate dehydrogenase where γ is in

[2] *sc* = syn-clinal ($\sim 60^o$) (*gauche*), *ac* = anticlinal ($\sim 120^o$), *ap* = antiperiplanar (*trans*)

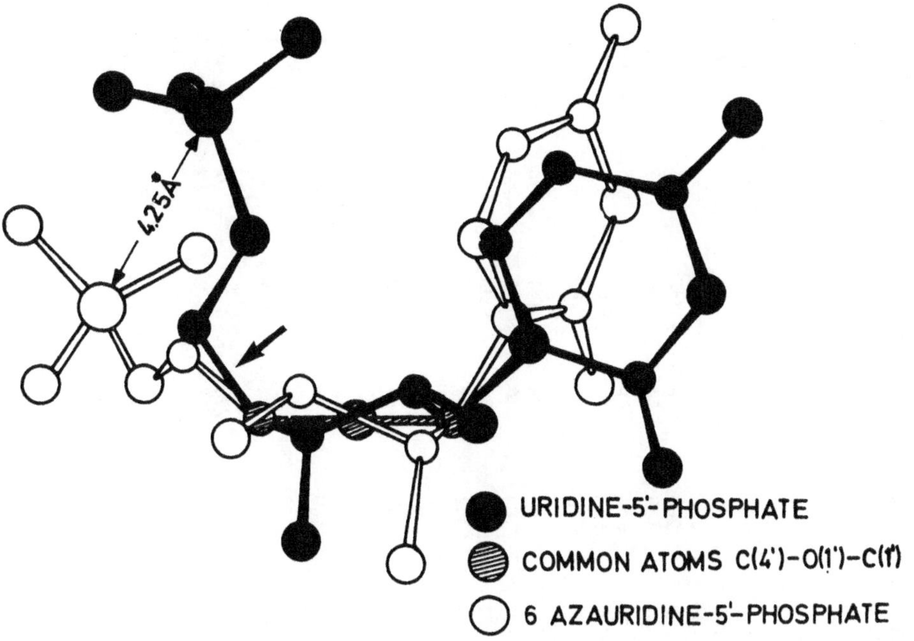

FIG. 3. Conformational change of a 5'-nucleotide with torsion angle γ ($O_{5'}$-$C_{5'}$-$C_{4'}$-$C_{3'}$) in the usual + *sc* form (structure with filled bonds = uridine-5'-phosphate) and in *ap* (open bonds, 6-azauridine-5'-phosphate). Note differences in position of the phosphate group relative to the nucleoside moiety. Central bond $C_{5'}$-$C_{4'}$ of torsion angle γ marked by arrow. Taken from (Suck *et al.*, 1979).

the " rigid " +*sc* range. In all other complexes, γ is in the unusual -*sc* or *ap* forms, consistent with maximum extension and optimum interactions with the protein surface.

If we compare the arrangements in the known protein-nucleic acid interactions, can we deduce any rules or recognize any recurrent motifs?

Macroscopically, many of the nucleotide binding proteins are composed of a certain sequence of α-helices and β-pleated sheets in (βαβ) topology. This kind of tertiary structure was first discovered by Rossmann in dehydrogenases. It seems to be a general feature that has been conserved evolutionarily and has been found in several other proteins as well (Rossmann *et al.*, 1975).

On a more detailed level, we can look at individual contacts between nucleotides and proteins. The question arises whether

TABLE I. *Conformations of Nucleotides when Associated with Binding Sites of Proteins. Taken from* (*Saenger*, 1983)

Protein	Nucleotide [a]		Torsion Angles (°) α	β	γ	ξ [b]	χ	sugar pucker
Lactate dehydrogenase	5′-AMP			177	178		—91	C_3'-*endo*
Lactate dehydrogenase	ADP		165	168	—168		—92	C_3'-*endo*
Lactate dehydrogenase	NAD^+	A	103	184	—72	156	—96	C_3'-*endo*
		N	109	178	—87	37	—90	C_3'-*endo*
Lactate dehydrogenase	NAD^+pyruvate	A	109	204	—78	156	—91	C_3'-*endo*
		N	87	189	—53	49	—74	C_3'-*endo*
Glyceraldehyde-3-phosphate dehydrogenase	NAD^+ green subunit	A	+ac	+ac	ap	ap	*syn*	C_3'-*endo*
		N	+sc	+ac	+sc	+ ac	*syn*	C_3'-*endo*
	red subunit	A	+ac	ap	—sc	ap	*anti*	C_2'-*endo*
		N	+sc	+sc	+sc	+ ac	*syn*	C_2'-*endo*
Malate dehydrogenase	NAD^+	A	58	193	—78	194	—70	C_3'-*endo*
		N	147	160	—40	88	—48	C_2'-*endo*
Liver alcohol dehydrogenase	ADP-ribose	A	177	167	159	39	—102	C_3'-*endo*
		R	51	152	154	112	—	
	8-Br-ADP-ribose	A	165	167	171	39	—80	C_2'-*endo*
		R	51	152	154	112	—	
Dehydrofolate reductase	NADPH	A	77	122	161	—112	—93	C_3'-*endo*
		N	—74	—156	—68	—74	—39	C_2'-*endo*
Tyrosyl tRNA synthetase	Tyrosil-5′-adenylate		170	140	—sc	30	*anti*	C_3'-*exo*
Elongation factor Tu	GDP		60	160	+sc	—	*anti*	C_3'-*endo*
Hexokinase	8-Br-ATP Cobalt		—	—	ap	—	*anti*	C_2'-*endo*
Ribonuclease S	UPcA	U	—	—	—	+ac	*anti*	C_3'-*endo*
		A	+sc	ap	ap	—	*anti*	C_3'-*endo*
Ribonuclease S	Cp (2′, 5′) A	C	—	—	270	—	—107	C_3'-*endo*
		A	—42	173	153	121	—82	C_3'-*endo*
Ribonuclease T_1	2′ GMP				+ ac		*syn*	C_3'-*endo*
Staphylococcal nuclease	p(dT)p				ap		*anti*	C_2'-*endo*
Tobacco mosaic virus	RNA (there nucleo-	1	145	220	170	70	—70	C_2'-*endo*
	tides per protein	2	40	185	195	275	—10	C_2'-*endo*
	subunit	3	150	135	270	290	—75	C_2'-*endo*

[a] A, C, N, R indicate adenosine, cytidine, nicotinamide ribose and ribose moieties.
[b] In NAD^+, NADPH and ADP, ξ denotes torsion angle O_5'-P-O-P.

there are some general rules and, in particular, are predictions possible as derived in Figure 1? If we look at Table 2 where protein-nucleic acid interactions found in several X-ray crystal structures are summarized, it is clear that this simple view cannot be true. There are many interactions between nucleotide phosphate, sugar, base moieties and protein main chain as well as side chain atoms which are unpredictable and cannot be assigned to a general scheme. The only more "logical" contacts in Table 2 are the pyrophosphate-Arg salt bridge in the complex between dihydrofolate reductase and NADPH (Matthews *et al.*, 1979) which could have been proposed on the basis of Figure 3, whereas the salt bridge between RNase S and the phosphate of Cp(2′, 5′)A involves the unpredictable combination of Glu11, His12, and peptide of Phe120 (Wodak *et al.*, 1977).

Moreover, there are "hydrophobic pockets" in protein tertiary structures which can accommodate bases. These pockets can be rather unspecific as, for instance, in case of the lactate dehydrogenase-NAD^+ pyruvate complex where adenine is not involved in any hydrogen bonding (Chandrasekhar *et al.*, 1973). A similar unspecific insertion is found in dihydrofolate reductase where adenine of NADPH is bound in an unspecific pocket in contrast to nicotinamide which is pocketed and recognized by hydrogen bonds to main chain CO and NH groups, Table 2 (Matthews *et al.*, 1979).

Staphylococcal nuclease is an enzyme that hydrolyzes RNA and DNA single strands at the $O_{5'}$-P bond and is not specific for bases. The latter are bound in a broad trough at the protein surface via hydrophobic forces and there are no hydrogen bonds involved in the recognition process. On the other hand, the phosphate group is very specifically and tightly bound by two arginines 35 and 87 in bidentate mode already discussed in the model structures in Figures 1 and 3; these interactions are augmented by coordination to Ca^{2+} ion and to a water molecule.

A completely different binding of the phosphate group is observed in ribonuclease A (RNase A) which claeves the $O_{3'}$-P bonds of single stranded RNA specifically at pyrimidine nucleotides via formation of a 2′, 3′-O-cyclophosphate intermediate. In con-

TABLE II. *Protein-Nucleic Acid Interactions in Two Complexes. Protein Side Chains are Designated Thr*64, *Ile*102 *etc., and Peptide Carbonyl as CO-Gly*17, *amide as NH-Ala*6 *etc. Salt Bridges and Hydrogen Bond Acceptors or Donors on the Nucleotide in Parentheses. Taken from* (*Saenger*, 1983)

	Nucleotide group	Salt Bridges	Bound to Protein via Hydrogen Bond	Hydrophobic
Dihydrofolate Reductase and NADPH	Adenine			Leu62, His64 Thr63, Ile102 His77, Asp78
	Adenine 2'-mononucleotide ribose	Arg 43 (phosphate)	Thr63 (phosphate) His64 (phosphate) Glu101 (O_3')	Ile102, Gly42
	Pyrophosphate	Arg44	NH-Thr63 NH-Ala 100 Thr45, Thr126	Gly99
	Nicontinamide mononucleotide ribose		CO-Gly17 (O_2') CO-His (O_3') Ser48 (O_2')	Ile13, Gly14
	Nicotinamide		NH-Ala6 (O_7) CO-Ala6 (N_7) CO-Ile13 (N_7)	Trp5, Ala6, Ile13 Leu19, Trp21 CO-Ala97
Ribonuclease S and Cp (2', 5') A	Cytidine		Thr45 (N_3)	Thr45, Phe120
	Ribose		Lys41 (O_4') His10 (O_5')	
	Phosphate		Gln11 (O) His12 (O) NH-Phe120 (O)	Phe120, His19
	Adenine		Glu111 (N_1) Gln69 (N_6) Asn71 (N_6)	Glu111, Ala 109
	Ribose		His119 (O_4')	His119

trast to staphylococcal nuclease, phosphate is anchored by His12, His119 and Glu11 in a rather " soft " mode without involvement of a metal ion, probably because the mechanism of phosphoester hydrolysis is completely different. Pyrimidine bases are recognized by main chain Thr45 NH···O_2(pyrimidine) and side chain Thr45-$O\gamma$-H···N_3(cytosine) and Thr45-$O\gamma$···H-N_3(uracil), i.e. hydroxyl $O\gamma$-H acts as H-bond donor and acceptor (Figure 4).

An even higher, more selective specificity is exerted by ribonuclease T_1 (RNase T_1) which hydrolyzes single stranded RNA exclusively at the $O_{3'}$-P bond of guanylic acid using a mechanism reminiscent of RNase A. In the complex RNase $T_1 \cdot 2'$-guanylic acid studied crystallographically, the 2′-phosphate group is hydrogen bonded to His40 and if the substrate with 3′-phosphate is bound, this hydrogen bond is probably broken and there is a hydrogen bond to His92 in adjacent position (Heinemann and Saenger, 1982). There is no apparent interaction between sugar and protein, but a number of different contacts between protein and guanine which can discriminate all other bases. It is surprising to find that there are none of the predictable interactions between guanine and amino acid side chains illustrated in Figure 1. Rather, main chain peptide groups form hydrogen bonds, viz. Asn43N-H···O_6(guanine) and (guanine)N_1-H···O(Asn44), (Figure 5). In addition, there is stacking of side chains of Tyr42 and Tyr45 with guanine sandwiched between these two groups.

Complex Between RNA Single Strand and Protein in Tobacco Mosaic Virus (TMV)

Tobacco mosaic virus (TMV) is a rod-like virus composed of 240 identical protein subunits, each of molecular weight 17,420 Dalton and arranged on a helix with 16 ⅓ subunits per turn. Along the " empty " interior of the helix, a single stranded RNA with 6400 nucleotides is wound. The individual protein subunits consist primarily of α-helical secondary structure and one of the helices formed by amino acids 108-129 interacts with three consecutive nucleotides, see Figure 6 (Holmes, 1980). The bases of these nucleotides embrace the α-helix and stick to it; nucleotide 2 is forced out on a larger radius because nucleotides 1 and

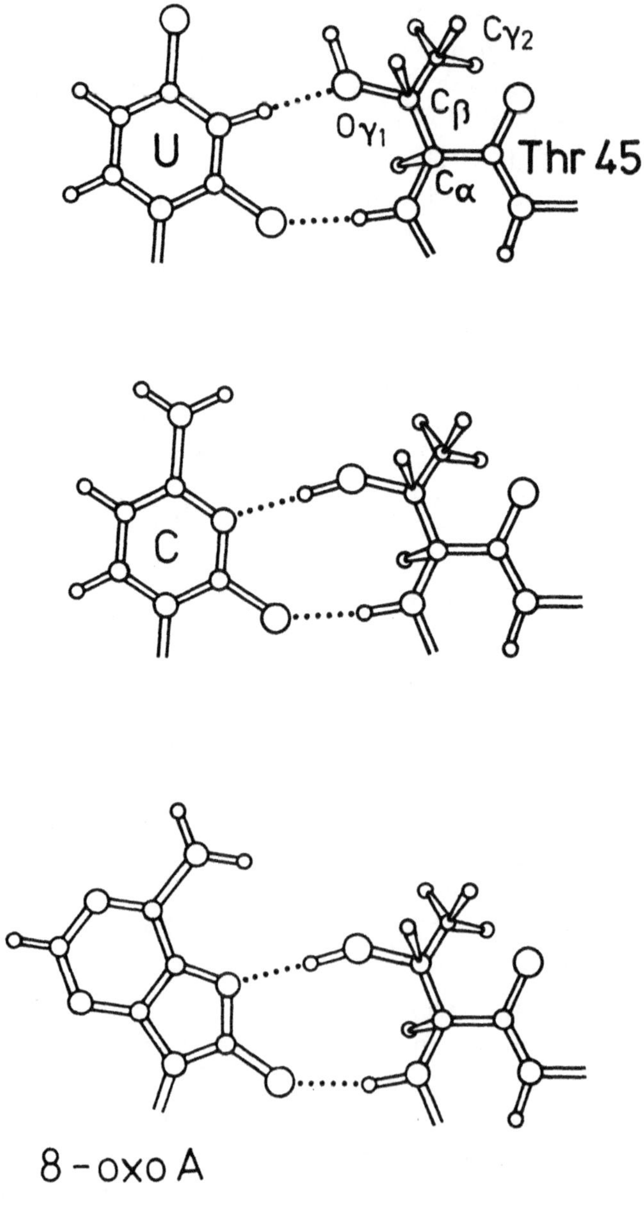

FIG. 4. Recognition between RNase A and pyrimidine base uracil and cytosine (top and center). In the bottom picture, interaction of 8-oxoadenosine (in *syn* form) and the protein active site is shown. In all cases, Thr 43 main chain NH and side chain $O_{\gamma}H$ act as hydrogen bond donors/acceptors. Of special interest is the dual function of $O_{\gamma}H$ which can act as donor (with cytosine) and as acceptor (with uracil). Taken from (Saenger, 1983).

3 are held so close together. All nucleotides are in extended conformation with torsion angle γ in *ap* and *-sc* range. The protein-base recognition has to be non-specific since RNA of " random " sequence binds with different triplets to the same protein subunit.

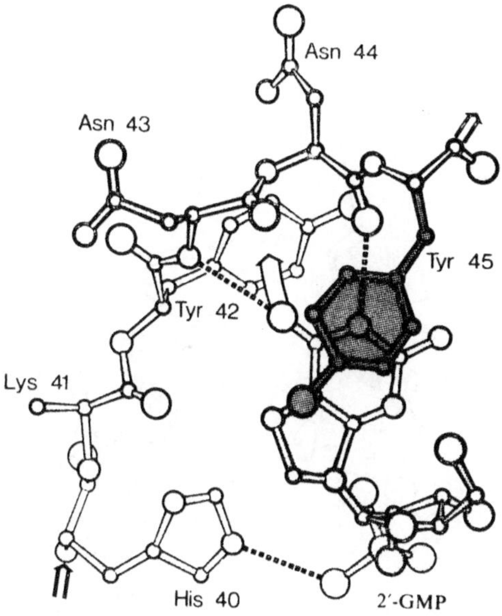

FIG. 5. Specific interactions of guanine with RNase T_1. Besides hydrogen bonds between guanine O_6 and N_1H, and main chain peptide groups NH and CO of Asn 43, the guanine system is sandwiched (stacked) between phenolic groups of Tyr 42 (in the rear) and of Tyr 45 (in front). Taken from (Heinemann & Saenger, 1982).

c) *Model Building of DNA Strand to Crystallographically Derived Structure of Phage fd Gene 5 Protein*

Expression of DNA is accompanied by unwinding of the double helix in order to facilitate recognition of DNA by several enzymes. There is a class of widely distributed proteins which serve to unwind and to stabilize the single stranded DNA. One of these proteins is the gene 5 product of phage fd, an 87 amino acids long polypeptide chain. It was crystallized and subjected to X-ray analysis (Brayer and McPherson, 1983).

The individual protein molecule is T-shaped and stabilized by β-pleated sheet structure. Two molecules form a dimer with the T-stems in antiparallel orientation and the T-bars (which are also antiparallel) form a shallow trough where, according to model building studies and NMR results, the DNA single strand is bound. As in the RNA··TMV complex, DNA-protein interactions have to be of general type because there is no sequence specificity. The contacts thus far identified for a pentanucleotide which just fills the trough are salt bridges between DNA phosphates and protein arginines and one lysine, as well as stacking interactions of all bases with three tyrosines and one phenylalanine. These interactions were model-built with only slight rearrangement of the protein side chains and are corroborated by spectroscopic data. The oligonucleotide fitted to the protein in an extended form, i.e. all torsion angles γ are not in the preferred $+sc$ orientation but rather $-sc$ or *ap*.

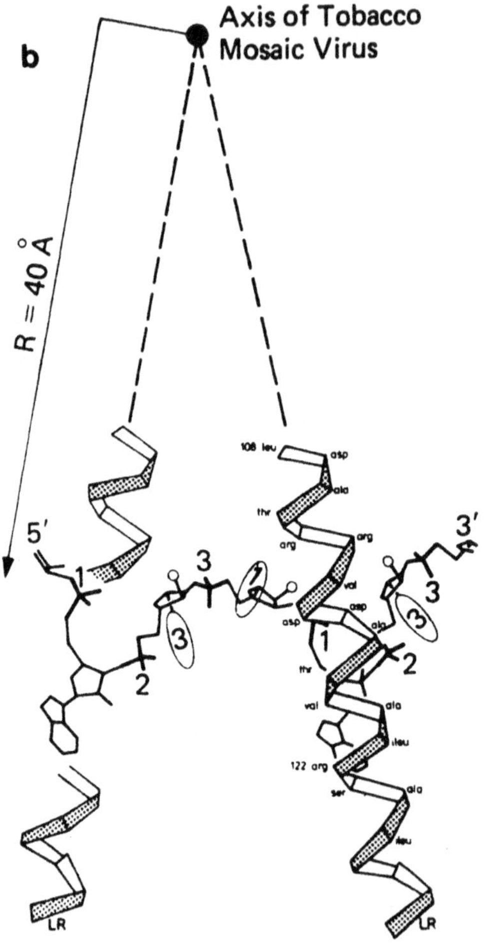

FIG. 6. Binding of an RNA single strand to protein in tobacco mosaic virus (TMV). In this virus, protein subunits are arranged helically around an axis (indicated by ●) and RNA screws at a radius of 40 Å. Shown are only two helices of the protein to which RNA in extended conformation is bound. Bases (drawn as ellipses numbered 1, 2, 3) are tightly attached to the α-helices; this binding has to be non-specific because RNA of "random" sequences is recognized by the TMV protein. Taken from (Saenger, 1983), redrawn from (Holems, 1980).

When binding to DNA double helix, gene 5 protein dimers (which also occur in solution) probably separate the antiparallel DNA strands initially and then aggregate linearly in a side-by-side manner. They build up a new helical arrangement with six gene 5 protein dimers per turn with 80-90Å pitch height and the DNA is inserted in a notch at a radius of about 23Å.

d) *Fitting of B-DNA to Crystal Structure Models of Netropsin and of* cro *and λ Repressors and Catabolite Activator Protein (CAP)*

The first " protein " to be considered here is netropsin. This antibiotic, in fact, contains peptide bonds but is not composed of amino acids. Its crystal structure revealed a bow-shaped molecule which could easily be fitted to the minor groove of B-DNA (Berman *et al.*, 1979) where it binds tightly, as has been demonstrated with other methods. The model suggests that interactions between netropsin and B-DNA comprise three netropsin N-H groups (as hydrogen bond donors) and O_2 of cytosine/thymine or N_3 of adenine/guanine which are in isomorphous locations in the Watson-Crick base pair. These groups belong to only one strand of the double helix, i.e. the netropsin molecule is aligned nearly parallel to the helix axis.

The crystal structures of three " real " proteins binding to DNA, namely catabolite activator protein (CAP), *cro* repressor and the N-terminal DNA binding domain of λ-repressor, have been determined. *Cro* is a 66 amino acids long polypeptide, binds to a near-palindromic sequence in DNA and forms a dimer or tetramer in solution, and the latter occurs in the crystal. Looking at a model of *cro* tetramer, it was striking that one half, a dimer, had two α-helices 34Å apart, in antiparallel orientation and tilted by 32°, a geometry just ideal for fitting the *cro* dimer to B-DNA double helix, Figure 7 (Anderson *et al.*, 1981). The model showed that the α-helices can be located in major groove sites and the minor groove in-between is filled by an antiparallel β-pleated sheet formed by the carboxy terminal stretch of the repressor dimer. The very last three amino acids at the carboxy terminus are not seen in electron density because, as has been proposed, they wag and act as " feelers " to recognize the DNA double helix.

What can be said about specific *cro*-operator interactions? As already outlined on a theoretical basis in a previous paragraph, protein-DNA recognition in the major groove is more favorable than in the minor groove, a picture consistent with the *cro*-DNA complex. The model indicates a large number of contact points which comprise hydrogen bonding and salt bridges and are in agreement with findings from chemical studies (Matthews *et al.*, 1983).

By and large, the situation in *cro*-DNA complex is comparable with the N-terminal DNA binding domain of λ repressor and with CAP (Pabo and Lewis, 1982; McKay and Steitz, 1981). They also occur as dimers and exhibit α-helices in suitable positions and orientations so that they can be model-fitted in major grooves of B-DNA. For CAP, binding to left-handed B-DNA was

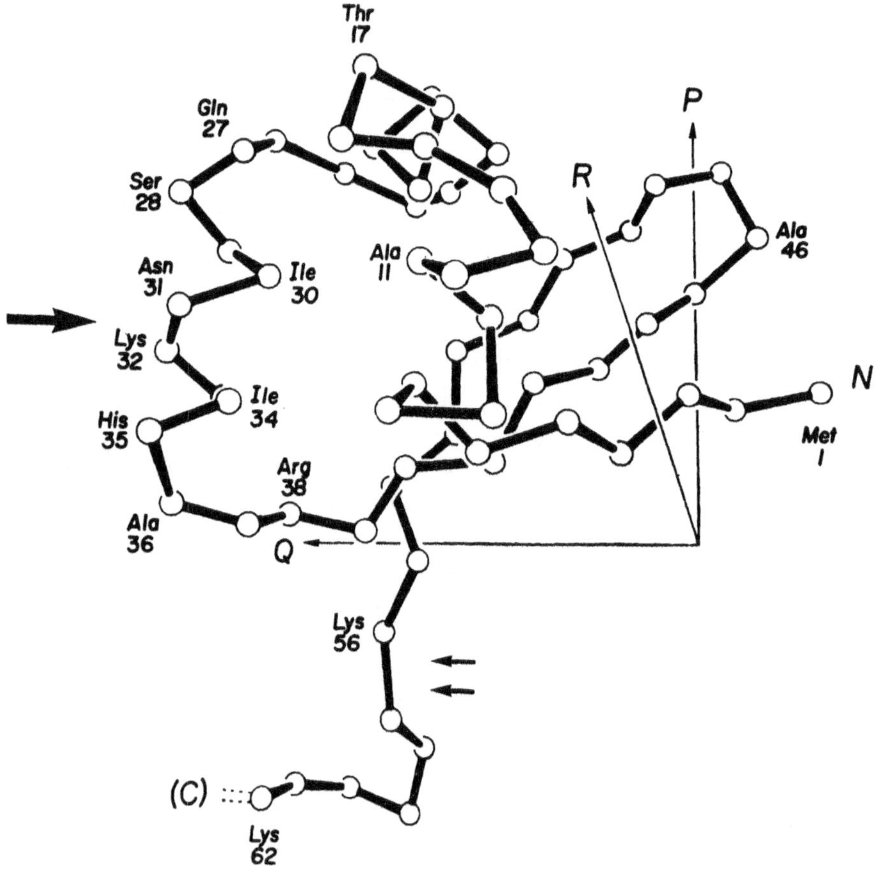

FIG. 7a. Model fitting of DNA to the crystallographically derived structure of *cro* repressor. The protein (Figure 7 a) exists as a tetramer with twofold symmetry and the DNA-*cro* arrangement is such that the complex exhibits dyad symmetry. Note that *cro* α-helices fit in DNA major grooves and β-pleated sheet formed by C-terminal part of *cro* inserts in DNA minor groove.

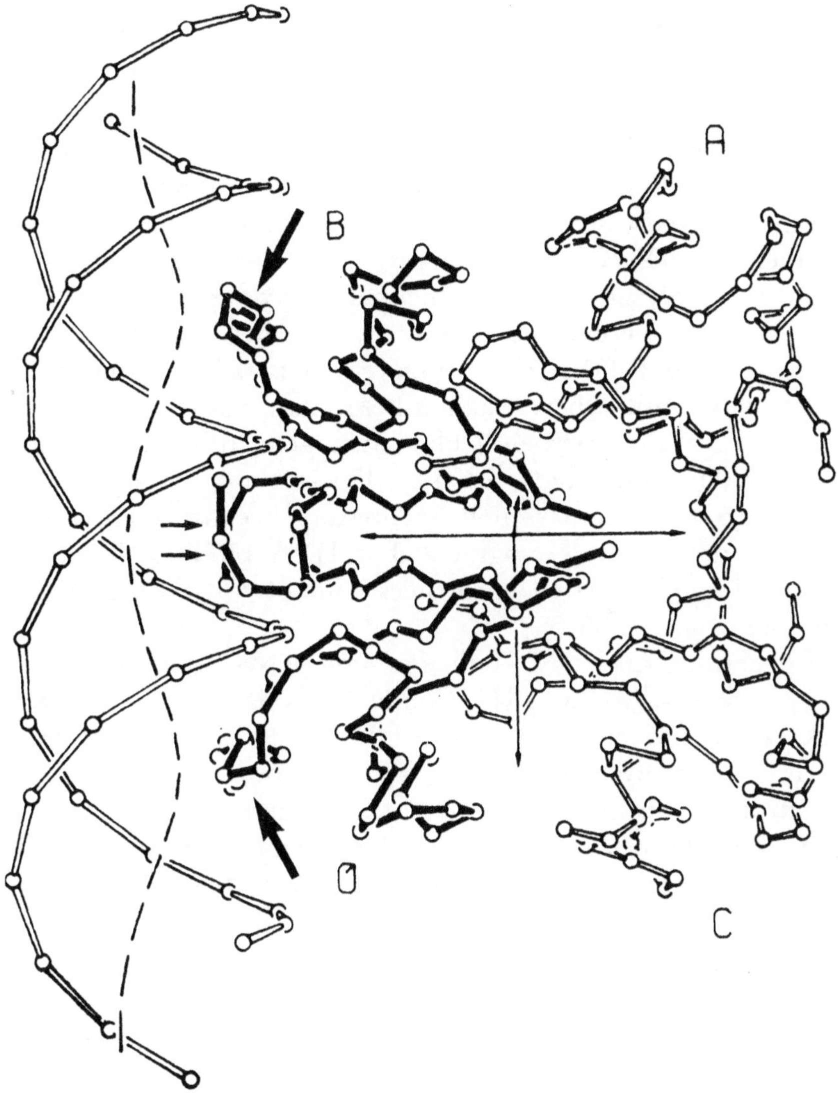

FIG. 7b. Illustrates the model-built DNA-*cro* complex tilted by 32° from the viewer to show fitting of *cro* α-helix (thick arrows) and β-plated sheet (thin arrows) in major and in minor grooves, resp. The same parts in *cro* monomer (Figure 7 a) are also indicated by thick and thin arrows. Taken from (Anderson *et al.*, 1981).

originally proposed but more recent data suggest that this interpretation might be incorrect (Salemme, 1982). In detail, of course, there are differences between these proteins. In CAP and λ-repressor, the β-pleated sheet is missing which fits so nicely into the minor groove of B-DNA in the *cro* repressor. There is no replacement for it in CAP but in λ-repressor, the N-terminus is not involved in secondary structure and model building suggests that it embraces the DNA double helix at the " backside " by insertion of the arms of the dimer in the major groove.

If the local geometries around the α-helices which are supposed to bind to B-DNA are compared, it is obvious that there is a two-helix-motif where each α-helix that inserts in the major groove is in near-perpendicular arrangement with another α-helix directly associated with it. This tertiary structure feature appears to be general among repressors because comparison of sequence data strongly suggests that DNA binding domains are conserved and consist of two helices joined by a short peptide link (Ptashne *et al.*, 1982).

Concerning the DNA-repressor complexes, the twofold symmetry inherent in the DNA double helix is conserved not only structurally but also nucleotide sequences are palindromic or nearly so and are matched by dimeric repressor structure. The palindromes are, in most cases, not ideal and present a good case where deviation from exact symmetry can lead to fine tuning of complex formation and association constants.

Conclusions

The examples of protein-nucleic acid recognition which were presented in this article cover a broad area from interactions between individual units to more complex systems where proteins associate with nucleotides, coenzymes or single and double stranded nucleic acids. It is clear that there is not a simple scheme by which the two partners recognize each other. The illustrations in Figure 1 are too naive and simplified and can only be valid in a few cases or they are augmented by other contacts or they are not observed at all.

One general rule is followed, however, because in almost all complexes where nucleotides, coenzymes or single stranded DNA or RNA interact with proteins, they adopt an extended form with torsion angle λ in the *-sc* or *ap* orientation. This facilitates recognition of phosphate-sugar-base by the protein, yet what protein side chain or main chain groups interact with phosphate, sugar, base, is not predictable.

On the other hand, recognition between proteins and nucleic acid double helices follows a more simple scheme. This is because double helices display and maintain a well defined structure to which the protein has to fit. Data available thus far on such systems suggest that protein α-helices insert into major grooves of B-DNA and in each case studied, the dyad symmetry of DNA is also preserved by the proteins which occur in dimeric forms.

ACKNOWLEDGEMENT

The author's own work mentioned in this paper was supported by Sonderforschungsbereich 9 of the Deutsche Forschungsgemeinschaft and by the Fonds der Chemischen Industrie.

REFERENCES

Anderson, W.F., Ohlendorf, D.H., Takeda, Y. and Matthews, B.W., 1981, Nature *290*, 754.
Berman, H.M., Hamilton, W.C. and Rousseau, R.J., 1973, Biochemistry *12*, 1809.
Berman, H.M., Neidle, S., Zimmer, Ch. and Thrum, H., 1979, Biochim. Biophys. Acta *561*, 124.
Brayer, G.D. and McPherson, A., 1983, J. Mol. Biol.
Chandrasekhar, K., McPherson, Jr., A., Adams, M.J. and Rossmann, M.G., 1973, J. Mol. Biol. *76*, 503.
Chothia, C., 1973, J. Mol. Biol. *75*, 295.
Church, G.M., Sussman, J.L. and Kim, S.H., 1977, Proc. Natl. Acad. Sci. USA. *74*, 1458.
Cotton, F.A., Hazen, E.A. and Legg, M.J., 1979, Proc. Natl. Acad. Sci. *76*, 2551.
Dickerson, R.E., 1983, J. Mol. Biol., *166*, 419.
Furberg, S. and Sølbakk, J., 1974, Acta Chem. Scand. *B 28*, 481.
Gabbay, E.J., Sanford, K., Baxter, C.S. and Kapicak, L., 1973, Biochemistry *12*, 4021.
Heinemann, U. and Saenger, W., 1982, Nature *299*, 27.
Hélène, C. and Maurizot, J.-C., 1981, CRS Crit. Rev. Biochem. *10*, 213.
Holmes, K.C., 1980, Trends Biochem. Sci. *5*, 4.
Ishida, T., Tomita, K.-I. and Inoue, M., 1980, Arch. Biochem. Biophys. *200*, 492.
Kyogoku, Y., Lord, R.C. and Rich, A., 1967, Proc. Natl. Acad. Sci. USA *57*, 250.
Matthews, B.W., Ohlendorf, D.H., Anderson, W.F., Fisher, R.G. and Takeda, Y., 1983, Trends Biochem. Sci. *8*, 25.

Matthews, D.A., Alden, R.A., Freer, S.T., Xuong, N.-H. and Kraut, J., 1979, J. Biol. Chem. *254*, 4144.
McKay, D.B. and Steitz, T.A., 1981, Nature *290*, 744.
Morita, F., 1974, Biochim. Biophys. Acta *343*, 674.
Narayanan, P. and Berman, H.M., 1977, Acta Crystallogr. *B 33*, 2047.
Ohki, M., Takenaka, A., Shimanouchi, H. and Sasada, Y., 1977 a, Bull. Chem. Soc. Japan *50*, 90; 1980, *53*, 2724.
Ohki, M., Takenaka, A., Shimanouchi, H. and Sasada, Y., 1977 b, Acta Crystallogr. *B 33*, 2954.
Ohki, M., Takenaka, A., Shimanouchi, H. and Sasada, Y., 1977 c, Acta Crystallogr. *B 33*, 2956.
Pabo, C.O. and Lewis, M., 1982, Nature *298*, 443.
Prescott, B., Chou, C.H. and Thomas, Jr., G.J., 1976, J. Phys. Chem. *80*, 1164.
Ptashne, M., Johnson. A.D. and Pabo, C.O., 1982, Sci. Amer. *245*, 106.
Rossmann, M.G., Liljas, A., Brändén, C.-I. and Banaszak, L.J., 1975, The Enzymes, XI, 61.
Saenger. W., 1983, Principles of Nucleic Acid Structure, Springer Verlag, New York.
Saenger, W., Suck, D., Knappenberg, M. and Dirkx, J., 1979, Biopolymers *18*, 2015.
Salemme, F.R., 1982, Proc. Natl. Acad. Sci. USA *79*, 5263.
Schulz, G.E. and Schirmer, R.H., 1979, Principles of Protein Structure, Springer Verlag, New York.
Seeman, N.C., Rosenberg. J.M. and Rich, A., 1976, Proc. Natl. Acad. Sci. USA. *73*, 804.
Sobell, H.M., 1976, Annu. Rev. Biophys. Bioeng. *5*, 307.
Suau, P. and Subirana, J.A., 1977, J. Mol. Biol. *117*, 909.
Suck, D., Saenger, W. and Rohde, W., 1974, Biochim. Biophys. Acta *361*, 1.
Sundaralingam, M. and Westhof, E., 1979, Int. J. Quant. Chem.: Quant. Biol. Symp. *6*, 115.
Takimoto, M., Takenaka, A. and Sasada, Y., 1981, Bull. Chem. Soc. Japan *54*, 1635.
Taylor, R. and Kennard, O., 1982, J. Amer. Chem. Soc. *104*, 5063.
von Hippel, P.H. and McGhee, J.D., 1972, Annu. Rev. Biochem. *41*, 231.
Warrant, R.W. and Kim, S.H., 1978, Nature *271*, 130.
Wodak, S.Y., Lin, M.Y. and Wyckoff, H.W., 1977, J. Mol. Biol. *116*, 855.

STUDYING ENZYME-SUBSTRATE INTERACTIONS BY SITE-DIRECTED MUTAGENESIS

ALAN R. FERSHT

Department of Chemistry
Imperial College of Science and Technology
London SW7 2AY, UK

and

GREG WINTER

MRC Laboratory of Molecular Biology
Hills Road
Cambridge CB2 2QH, UK

ABSTRACT

It is now possible to change at will amino acid residues in proteins by site-directed mutagenesis of their genes and hence perform structure-activity studies on molecular interactions. Residues in the tyrosyl-tRNA synthetase that are known from protein crystallographic studies to interact with the substrate are being systematically altered in this manner and the effects on binding and catalysis analysed. It is shown how the strengths of molecular interactions may be measured and how site-directed mutagenesis can probe such interactions. The results indicate how hydrogen bonding in enzyme-substrate complexes must be analysed in terms of an exchange process with solvent. This concept is used to predict a structural modification that increases the affinity of the enzyme for ATP. It is also seen how site-directed mutagenesis can provide evidence on theories of enzyme catalysis.

I. INTRODUCTION

a) *Oligodeoxynucleotide-directed mutagenesis*

Until recently, the only general experimental procedure for studying structure-activity relationships in protein-ligand interactions was to modify the structure of the ligand. But it is now possible to study such structure-activity relationships by systematically altering the amino acid residues of a protein by site-directed mutagenesis of its gene. For example, amino acid residues in the tyrosyl-tRNA synthetase from *Bacillus stearothermophilus* have been modified as follows — see Fig. 1 (Winter *et al.*, 1982; Wilkin-

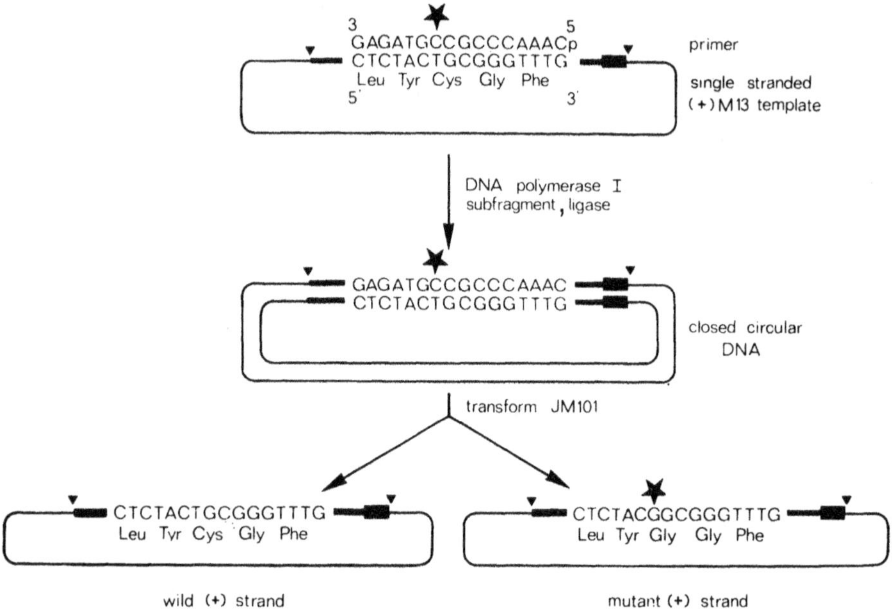

FIG. 1. Scheme for site-directed mutagenesis (Winter *et al.*, 1982).

son *et al.*, 1983 a). The gene coding for the enzyme was cloned into the bacteriophage vector M 13; a short oligodeoxynucleotide primer was synthesized that is complementary to the sequence of the DNA to be mutated apart from a single mismatch; the mismatch was designed to convert the codon for the target amino acid into that for the desired mutant amino acid residue; the oligodeoxynucleotide was annealed to the single-stranded DNA and used as a primer for replication catalysed by a DNA polymerase; the new strand was ligated using DNA ligase to produce a heteroduplex of wildtype and mutant DNA; host cells were transformed by the heteroduplex to give cells containing mutant or wildtype vector. Mutant plaques were identified using the ^{32}P-labelled mutagenic primer as a probe in a hybridisation assay since it binds preferentially to the mutant DNA (Wallace *et al.*, 1981).

Site-directed mutagenesis has been used in the past few years to produce mutations at the nucleotide level for studying promoters, for example, or for modifying genes in regions that do not correspond to the final structural regions of proteins. It is only recently that the technique has been successfully applied to studying the effects of mutation on an enzyme of known three dimensional structure. When combined with a high resolution crystal of a protein, site-directed mutagenesis provides a means of measuring the contribution of all residues at the active site to binding and catalysis. It may also allow an assessment of the roles of amino acid residues in the folding of the protein and the assembly of subunits where involved.

b) *The target — the tyrosyl-tRNA synthetase (TyrTS)*

The first such systematic study has been on the tyrosyl-tRNA synthetase from *Bacillus stearothermophilus* (Winter et al, 1982).

This enzyme catalyses the aminoacylation of $tRNA^{Tyr}$ in a two-step reaction (eq 1 and 2) (Fersht and Jakes, 1975). It is a

$$E + Tyr + ATP \rightleftharpoons E.Tyr\text{-}AMP + PP_i \qquad (1)$$

$$E.Tyr\text{-}AMP + tRNA^{Tyr} \rightleftharpoons Tyr\text{-}tRNA^{Tyr} + AMP + E \quad (2)$$

symmetrical dimer of M_r 2 × 47.5k (Koch, 1974; Irwin *et al.*, 1976). The nucleotide sequence of the gene has been determined from a clone in the vector pBR322 (Winter *et al.*, 1983). X-ray crystallographic studies on the enzyme at 0.3 nm resolution have been published (Bhat *et al.*, 1982) and in subsequent work refinement has been extended to a nominal 0.21 nm (T.N. Bhat, P. Brick & D.M. Blow, unpublished). Most importantly, the crystal structure of the enzyme-bound tyrosyl-adenylate complex has also been solved (Rubin and Blow, 1981) so that there is the rare opportunity of directly knowing the interactions of the enzyme with a substrate (Fig. 2). Aside from this, the tyrosyl-tRNA synthetase has convenient kinetic properties that render it a suitable target for mutagenesis studies: the enzyme is readily assayed by active site titration so that accurate and reproducible steady state kinetic measurements may be made (Fersht *et al.*, 1975 a); the partial reactions of activation (eqn 1) and the overall charging reaction are readily determined by easy assays; pre-steady state kinetics of activation may be followed by stopped flow fluorescence

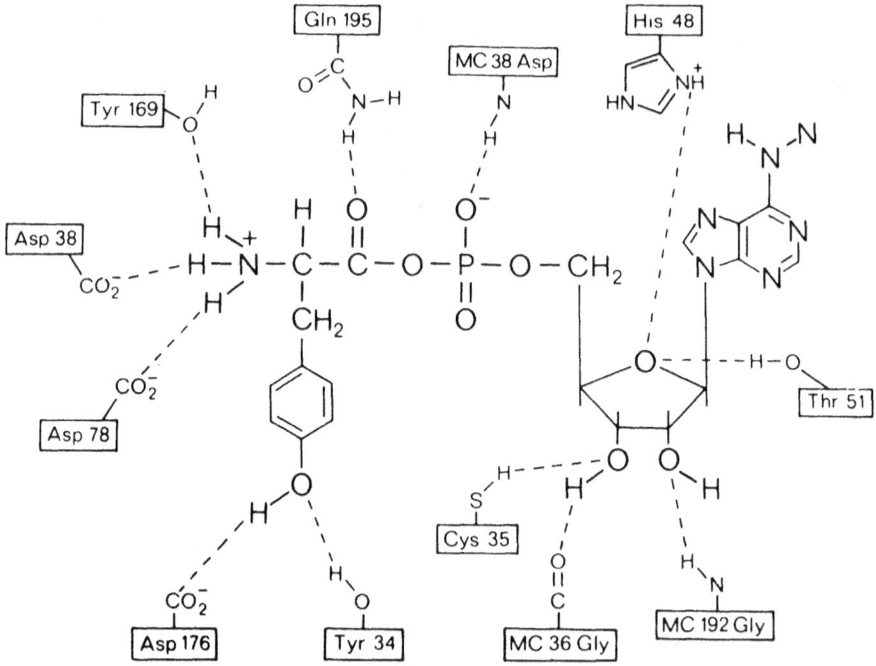

FIG. 2. Possible hydrogen bonds between tyrosyl-tRNA synthetase and tyrosyl-adenylate (courtesy of D.M. Blow).

studies (Fersht *et al.*, 1975 b) and of the transfer step (eqn 2) by rapid quenching methods (Fersht and Jakes, 1975). There are also interesting properties of the enzyme as oligomer-kinetic studies indicate that the two active sites interact: only one mol of tyrosine is bound (tightly) per dimer in solution (Fersht, 1975) as is only one mol of $tRNA^{Tyr}$ (Fersht, and Jakes 1975: Dessen *et al.*, 1982); one mol of tyrosyl adenylate is formed far more rapidly than the other at the two active sites (Mulvey and Fersht, 1977). Under the conditions of pyrophosphate exchange, only one active site is primarily responsible for the chemical catalysis (Fersht *et al.*, 1975 b).

c) *Strategy*

Structure-activity relationships. The utilization of binding energy is at the heart of enzyme catalysis and specificity. There is a simple equation relating the rate constant k_{cat}/K_M of the Michaelis-Menten equation to the binding energy of the enzyme and substrate (eqn 1; Fersht, 1974; Fersht, 1977).

$$RT \ln(k_{cat}/K_M) = RT \ln(kT/h) - \Delta G^{\neq} - \Delta G_s \quad (3)$$

(where ΔG_s is the binding energy between the enzyme and substrate, $\Delta G^{\neq}$ is the energy of activation of the chemical steps, and k and h are the constants of Boltzmann and Planck respectively.)

Using site-directed mutagenesis, we are able to alter the binding energy term, ΔG_s, by changing the side chains of the amino acid residues that interact with the substrate. The changes of binding energy may be calculated from k_{cat}/K_M using eqn 3 and the importance of the interaction evaluated. This provides an experimental means of measuring the binding energies of the interactions. Theoretical considerations (Fersht, 1974) show that reaction rate is optimized when the binding energies of these interactions are realized in the enzyme-transition state complex rather than the enzyme-substrate complex — the concept of transition state stabilization. Further, hydrogen bonds should be particularly important in mediating such differential binding effects because the strength of hydrogen bonding varies strongly with interatomic distance and so is sensitive to the movement of atoms during the reaction. Accordingly, the initial strategy of our experiments is to alter residues that form hydrogen bonds with the substrates.

2. Results

a) *Nature of hydrogen bonding of enzymes with substrates*

The mutation of Cys-35 to Ser-35 was the very first experiment in which a residue at the active site of an enzyme of known three-dimensional structure was specifically mutated (Winter *et al.*, 1982; Wilkinson *et al.*, 1983 a). This residue is seen in Fig. 2 to make a possible hydrogen bond with the 3'-hydroxyl of the ribose moiety of Tyr-AMP. The mutation experiment changes -SH to -OH in the amino acid side chain. Although the *absolute* strength of the hydrogen bond -OH···O- is greater than that of -SH···O- in simple intermolecular associations, it is found on comparing the wild-type TyrTS with the mutant TyrTS(Ser-35) (Table 1) that

TABLE 1. *Activation of Tyr*

Mutant	k_{cat} s^{-1}	K_M (ATP) mM	K_M (Tyr) μM	k_{cat}/K_M (ATP) s^{-1} M^{-1}
Native	7.6	0.9	2.4	8,400
Cys–35→Gly	2.8	2.6	2.7	1,120
Cys–35→Ser	2.4	2.4	2.6	1,000

From Wilkinson *et al.* (1983a)

the wild type enzyme has a higher affinity for the substrate (calculated from eqn 3). The difference between the enzymic and non-enzymic hydrogen bonding strengths stems from two reasons. *i*, it is not the absolute strength of a hydrogen bond in an enzyme-substrate complex that affects the enzyme-substrate dissociation constant (Jencks, 1969; Fersht, 1977). It is the *difference* in energies of the hydrogen bond donor/acceptor in the free enzyme and the acceptor/donor in the free substrate when they are separately hydrogen-bonded to solvent water compared when hydrogen bonded to each other that is important (eqn 4).

$$E\cdots OH_2 + S\cdots HOH = E\cdots S + HOH\cdots OH_2 \qquad (4)$$

ii, there are geometrical constraints on hydrogen bond formation in enzyme-substrate complexes because the donor and acceptor groups are held at fixed distances from each other. For example, the S-O distance in SH···O bonds is 0.5 nm longer than the O-O in OH···O. The substitution of the side chain hydroxyl of a serine residue for a cysteinyl is thus energetically unfavourable because the O-O distance is too long (Fersht and Dingwall, 1979). Comparison of TyrTS(Ser-35) and TyrTS(Gly-35) (Table 1) shows that the serine side chain contributes no apparent binding energy with the substrate and, if anything, the presence of the side chain actually weakens binding.

It is concluded that the presence of a poor (too long) hydrogen bond in an enzyme-substrate complex can increase the dissociation constant and lower k_{cat}/K_M. Perhaps the removal of a weak hydrogen bond by site-directed mutagenesis could be used to increase the affinity of an enzyme for its substrate?

b) *Improving the enzyme*

Preliminary indications suggested that Thr-51 makes a long hydrogen bond with the adenylate (Fig. 2). Further, although all the other side chains illustrated in Fig. 2 are conserved in the *E. coli* enzyme, it has a proline residue in this position that cannot possibly make a hydrogen bond with the substrate and will also distort the secondary structure in this region. Accordingly, in light of the conclusion of the last section, we changed Thr-51 to first Ala-51 to note the effect of deleting the hydrogen bond, and then to Pro-51 to examine the results of causing a larger structural change (Wilkinson *et al.*, 1983 b).

The results for activation of tyrosine by these mutants are listed in Table 2. It is seen that the K_M for ATP of TyrTS(Ala-51) is two times lower than that of wild type. It does appear in fact that the interaction of Thr-51 with the substrate is very weak and so lowers the affinity of the enzyme for the substrate. Site-directed mutagenesis is thus a good fine structure probe for hydrogen bonding.

TABLE 2. *Improving the Enzyme*

Mutant	Activation of Tyr			
	k_{cat} s^{-1}	K_M (ATP) mM	K_M (Tyr) μM	k_{cat}/K_M (ATP) s^{-1} M^{-1}
Native	7.6	0.9	2.4	8,400
Thr–51→Ala	8.6	0.54	2.0	15,900
Thr–51→Pro	12.0	0.058		208,000

From Wilkinson *et al.* (1983b)

Mutation to TyrTS(Pro-51) causes a massive increase in k_{cat}/K_M, involving both an increase in k_{cat} and a decrease in K_M. Thus, engineering a single point mutation has caused a significant improvement in enzymic activity and so shows that there are real biotechnological possibilities of engineering the activities of enzymes *in vitro*.

c) *Evidence for transition state stabilization in the activation reaction*

How is the attack of the tyrosyl carboxyl group on ATP (eqn 1) catalysed? General base catalysis of the nucleophilic attack is not possible since the carboxylate is fully ionized. Nucleophilic catalysis by a group on the enzyme forming an intermediate with AMP is unlikely since ATP is already an activated compound containing a good leaving group (pyrophosphate). The remaining possibilities are electrostatic catalysis and transition state stabilization whereby certain groups on the enzyme make stronger interactions with the substrate when it is in the form of its transition state. A predicted characteristic of transition state stabilization is the following. When a group on the enzyme is removed, as in a site-directed mutagenesis experiment, the affinity of the enzyme for the substrate will be decreased to some extent, but the affinity for the transition state will be decreased to a much greater extent if the transition state is stabilized. This means that the value of K_S may be increased to a greater or lesser extent but the value of k_{cat} will fall. This is seen to be happening in the data in in Table 3. Particularly noteworthy is the change from His-45 to Asn-45 leading to a 2000-fold drop in k_{cat} whilst the values of K_M are hardly affected. Similarly, the rate of attack of pyrophosphate on the tyrosyl adenylate complex is lowered 2000-fold in TyrTS(Asn-45). His-45 is not seen to make an interaction with the tyrosyl adenylate complex (Fig. 2). It is likely that an electrostatic interaction occurs between the positively charged side chain and the transition state structure of the substrate as the α-phosphoryl group of ATP becomes pentacovalent.

Changes in k_{cat} and K_M as found in Table 3 may also arise from mechanisms such as ' induced fit ' and ' non-productive binding.' Because of this, the results of previous kinetic studies on structure-reactivity relationships from studies where the structure of the substrate has been altered have usually failed to provide unambiguous evidence for 'strain' mechanisms (Fersht, 1977). However, the changes induced by site-directed mutagenesis of the

TABLE 3. *Binding Energies Altering* k_{cat} *and* K_M

Mutant	k_{cat} s^{-1}	K_M (ATP) mM	k_{cat}/K_M s^{-1} M^{-1}
Native	7.6	0.9	8,400
Cys–35→Gly	2.8	2.6	1,120
Cys–35→Ser	2.4	2.4	1,000
His–48→Gly	1.6	1.4	1,140
His–45 →Asn	0.003	1.0	3

Data from Wilkinson *et al.* (1983a) and unpublished.

tyrosyl-tRNA synthetase are more amenable for producing unambiguous analysis for the following reasons: *i*, it will be possible to solve the structures of the native enzyme and enzyme-substrate complexes for the mutants by X-ray crystallography; *ii*, many of the structural changes will be sufficiently small that the binding energetics may be calculated. Even in the absence of precise structural data, the combination of chemical intuition and inspection of the enzyme structure by computer graphics strongly suggests that transition state stabilisation is operating.

REFERENCES

Bhat, T.N., Blow, D.M., Brick, P., and Nyborg, J. (1982) *J. Mol. Biol.* **158**, 699-709.
Fersht, A.R. (1974) *Proc. R. Soc. London, Ser. B* **187**, 397-407.
Fersht, A.R. (1975) *Biochemistry* **14**, 5-12.
Fersht, A.R. (1977) *Enzyme Structure and Mechanism*, W.H. Freeman, San Francisco, CA and Oxford U.K.
Fersht, A.R. and Jakes, R. (1975) *Biochemistry* **14**, 3350-3356.
Fersht, A.R., Ashford, J.S., Bruton, C.J., Jakes, R., Koch, G.L.E. and Hartley, B.S. (1979a) *Biochemistry* **14**, 1-4.
Fersht, A.R., Mulvey, R.S.M., and Koch, G.L.E. (1975b) *Biochemistry* **14**, 13-18.
Koch, G.L.E. (1974) *Biochemistry* **13**, 2307-2312.

Rubin, J. and Blow, D.M. (1981) *J. Molec. Biol.* **145**, 489-500.

Wallace, R.B., Schold, M. Johnson, M.J., Dembek, P., and Itakura, K. (1981) *Nucleic Acids Res.* **9**, 3647-3657.

Wilkinson, A.J., Fersht, A.R., Blow, D.M. and Winter, G. (1983a) *Biochemistry* **22**, 3581-3586.

Wilkinson, A.J., Fersht, A.R., Blow, D.M., Carter, P. and Winter, G. (1983b) *Nature* (*London*) *in press.*

Winter, G., Fersht, A.R., Wilkinson, A.J., Zoller, M., and Smith, M. (1982) *Nature* (*London*) **299**, 756-758.

Winter, G., Koch, G.L.E., Hartley, B.S., and Barker, D.G. (1983) *Eur. J. Biochem.* **132**, 383-387.

SEQUENCE DEPENDENCE OF DNA CONFORMATION, DYNAMICS AND INTERACTIONS IN SOLUTION

DINSHAW J. PATEL, SHARON A. KOZLOWSKI
Bell Laboratories
Murray Hill, New Jersey 07974
ARTHUR PARDI
Chemistry Department, University of California
Berkeley, California 94720
RAM BHATT
Molecular Genetics Department, Hoffman-LaRoche
Nutley, New Jersey 07110
SATOSHI IKUTA AND KEIICHI ITAKURA
Molecular Genetics Department, City of Hope National Medical Center
Duarte, California 91010

ABSTRACT

We have applied both distance-dependent nuclear Overhauser effect measurements to elucidate structural information and saturation recovery methods to evaluate hydrogen exchange kinetics at the individual base pair level for nucleic acids in aqueous solution. These experiments demonstrate a sequence dependence of the local conformation and dynamics of DNA in solution. We show that intermolecular nuclear Overhauser effect measurements can map out the complexation sites of antibiotics on the DNA helix and hydrogen exchange measurements can estimate the extent of kinetic stabilization at and adjacent to the ligand binding site.

INTRODUCTION

There has been a significant increase in the last few years in our knowledge of DNA structure and its interplay with various other components of the metabolic apparatus. These developments have been sparked by recent advances in the synthesis, sequencing and cloning of nucleic acids.

Our efforts are focused on the conformation and dynamics of DNA and their complexes with low molecular weight ligands and proteins. We have used high resolution NMR methods to investigate fragments containing one to two turns of helix of known defined sequence and their antibiotic complexes (Patel, *et al.*, 1982; Patel, *et al.*, 1983; Feigon, *et al.*, 1983) in an attempt to correlate with and add to our current knowledge of structure based on single crystal X-Ray analysis of related DNA sequences. (Dickerson, *et al.*, 1983; Viswamitra, 1983; Wang, *et al.*, 1983).

This contribution outlines the application of NMR methods to characterize conformational and dynamic features of the d(CGATTATAATCG) self-complementary duplex (henceforth called Pribnow 12-mer, Scheme I). These results are compared where appropriate with two related sequences d(CGCGTATA-CGCG) (henceforth called TATA 12-mer, Scheme II) and d(CGCGAATTCGCG) (henceforth called AATT 12-mer, Scheme III). The TATA sequence in the TATA 12-mer and the TATAAT sequence in the Pribnow 12-mer are common features of DNA promotor regions which are the sites of initiation of the transcription of nucleic acids.

We have also elucidated structural aspects of the binding site of netropsin on the AATT 12-mer duplex and monitored the dynamics by measuring hydrogen exchange characteristics at and adjacent to the antibiotic complexation site.

1. *Spectral Resolution*

The excellent spectral resolution at high magnetic fields (Patel and Canuel, 1979; Early, *et al.*, 1980; Patel, *et al.*, 1982 a; Kan, *et al.*, 1982; Tran-Dinh, *et al.*, 1982; Sanderson, *et al.*, 1983) is demonstrated by the 498 MHz imino proton NMR spectra of the Pribnow 12-mer in 0.1M phosphate, H_2O solution between 12 and 15 ppm at -5°C (Figure 1A). We observe six well resolved imino resonances (thymidine H-3 from an A·T base pair and guanosine H-1 from a G·C base pair, Scheme IV) so that NMR at high magnetic fields provides markers of hydrogen-bonding at each base pair along the Pribnow, 12-mer sequence (Figure 1A).

C	G	A	T	T	A	T	A	A	T	C	G
1	2	3	4	5	6	6	5	4	3	2	1
G	C	T	A	A	T	A	T	T	A	G	C

C	G	C	G	T	A	T	A	C	G	C	G
1	2	3	4	5	6	6	5	4	3	2	1
G	C	G	C	A	T	A	T	G	C	G	C

C	G	C	G	A	A	T	T	C	G	C	G
1	2	3	4	5	6	6	5	4	3	2	1
G	C	G	C	T	T	A	A	G	C	G	C

The nonexchangeable base and sugar proton NMR spectrum of the Pribnow 12-mer in 0.1 M phosphate solution between 5.5 and 8.5 ppm at 5°C is presented in Figure 1B. The purine H-8 and H-2 and pyrimidine H-6 protons (Scheme IV) are partially resolved from each other (Figure 1B) as are the pyrimidine CH_3–5 resonances (Scheme IV) between 1.2 and 1.6 ppm. These resolved proton resonances provide additional markers for monitoring base stacking and the duplex to strand transition (Patel, *et al.*, 1982a).

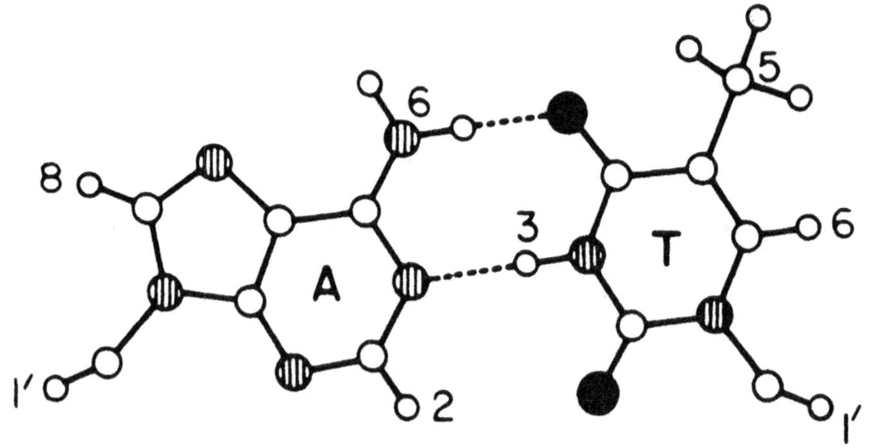

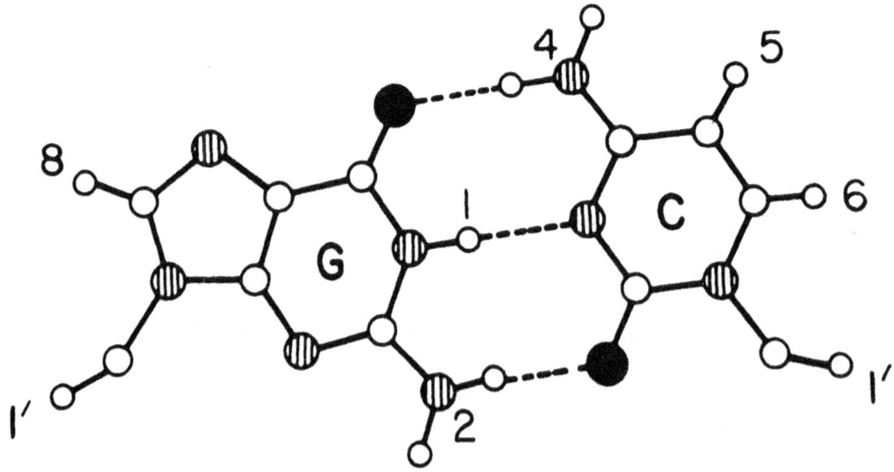

We next describe distance dependent nuclear Overhauser effect measurements (Bothner-By, 1979; Schimmel and Redfield, 1980; Redfield, *et al.*, 1981) which permit assignment of the exchangeable imino and nonexchangeable base protons to specific positions in the Pribnow 12-mer sequence.

2. *Inter-base Pair NOEs — Imino Protons*

Redfield and coworkers have pioneered the application of nuclear Overhauser effects (NOE) to nucleic acids in their seminal

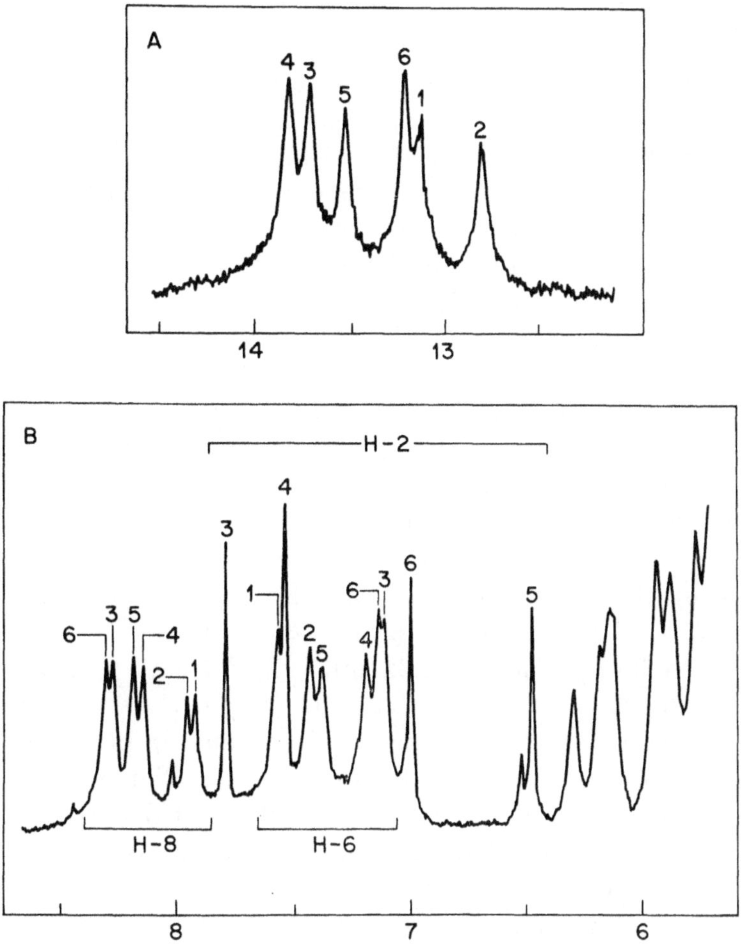

FIG. 1. The 498 MHz imino proton NMR spectra of the Pribnow 12-mer in 0.1M phosphate, 2.5 mM EDTA solution. The signal to noise of the spectra were improved by applying a 1 Hz line broadening contribution. The proton assignments to specific base pairs are designated over the resonances. (A) The imino spectral region (12.5 to 14.5 ppm) in 5:1 H_2O:D_2O, pH 6.90 at -5°C. (B) The aromatic spectral region (5.5 to 8.5 ppm) in D_2O, pH 7.17 at 5°C.

studies on transfer RNA (Roy and Redfield, 1981; Tropp and Redfield, 1981; Hare and Reid, 1982; Heerschap, *et al.*, 1982; Kime and Moore, 1983). The magnitude of the NOE between proton spins depends on the inverse sixth power of the inter-proton separation and on the frequency of motion of the spins. We observe negative NOEs corresponding to a loss of resonance intensity for DNA fragments at low temperature and their magnitude varies from a few percent at inter-base pair separations of 3–4 Å to much larger effects at shorter proton separations within the same base pair. (Patel, *et al.*, 1982a,b; Kan, *et al.*, 1982; Broido, *et al.*, 1982; Patel, *et al.*, 1983c; Reid, *et al.*, 1983; Chou, *et al.*, 1983). The assignment procedure is based on starting with an independently established imino proton at either the terminal or central base pair and using NOEs between imino protons on adjacent base pairs to walk down the duplex.

We can distinguish dA·dT from dG·dC base pairs in the Pribnow 12-mer duplex at -5°C (Figure 1A) since saturation of the thymidine imino protons (12 to 14 ppm) results in strong intrabase pair NOEs to the narrow nonexchangeable adenosine H-2 protons (6.5 to 8 ppm) on the same base pair (Scheme IV). By contrast, saturation of the guanosine imino protons (12 to 14 ppm) results in intra-base pair NOEs to broad amino resonances in the same pair (Scheme IV) (Redfield, *et al.*, 1981; Patel, *et al.*, 1982a). This permits assignment of the two imino protons to highest field (12.82 and 13.13 ppm) to dG·dC base pairs in the Pribnow 12-mer at -5°C (Figure 1A) with the 13.13 ppm resonance assigned to base pair 1 since it broadens out on raising the temperature to 5°C (Figure 2A). This reflects fraying at the ends of the duplex (Patel and Hilbers, 1975; Hilbers, 1979) with the imino proton broadening out when its lifetime in the paired state becomes <5 msec.

The thymidine imino protons from the four dA·dT base pairs resonate at 13.18, 13.48, 13.65 and 13.75 ppm in the Pribnow 12-mer at 5°C (Figure 2A). The presence of a two-fold center of symmetry in the Pribnow 12-mer results in dA·dT base pair 6 being flanked by an identical base pair in one direction and dA·dT base pair 5 in the other direction (Scheme I). We observe a single inter-base pair NOE at the 13.48 ppm imino proton on saturation of the 13.18 ppm imino proton (Figure 2C) permitting assignment of these resonances to dA·dT base pairs 5 and 6 respectively. By contrast, the remaining dA·dT base pairs 3, 4 and 5 have different base pairs flanking them in the Pribnow 12-mer sequence (Scheme I). Thus, saturation of the 13.48 ppm thymi-

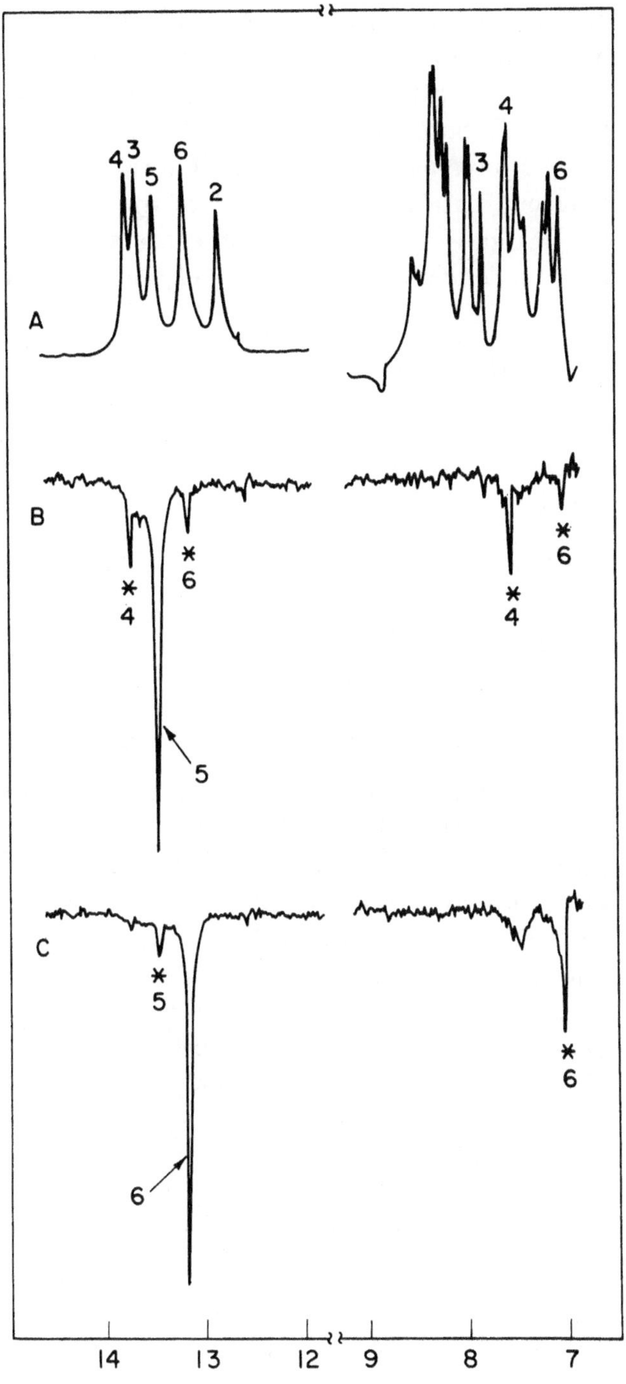

FIG. 2. (A) The 498 MHz proton Fourier transform spectrum of the imino (12.5 to 14.5 ppm) and the aromatic (7 to 9 ppm) of the Pribnow 12-mer duplex in 0.1M phosphate, 5:1 $H_2O:D_2O$, pH 6.90 at 5°C. The imino proton assignments to specific base pairs are designated over the resonances. (B) The difference spectrum following 1 second saturation of the 12.83 ppm guanosine imino proton (designated by an arrow). We observe negative NOEs at the 8.51 ppm cytidine amino protons of the same base pair and at the 7.80 ppm adenosine H-2 proton and the 13.65 ppm thymidine imino proton of an adjacent base pair (designated by asterisks). (C) Saturation of the 13.18 ppm thymidine imino proton results in negative NOEs at the 7.01 ppm adenosine H-2 proton of the same base pair and the 13.48 ppm thymidine imino proton of an adjacent base pair.

dine imino proton assigned to dA·dT base pair 5 results in inter-base pair NOEs at the 13.18 ppm imino proton assigned to dA·dT base pair 6 and the 13.75 ppm imino proton which must be assigned to dA·dT base pair 4 in the Pribnow 12-mer sequence (Figure 2B).

We have adopted this procedure of monitoring NOEs between adjacent base pairs to complete the imino proton assignments in the Pribnow 12-mer and these are shown in Figures 1A and 2A and the chemical shifts at 5°C and listed in Table I.

TABLE 1. *The imino and nonexchangeable base proton chemical shifts in the Pribnow 12-mer duplex in 0.1M phosphate solution at 5° C.*

Pair	Position		Purine			Pyrimidine	
		(H-1)	(H-8)	(H-2)	(H-3)	(H-6)	(H-5)
G·C	1	∾13.1	7.93			7.58	5.81
G·C	2	12.83	7.95			7.45	5.60
A·T	3		8.28	7.80	13.65	7.12	1.26
A·T	4		8.14	7.54	13.75	7.18	1.33
A·T	5		8.19	6.50	13.48	7.38	1.58
A·T	6		8.31	7.02	13.18	7.12	1.34

3. *Inter-Base Pair NOEs — Nonexchangeable Protons*

It is quite straightforward to differentiate between the narrow adenosine H-8 (8.1 to 8.3 ppm) and guanosine H-8 (7.9 to 8.0 ppm) singlets, the cytidine H-6 (7.4 to 7.6 ppm) and cytidine H-5 (5.6 to 5.8 ppm) doublets, the broad thymidine H-6 (7.1 to 7.4 ppm) and thymidine CH_3—5 (1.2 to 1.6 ppm) singlets and the very narrow adenosine H-2 (6.5 to 7.8 ppm) singlets based on chemical shifts, coupling constants and deuteration rates (Figure 3A). We outline procedures for assignment of the nonexchangeable protons to specific positions in the Pribnow 12-mer duplex based on NOEs between nonechangeable protons on adjacent base pairs.

It can be shown that the thymidine CH_3—5 and adenosine H-8 protons are < 4Å from each other in the dA(3′—5′)dT segment but < 4Å from each other in the dT(3′—5′)dA segment based on an examination of molecular models of right-handed B-DNA (Patel, *et al.*, 1983a; Feigon, *et al.*, 1983; Reid, *et al.*,

1983). Thus, we observe an NOE at the 1.33 ppm thymidine CH_3—5 proton on saturation of the 8.28 ppm adenosine H-8 proton in the Pribnow 12-mer at 5°C (Figure 3B). Such an NOE would be predicated between adjacent bases in the dinucleotide segments dA_3—dT_4, dA_6—dT_6 and dA_4—dT_3 but not for the dinucleotide segments dT_5—dA_6 and dT_6—dA_5 in the Pribnow 12-mer sequence (Scheme I).

We also observe an NOE at the 7.95 ppm guanosine H-8 proton on saturation of the 8.28 ppm adenosine H-8 proton in the difference spectrum in Figure 3B. This permits assignment of the 7.95 guanosine H-8, 8.28 ppm adenosine H-8 and 1.33 ppm thymidine

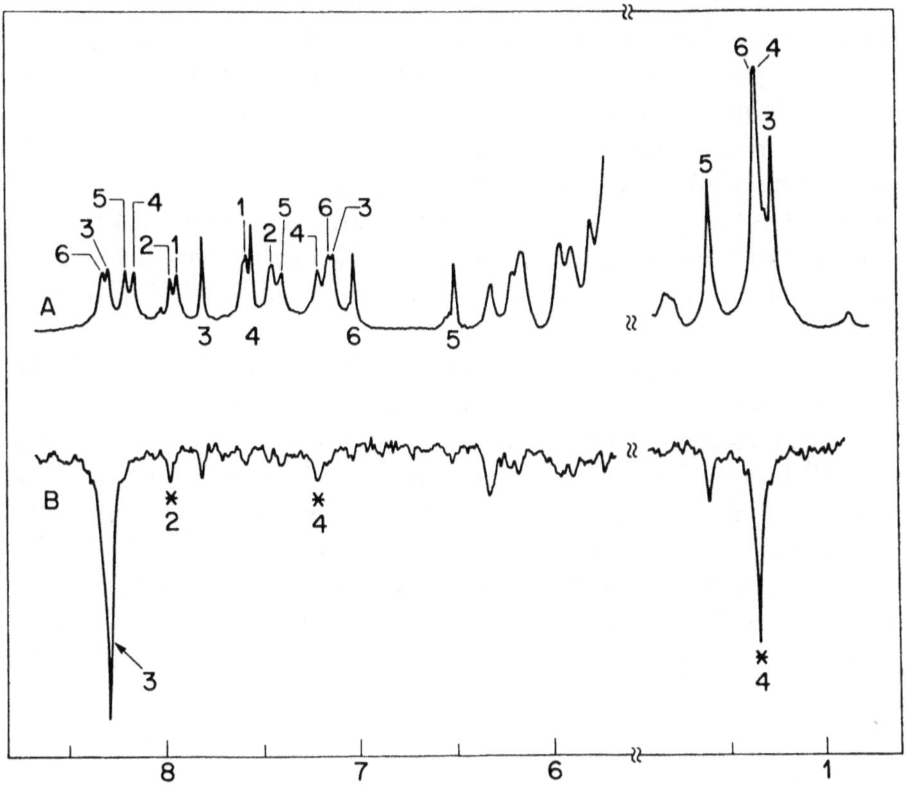

FIG. 3. (A) The 498 MHz proton NMR spectrum (5.5 to 8.5 ppm; 1 to 2 ppm) of the Pribnow 12-mer in 0.1M phosphate, 2.5 mM EDTA, D_2O, pH 7.17 at 5°C. (B) The difference spectrum following 1 second saturation of the adenosine H-8 proton at 8.28 ppm. The saturated resonance is designated by an arrow and the observed NOEs are designated by asterisks.

CH_3—5 to the trinucleotide segment dG_2—dA_3—dT_4 in the Pribnow 12-mer sequence.

We have recorded the one-dimensional NOEs by saturating all the resolvable base protons in turn and this has permitted us to complete the base proton assignments in the Pribnow 12-mer duplex. These are indicated over the spectra in Figures 1B and 3A and their chemical shifts at 5°C are summarized in Table I.

4. *Premelting Transition*

The resolved and assigned proton chemical shifts in the Pribnow 12-mer duplex have been monitored between 0° and 40°C. We observe pronounced premelting chemical shift changes at the dA·dT base pair protons which are not detected at the dG·dC base pair protons in this temperature range (Figures 4A and 4B).

This is most dramatically demonstrated for the four thymidine imino protons in the Pribnow 12-mer which undergo large upfield premelting shifts with increasing temperature (Figure 4A). The observed ~0.3 ppm premelting shift change for the imino protons over a 40°C temperature range is ~10% of the overall chemical shift difference of ~2.5 ppm between the thymidine imino protons in the paired and unpaired states.

The single crystal X-Ray analysis of the d(CGCGAATTCGCG) duplex reveals several interesting features for the A—A—T—T tetranucleotide core of the helix (Dickerson, *et al.*, 1983). The dA·dT base pairs are propeller-twisted and there is an organized spine of hydration in the minor groove in the crystalline state. The premelting transition monitored at the dA·dT rich ATTA-TAAT segment of the Pribnow 12-mer probably reflects changes in the propeller-twists and/or changes in the minor groove dimensions resulting from disruption of bound water with increasing temperature.

5. *Melting Transition*

The majority of the nonexchangeable base protons shift as average peaks during the duplex to strand transition of the Pribnow 12-mer with increasing temperature (Figure 4B). We mea-

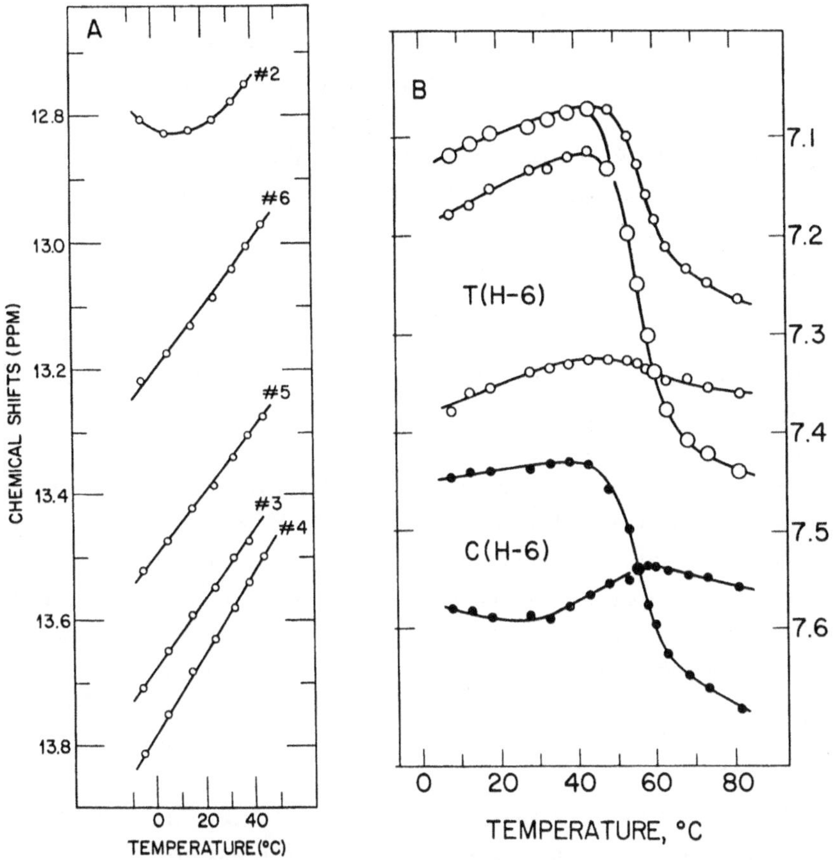

FIG. 4. (A) The temperature dependence of the imino proton chemical shifts in the Pribnow 12-mer duplex in 0.1M phosphate, 2.5 mM EDTA, 5:1 H_2O:D_2O, pH 6.90 in the premelting region between 24° and 44°C. (B) The temperature dependence of the nonexchangeable thymidine H-6 and cytidine H-6 protons in the Pribnow 12-mer in 0.1M phosphate, 2.5 mM EDTA, D_2O in the premelting and melting transition regions between 5° and 85°C.

sure a transition midpoint of 55±1°C at the five nonterminal base pairs 2, 3, 4, 5 and 6 for the Pribnow 12-mer in 0.1M phosphate solution. This observation requires that a minimum of ten non-terminal base pairs in the Pribnow 12-mer melt in a cooperative manner during the helix-coil transition.

6. *Inter-Base Pair NOEs — Adenosine H-2 Protons*

Recent single crystal X-Ray investigations of oligonucleotide duplexes (reviewed by Neidle and Berman, 1983) strongly demonstrate that the double helix is not built up by rotational and translational transformations of the same mononucleotide repeat as initially suggested from X-Ray fiber diffraction analysis. Dickerson and coworkers observed propeller-twisting of base pairs in their single crystal X-Ray analysis of the AATT 12-mer (Scheme III) which results in subtle sequence dependent effects along the length of the dodecanucleotide duplex (Dickerson, *et al.*, 1983). Specifically, propeller-twisting results in close contacts between adjacent purines in the minor groove in pyrimidine (3′-5′) purine steps and in the major groove in purine (3′-5′) pyrimidine steps. By contrast, there are no close steric contacts in pyrimidine (3′-5′) pyrimidine steps when the purines are on the same strand (Calladine, 1982; Dickerson, 1983).

We set out to deduce whether the striking sequence dependent observations observed for right-handed B-DNA fragments in crystal also holds in solution. We have used the adenosine H-2 proton located in the minor groove of dA·dT base pairs to differentiate between the three sequence dependent geometries for the dinucleotide segments (dA_3—dT_4)·(dA_4—dT_3) (Scheme V-A), (dT_4—dT_5)·(dA_5—dA_4) (Scheme V-B), and (dT_5—dA_6)·(dT_6—dA_5) (Scheme V-C) in the Pribnow 12-mer duplex (Scheme I) (Patel, *et al.*, 1983c).

The adenosine H-2 protons of dA·dT base pairs 3, 4, 5 and 6 in the Pribnow 12-mer duplex have been assigned from intra- and inter-base pair NOEs between the assigned thymidine imino protons and the adenosine H-2 on the same and adjacent base pairs. Three of the four adenosine H-2's in the Pribnow 12-mer resonate between 7 to 8 ppm with the remaining proton at 6.5 ppm resonating to high field of the spectra in Figure 2A.

Thus, saturation of the 13.18 ppm thymidine imino proton of base pair 6 in the Pribnow 12-mer at 5°C results in an intra-base pair NOE at the 7.01 ppm adenosine H-2 proton of the same base pair 6 (Figure 2C) and an inter-base pair NOE at the 6.50 ppm adenosine H-2 of adjacent base pair 5 (data not shown). Saturation of the 13.48 ppm thymidine imino proton of base pair 5 results in an intra-base pair NOE at the 6.50 ppm adenosine H-2 of the same base pair (data not shown) as well as inter-base pair

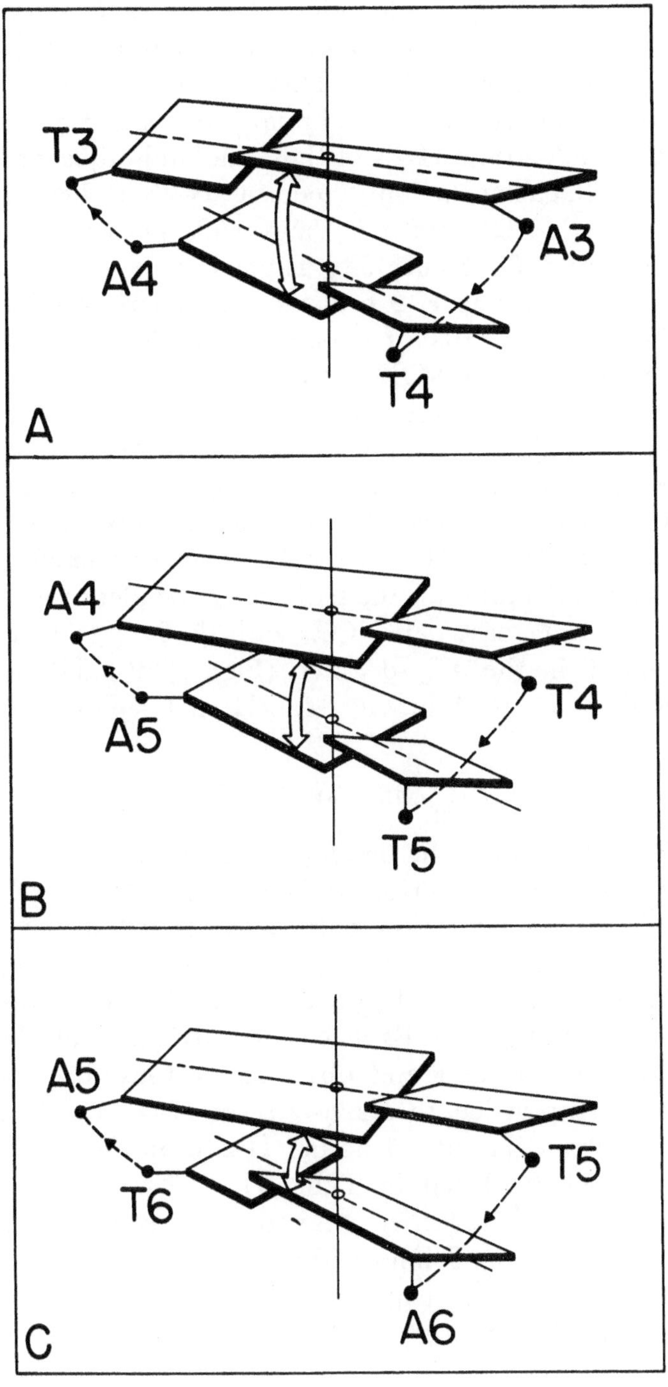
T3
A3
A4
T4
A
A4
T4
A5
T5
B
A5
T5
T6
A6
C

NOEs at the 7.02 ppm adenosine H-2 of adjacent base pair 6 and the 7.54 ppm adenosine H-2 of adjacent base pair 4 in the Pribnow at 5°C (Figure 2B). This approach of saturating the assigned and resolved thymidine imino protons and monitoring the NOEs at the adenosine H-2's of the same and adjacent base pairs permits assignment of the four adenosine H-2's at 7.80 ppm (dA_3), 7.54 ppm (dA_4), 7.02 ppm (dA_6) and 6.50 ppm (dA_5) in the Pribnow 12-mer at 5°C (Figure 5A, Table I).

The four adenosine H-2 protons are well resolved from each other (Figure 5A) so that it is possible to selectively saturate the resonances and monitor NOEs between adenosine H-2's on adjacent base pairs in the minor groove of the d(ATTA)·d(TAAT) segment of the Pribnow 12-mer duplex. We demonstrate this with two examples involving 1 second saturation of the 6.50 ppm adenosine H-2 of base pair 5 (Figure 5B) and the 7.54 ppm adenosine H-2 of base pair 4 (Figure 5C) in the Pribnow 12-mer at 5°C. The most striking observation is the sequence dependence of the magnitude of these inter-base pair NOEs with the largest effect between pairs 5 and 6 in the dT—dA step (Figure 5B) and the smallest between pairs 3 and 4 in the dA—dT step (Figure 5C).

7. *Sequence Dependence of Base Pair Stacking*

We have monitored the buildup of the NOEs between adjacent adenosine H-2's in the Pribnow 12-mer at 5°C as a function of the length of the saturation pulse between 50 msec and 3 sec. The results on saturation of the adenosine H-2's of base pairs 5, 6, 4 and 3 are presented in Figures 6A, 6B, 6C and 6D respectively. The magnitude of the NOEs reach a steady state value η_{ij} (where i and j are the satured and observed protons, respectively) for saturation pulses of length above 2 seconds (Figure 6) and these values between adjacent adenosine H-2 protons in the Pribnow 12-mer at 5°C are listed in Table II.

The selective spin lattice relaxation times for the individual adenosine H-2's in the Pribnow 12-mer at 5°C have been measured from saturation recovery experiments. This approach monitors the recovery of magnetization as a function of the delay between the

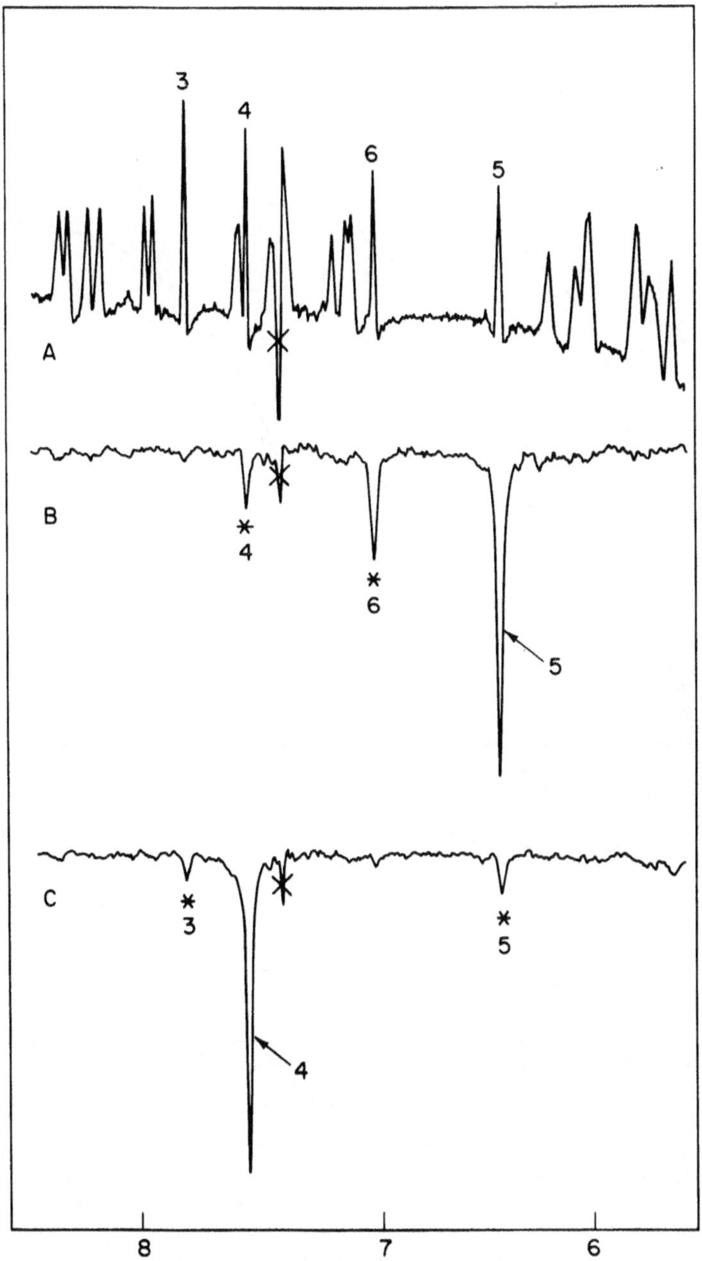

FIG. 5. (A) The resolution enhanced 498 MHz proton Fourier transform spectrum of the aromatic (6 to 8.5 ppm) protons of the Pribnow 12-mer duplex in 0.1M phosphate, D_2O, pH 7.17 at 5°C. (B) The difference spectrum following 1 second saturation of the 6.50 ppm adenosine H-2 proton of dA·dT base pair 5. Steady state negative NOEs are observed at7.55 adenosine H-2 of dA·dT base pair 4 and the 7.03 ppm adenosine H-2 of dA·dT base pair 6. (C) The difference spectrum following 1 second saturation of the 7.55 ppm adenosine H-2 proton of dA·dT base pair 4. Steady state negative NOEs are observed at 7.81 ppm adenosine H-2 of dA·dT base pair 3 and the 6.50 ppm adenosine H-2 proton of dA·dT base pair 5. The NMR spectra contain a spike at 7.45 ppm designated by an x.

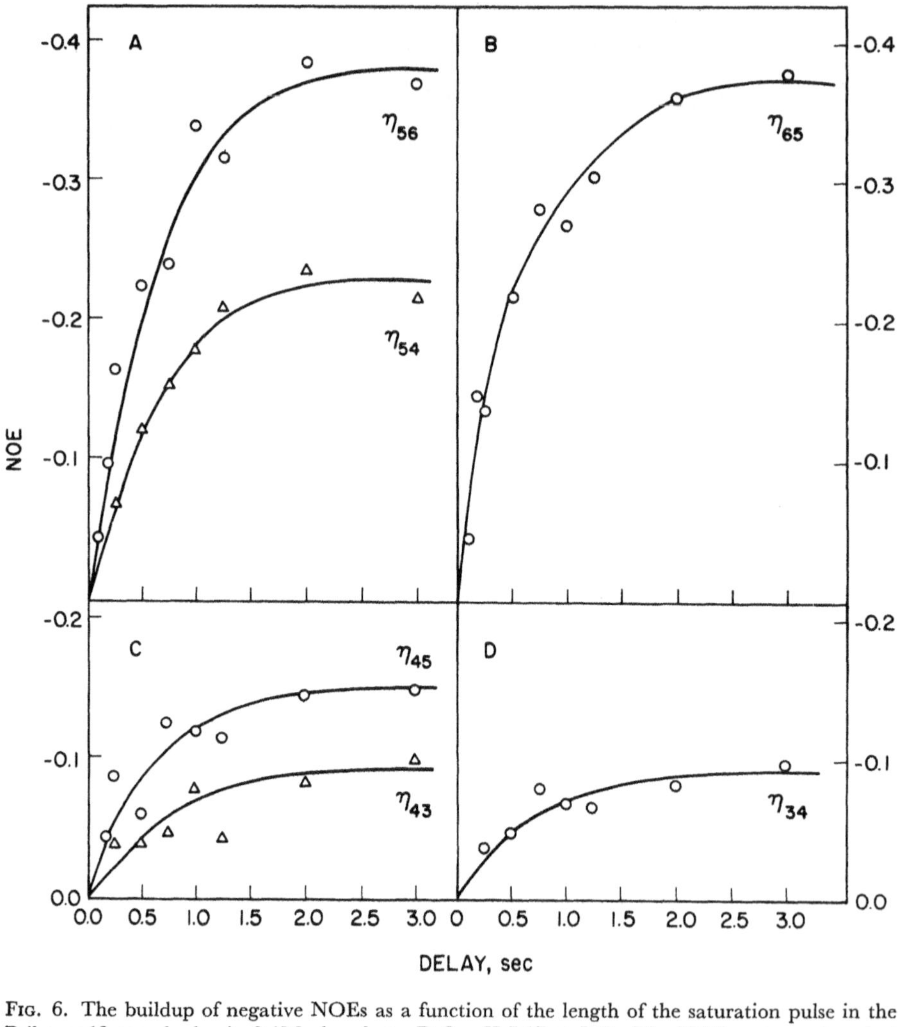

FIG. 6. The buildup of negative NOEs as a function of the length of the saturation pulse in the Pribnow 12-mer duplex in 0.1M phosphate, D_2O, pH 7.17 at 5°C. The NOEs are monitored at adjacent adenosine H-2 protons on saturation of (A) the 6.50 ppm adenosine H-2 of dA · dT base pair 5, (B) the 7.03 ppm adenosine H-2 of dA ·dT base pair 6, (C) the 7.55 ppm adenosine H-2 of dA · dT base pair 4 and (D) the 7.81 ppm adenosine H-2 of dA · dT base pair 3. The symbol η_{ij} corresponds to the magnitude of the NOE on saturation of i and observation of j.

saturation and observation pulses. The selective spin-lattice relaxation rates ϱ_j for the four adenosine H-2 protons in the Pribnow 12-mer at 5°C are listed in Table II.

The cross relaxation rate $\sigma_{ij} = \eta_{ij}\varrho_j$ between adenosine H-2

TABLE 2. *Cross-relaxation rates and inter-proton distances between adenosine H-2 protons on adjacent dA·dT base pairs in the Pribnow 12-mer at 5° C.*

Spin *i*	*j*	η_{ij}	ρ_j	σ_{ij}	r_{ij}
3	4	—09	1.29	—.12	4.0
4	3	—.09	.97	—.09	4.2
4	5	—.15	1.63	—.25	3.6
5	4	—.23	1.29	—.30	3.5
5	6	—.38	1.31	—.50	3.2
6	5	—.37	1.63	—.60	3.1

protons on adjacent base pairs can be determined from the measured values of the steady state NOE, η_{ij}, and the selective spin-lattice relaxation rate constant, ϱ_j. The assumption of an isolated two spin (i,j) system is appropriate in this case since the adenosine H-2 protons are isolated from other non-exchangeable base and sugar protons in the duplex state. The cross relaxation rate constants σ_{ij} between the adenosine H-2 protons on adjacent base pairs in the Pribnow 12-mer at 5°C are summarized in Table II.

The inter-proton distance r_{ij} can be estimated from the measured value of σ_{ij} and τ_c, the correlation time for isotropic tumbling of the dodecanucleotide duplex, by

$$\sigma_{ij} = \frac{-5.68 \times 10^{10}}{r_{ij}{}^6} \tau_c$$

This equation (Tropp and Redfield, 1981) is applicable when $\omega\tau_c > 1$ and holds true at 498 MHz and a τ_c value of 9 ns for the dodecanucleotide at 5°C.

The experimental data permits two estimates each of the inter-proton distance between adenosine H-2's on adjacent pairs 3, 4, pairs 4, 5 and pairs 5, 6 in the Pribnow 12-mer duplex (Table II). These values are ~4.1Å for the $(dA_3—dT_4) \cdot (dA_4—dT_3)$ step (Scheme V-A), ~3.55Å for the $(dT_4—dT_5) \cdot (dA_5—dA_4)$ step (Scheme V-B) and ~3.15Å for the $(dT_5—dA_6) \cdot (dT_6—dA_5)$ step (Scheme V-C).

These experimentally measured inter-proton distances between adjacent adenosine H-2's in the $d(A_3—T_4—T_5—A_6) \cdot d(T_6—A_5—A_4—T_3)$ fragment of the Pribnow 12-mer can be compared with the corresponding values computed for DNA conformations based on an analysis of fiber diffraction patterns (Arnott and Hukins, 1972; Arnott, *et al.*, 1983). These computed values are summarized in Table III in the presence of a small (4°) propeller-twist of opposite handedness as in Scheme V (Arnott-Hukins model)

TABLE 3. *Interproton distances between adenosine H-2 protons on adjacent dA · dT base pairs in the* $d(A_3—T_4—T_5—A_6) \cdot d(T_6—dA_5—dA_4—dT_3)$ *segment based on fiber diffraction models of right handed DNA with propeller-twists of different magnitude and handedness*

Pair	r_{ji}, Å	
	Arnott-Hukins[1]	Arnott-Chandrasekaran[2]
3,4	4.12	4.36
4,5	3.76	3.63
5,6	3.57	2.94

[1] Computations undertaken by Dr. R. Chandrasekaran.
[2] The propeller twist is 4° and has an opposite sense of rotation from Scheme V.
[3] The propeller twist is 13° and has the same sense of rotation as in Scheme V.

and in the presence of a large (13°) propeller-twist of the same handedness as in Scheme V (Arnott-Chandrasekaran model).

We note that the experimental NMR observation that r_{34} is ~0.55Å longer than r_{45} which in turn is ~0.4Å longer than r_{56} (Table II) is in good agreement with the Arnott-Chandrasekaran model which predicts that r_{34} is ~0.7Å longer than r_{45} which in turn is ~0.65Å longer than r_{56}. Thus, the NOE measurements on the Pribnow 12-mer duplex demonstrate that the dA · dT base pairs are propeller-twisted with the same handedness as observed in the solid state (Patel, *et al.*, 1983c).

8. *Saturation Recovery Measurements of Hydrogen Exchange Kinetics*

It has been shown from hydrogen-deuterium exchange stop flow measurements that the exchange kinectics of the Watson-Crick imino protons are sensitive markers of the dynamics of base pair opening in the interior of nucleic acid duplexes (Mandal, *et al.*, 1979; Mandal, *et al.*, 1980; Hayashi, *et al.*, 1981). Hydrogen exchange can also be monitored by the saturation recovery NMR method first applied by Redfield and coworkers to probe the dynamics of transfer RNA in solution (Johnston and Redfield, 1977, 1981; Hurd and Reid, 1980). The method consists of saturating the imino protons of the nucleic acid in H_2O and monitoring the kinetics of magnetization recovery as a function of the delay between the saturation and observation pulses. The contribution to the recovery due to exchange of the imino protons with unperturbed solvent can be partitioned out to yield the hydrogen exchange kinetics at individual base pairs in the nucleic acid sequence (Schimmel and Redfield, 1980; Early, *et al.*, 1981a,b). We have used this approach to evaluate the kinetics of base pair opening reactions in DNA duplexes (Pardi and Tinoco, 1982; Pardi, *et al.*, 1982, 1983; Patel, *et al.*, 1983b).

We demonstrate the application of the saturation recovery method to DNA fragments with studies on the Pribonw 12-mer in 0.1M phosphate, 5:1 H_2O:D_2O, pH 6.90 solution. The recovery of imino proton magnetization as a function of the delay between saturation and observation pulses at 22.5°C is presented in Figure 7. The individual imino protons are well resolved so that the recovery from saturation can be monitored at base pairs 2, 3, 4, 5 and 6 in the Pribnow 12-mer duplex. The magnetization recovers by a first order process and the saturation recovery lifetimes can be deduced from the slope of a plot of the natural logarithm of the fractional saturation as a function of the delay between saturation and observation pulses. We estimate lifetimes ranging from 120 to 180 msecs for the four dA·dT base pairs in the Pribnow 12-mer sequence in 0.1M phosphate, pH 6.90 at 22.5°C (Figure 8).

The saturation recovery lifetimes have been evaluated as a function of temperature and these values for the Pribnow 12-mer in 0.1M phosphate, pH 6.90, between -4° and 34°C are listed in Table IV. The corresponding Arrhenius plots of the natural logarithm of the saturation recovery rates as a function of the inverse of the absolute temperature are plotted in Figure 9 for the four dA·dT base pairs in the Pribnow 12-mer duplex in 0.1M phosphate, pH 6.90 solution. The saturation recovery rate at any given temperature is the sum of the magnetic (spin-lattice relaxation) and chemical (exchange) contributions which can be

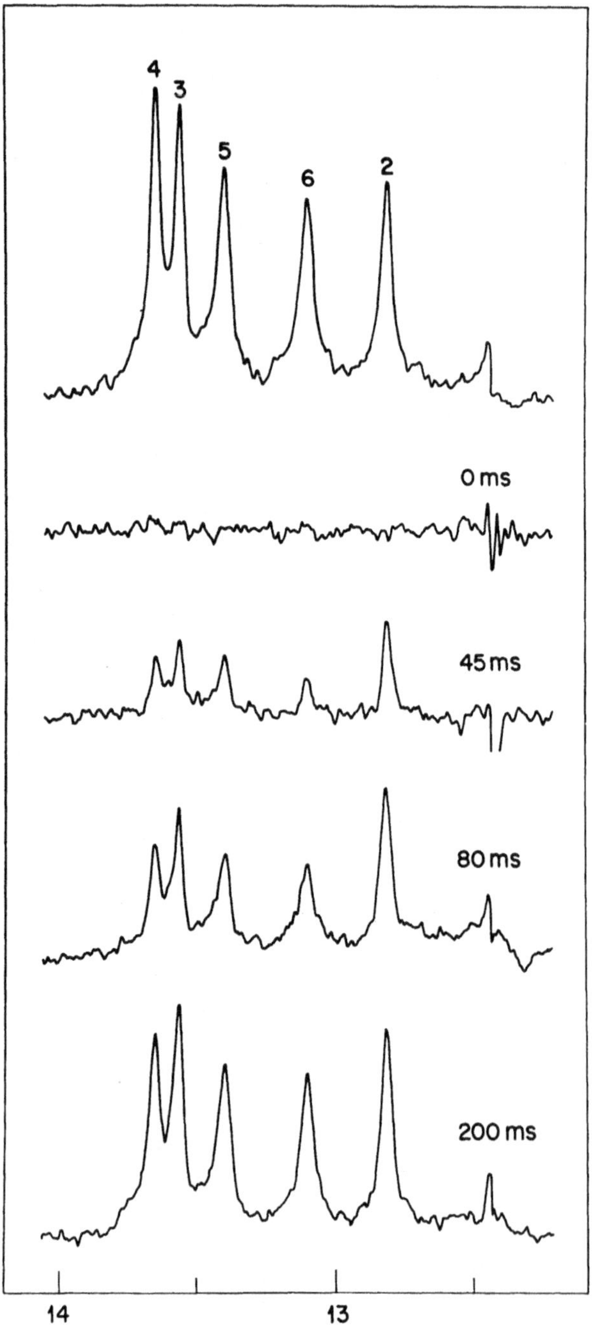

FIG. 7. The 498 MHz proton Fourier transform spectra (12.5 to 13.5 ppm) of the Pribnow 12-mer in 0.1M phosphate, 5:1 H_2O:D_2O, 2.5 mM EDTA, pH 6.90 at 22.5° C. A variable delay was introduced between the broad band saturation and observation pulses with a 1500 msec value for the top spectrum and the various values for the remaining spectra as noted in the Figure. A repetition delay of 1.5 seconds was introduced after each accumulation.

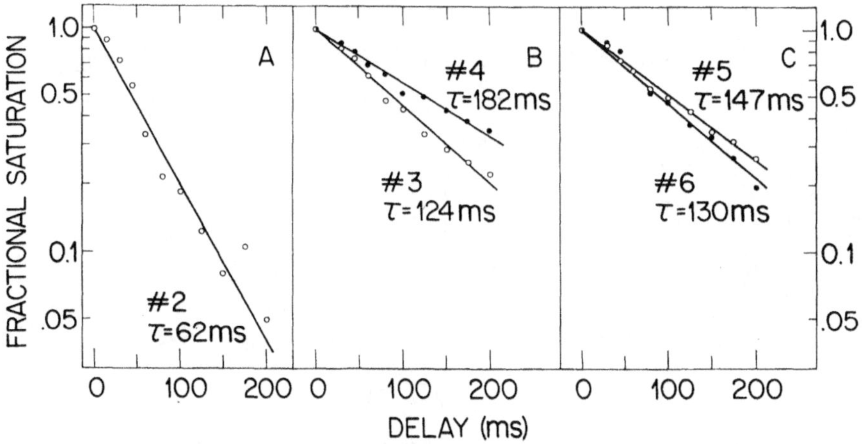

FIG. 8. Plots of the fractional saturation of the imino protons of (A) dG·dC base pair 2, (B) dA·dT base pairs 3 and 4 and (C) dA·dT base pairs 5 and 6 in the Pribnow 12-mer duplex as a function of the delay between the saturation and observation pulses in msec at 22.5°C. The buffer was 0.1M phosphate, 5:1 H_2O:D_2O, 2.5 mM EDTA, pH 6.90.

TABLE 4. *Temperature and pH dependence of the 498 MHz saturation recovery lifetimes (msecs) of the imino protons in the Pribnow 12-mer duplex in 0.1 phosphate solution*

Temp. °C	Imino proton lifetimes in msec				
	#2	#3	#4	#5	#6
pH 6.90					
— 4°	154	216	215	224	194
3°	177	276	273	294	247
10°	167	260	299	273	251
15°	142	224	274	230	240
18.7°	101	170	239	193	187
22.5°	62	124	182	147	130
26.2°	38	72	128	80	84
30°	32	46	77	55	60
33.7°		32	51	39	44
pH 8.51					
18.7°		73	133	63	85
22.5°		35	49	31	47
26.2°			37	30	34

partitioned as outlined below. The saturation recovery rates are essentially temperature independent between -4° and +15°C and are predominantly a measure of the magnetic spin-lattice relaxation rate in this low temperature region. By contrast, the saturation

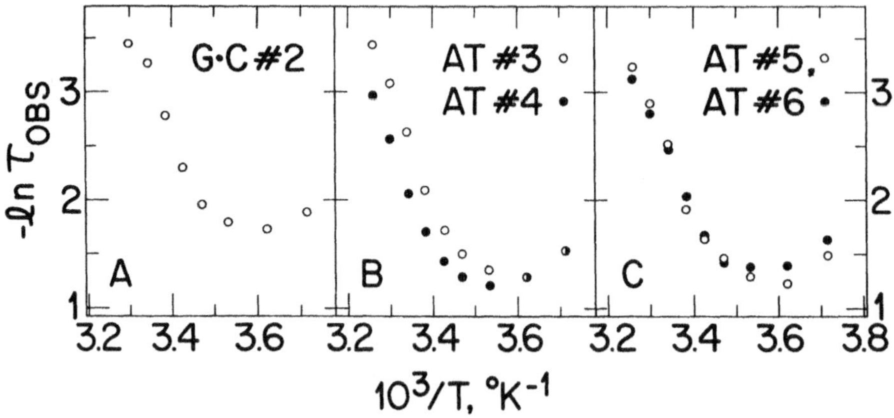

FIG. 9. Arrhenius plots for the inverse of the imino proton lifetimes of (A) dG·dC base pair 2, (B) dA·dT base pairs 3 and 4 and (C) dA·dT base pairs 5 and 6 in the Pribnow 12-mer duplex in 0.1M phosphate, 2.5 mM EDTA, 5:1 $H_2O:D_2O$, pH 6.90.

recovery rates increase dramatically between 15° and 34° and are predominantly a measure of the chemical exchange rate in this elevated temperature region.

The slopes of the saturation recovery data between 15° and 34°C yield activation barriers of 18 to 22 kcal for the Pribnow 12-mer base pairs 3, 4, 5 and 6 (Table V). Subtraction of the constant magnetic contribution (saturation recovery rate at 0°C) from the saturation recovery data at elevated temperatures yields a corrected activation barrier of 29 to 34 kcal for the chemical hydrogen exchange contribution for Pribnow 12-mer base pairs 3, 4, 5 and 6 between 15° and 34°C (Table V).

TABLE 5. *Activation barriers (18.7° to 33.7° C) in the Pribnow 12-mer Duplex* [1]

Contribution	Activation Barrier (kcal)			
	#3	#4	#5	#6
Saturation Recovery[2]	22	20	21	18
Exchange[3]	30	34	33	29

[1] Buffer is 0.1M phosphate, 5:1 $H_2O:D_2O$, pH 6.90.

[2] The saturation recovery rates between 18.7° and 33.7°C are directly used to determine the activation energy.

[3] The magnetic contribution (saturation recovery rate at 0°C) is subtracted from the saturation recovery rates between 18.7 and 33.7° to evaluate the exchange contribution.

9. *Sequence Dependence of Transient Base Pair Opening*

It was of interest to evaluate the hydrogen exchange kinetics in DNA fragments containing the TATA and TATAAT promotor sequences with the corresponding values for non-promotor sequences. We therefore set out to evaluate the sequence dependence of hydrogen exchange kinetics by comparing the saturation recovery parameters for the AATT 12-mer (Scheme III) and TATA 12-mer (Scheme II) duplexes which have common d(CGCG) segments flanking a sequence dependent dA · dT tetranucleotide core (Patel, *et al.*, 1983b).

We have evaluated the saturation recovery lifetimes for these two dodecanucleotide duplexes in 0.1M phosphate, neutral pH as a function of temperature. The lifetimes at the individual base pairs in the AATT 12-mer and the TATA 12-mer at 22.5°, 37.5° and 52.5°C are summarized in Table VI under temperature conditions where hydrogen exchange makes the predominant contribution to the recovery of magnetization.

It is readily apparent that faster hydrogen exchange rates are observed in the TATA 12-mer compared to the AATT 12-mer at dA · dT base pair 5 and at dA · dT base pair 6 between 22.5° and 52.5°C (Table VI). We also observe faster hydrogen exchange rates for the TATA 12-mer duplex at dG·dC base pair 4 but the exchange rates are similar for both duplexes at dG · dC base pair 2 (Table VI). These studies establish a sequence dependence

TABLE 6. *Temperature dependence of the 498 MHz Saturation recovery lifetimes (msec) of the imino protons in the AATT 12-mer (pH 6.95) and the TATA 12-mer (pH 6.54) in 0.1M phosphate, 4:1 H_2O:D_2O solution*

		Imino proton Lifetime in msecs				
		#2	#3	#4	#5	#6
22.5°C	AATT 12-mer	139	336	418	373	383
	TATA 12-mer	132	339	377	157	181
37.5°C	AATT 12-mer	33	170	283	114	172
	TATA 12-mer	46	177	228	51	70
52.5°C	AATT 12-mer		38	145	30	50
	TATA 12-mer		17	46	20	23

to the hydrogen exchange rates of imino protons in the interior of DNA fragments.

Imino protons in nucleic acids are postulated to undergo hydrogen exchange from the unpaired state as depicted schematically in Scheme VI. The exchange rate constant (k_{EX}) is given by the expression

$$k_{EX} = \frac{k_{OP} k_{CH}(BH)}{k_{CL} + k_{CH}(BH)}$$

where k_{OP} and k_{CL} are the opening and closing rate constants and $k_{CH}(BH)$ is the base and buffer catalyzed rate from the open form with solvent. (Teitelbaum and Englander, 1975a,b).

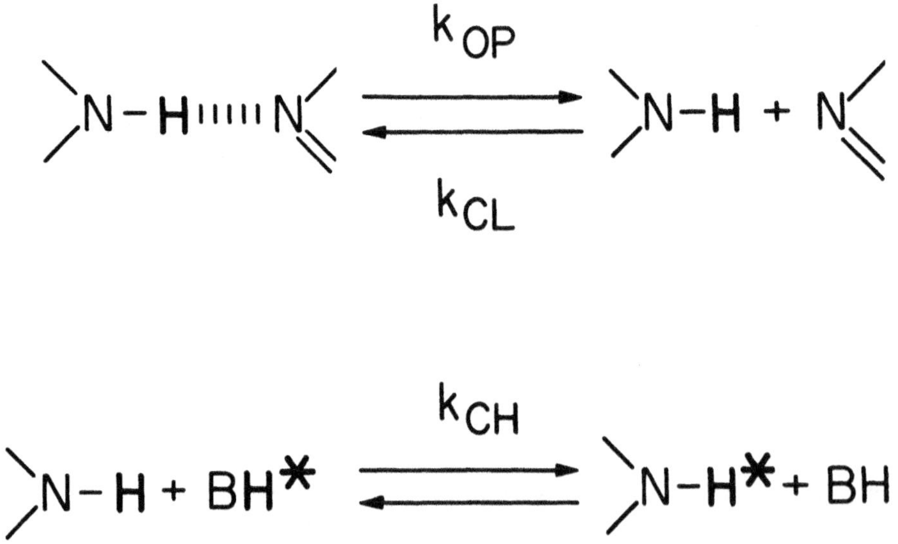

When $k_{CH}(B) >> k_{CL}$, exchange occurs every time the duplex opens and is a direct measure of the local duplex opening rate constant.

$$k_{EX} = k_{OP}$$

The hydrogen exchange rate constant is independent of base and general buffer catalysts and this mechanism is found to predominate for base pairs in the interior of duplexes (Patel and Hilbers, 1975).

By contrast, when $k_{CL} >> k_{CH}(BH)$, the paired and open states are in rapid equilibrium with occasional leakage from the open state with solvent H_2O

$$k_{EX} = \frac{k_{OP}}{k_{CL}} k_{CH}(BH)$$

The hydrogen exchange rate is dependent on the concentration of base and general catalysts and this mechanism is found to predominate for base pairs at the ends of duplexes (Patel and Hilbers, 1975).

We have measured the hydrogen exchange lifetimes for the AATT 12-mer and TATA 12-mer duplexes between pH values ranging from 7 to 8.7 and observe that the lifetimes for internal base pairs 4, 5 and 6 decrease by a factor of < 1.5. The exchange rates are therefore essentially independent of base concentration and hence are a direct measure of the duplex opening rate constants at internal base pairs 4, 5 and 6 in the two dodecanucleotide duplexes. Thus, we observe much faster transient base pair opening rates at the GTATAC segment of the TATA 12-mer (Scheme II) compared to the GAATTC segment in the AATT 12-mer duplex (Scheme III).

By contrast, hydrogen exchange lifetimes decrease by a factor of 2 to 4 on raising the pH from 6.9 to 8.5 in the Pribnow 12-mer in 0.1M phosphate solution (Table IV). This indicates that the prequilibrium pathway also contributes to hydrogen exchange for the Pribnow 12-mer sequence which contains a stretch of eight internal base pairs.

10. *Antibiotic — DNA Complex Formation*

Conformational aspects of ligand-DNA interactions can be approached by searching for and measuring intermolecular NOE effects between protons on the ligand and those on the DNA. A few examples of such complexes are available where the ligand is an antibiotic (Patel, 1982; Reid, *et al.*, 1983b) and a protein (Alma, *et al.*, 1981, 1983; Nick, *et al.*, 1982; Kime and Moore, 1983; Scheek, *et al.*, 1983). We report below on such an investigation of the netropsin·AATT 12-mer complex where we have successfully observed intermolecular NOEs between antibiotic and nucleic acid protons (Patel, 1982).

Previous studies have demonstrated that the peptide antibiotic netropsin (Scheme VII) binds in the minor groove at dA · dT rich sites on double helical DNA (for reviews see, Zimmer, 1975;

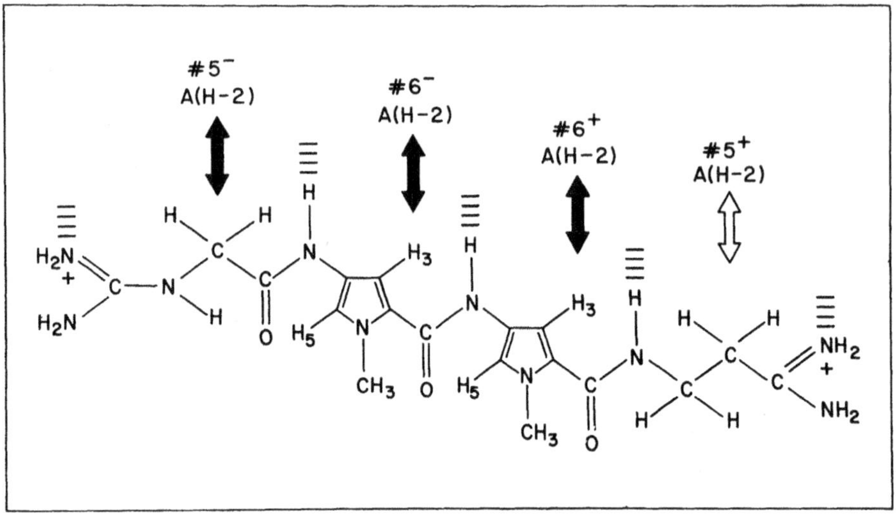

Hahn, 1975; Krey, 1980). The N-methylpyrrole ring, the peptide groups and the charged ends of the antibiotic can form van der Waals' contacts, intermolecular hydrogen bonds and electrostatic interactions with the nucleic acid base pair edges in the complex. (Wartell, *et al.*, 1974; Patel and Canuel, 1977; Patel, 1979; Dattagupta, *et al.*, 1980; Patel, *et al.*, 1981; Patel, 1982). The crystal structure of netropsin (Berman, *et al.*, 1979) and the AATT 12-mer (Dickerson *et al.*, 1983) have been individually determined but no crystallographic information is available on the antibiotic complex with DNA.

The proton NMR spectrum of the complex formed on addition of 1 equivalent of netropsin to the AATT 12-mer in H_2O at 30°C is presented in Figure 10A. The exchangeable resonances are designated by asterisks and include the imino (12.5 to 14.5 ppm) and amino (6 to 8.5 ppm) protons of the dodecanucleotide and the peptide and amino end groups (9 to 10 ppm; 6 to 8.5 ppm)

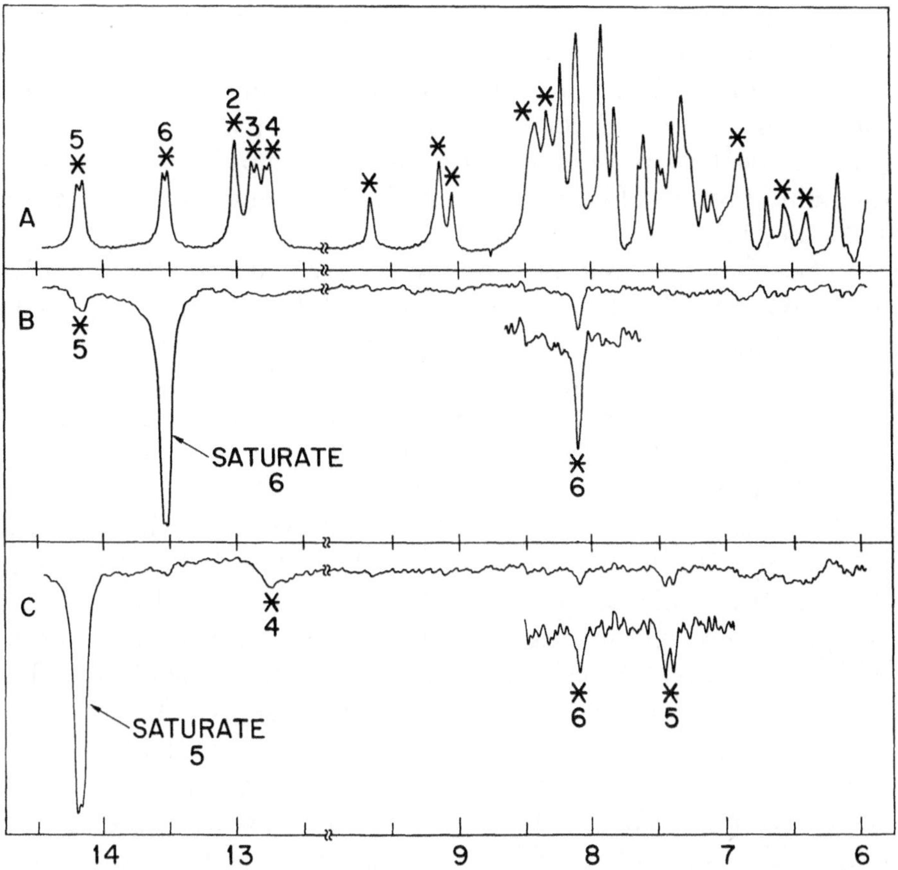

FIG. 10. (A) The 498 MHz proton NMR spectrum (12.5 to 14.5 ppm; 6.0 to 10.0 ppm) of the 1 netropsin per d(CGCGAATTCGCG) complex in 0.2M phosphate, 5 mM EDTA, 4:1 H_2O: D_2O, pH 6.9 at 30°C. The exchangeable resonances are designated by asterisks. (B) Difference spectrum following saturation of the 13.53 ppm thymidine imino proton in the complex. Negative NOE effects designated by asterisks are observed at the 8.09 ppm adenosine H-2 proton of the same dA · dT base pair and 14.17 ppm thymidine H-3 imino protons of an adjacent dA · dT base pair. (C) Difference spectrum following saturation of the 14.17 ppm thymidine imino proton in the complex. Negative NOE effects designated by asterisks are observed at the 7.44 and 7.37 ppm adenosine H-2 protons of the same dA · dT base pair, the 8.09 ppm adenosine H-2 proton of an adjacent dA · dT base pair and the 12.75 ppm guanosine imino proton of an adjacent dG · dC base pair. The saturation pulse was on for 250 msec during each accumulation.

of the antibiotic in the complex. Netropsin (Scheme VII) lacks an element of symmetry and hence removes the two fold symmetry of the AATT 12-mer on formation of a tight complex. Thus, the symmetry related imino protons of base pairs 3, 4, 5 and 6 give resolved pairs of resonances in the complex in contrast to the imino protons of base pair 2 which give a single peak. (Figures 10A and 11B).

The imino protons of individual base pairs in the complex were assigned from inter-base pair NOE measurements with typical data used to differentiate dA · dT base pair 5 from dA · dT base pair 6 shown in Figures 10B and 10C. These studies demonstrate that the thymidine imino proton of dA · dT base pair 5 resonates to low field from that of dA · dT base pair 6 in the complex (Figure 10A).

We observe intra- and inter-base pair NOEs in the aromatic region on saturation of the assigned thymidine imino protons of dA · dT base pairs 6 (Figure 10B) and 5 (Figure 10C) in the complex. This permits assignment of the adenosine H-2 protons with that of dA · dT base pair 6 resonating to low field from that of dA · dT base pairs 5 in the complex (Figures 10B and 10C).

The large selective complexation shifts observed at the thymidine imino protons (Figure 11) and adenosine H-2 protons of AATT 12-mer (Table VII) suggests that netropsin binds in the dA · dT rich core of the DNA duplex. The chemical shift information cannot be directly analyzed in structural terms since large downfield complexation shifts are observed at the thymidine imino proton of base pair 5 and adenosine H-2 of base pair 6 while the thymidine imino proton of base pair 6 and the adenosine H-2 of base pair 5 are essentially unperturbed on complex formation (Table VII).

We have differentiated between the pyrrole H-3 and H-5 protons of netropsin in the AATT 12-mer complex since the latter show intramolecular NOEs to the pyrrole NCH_3 protons. The pyrrole H-3 protons resonate to high field of the pyrrole H-5 protons for netropsin in the absence and presence of AATT 12-mer with large downfield shifts observed at the pyrrole H-5 protons on complex formation (Table VII). Though this chemical shift

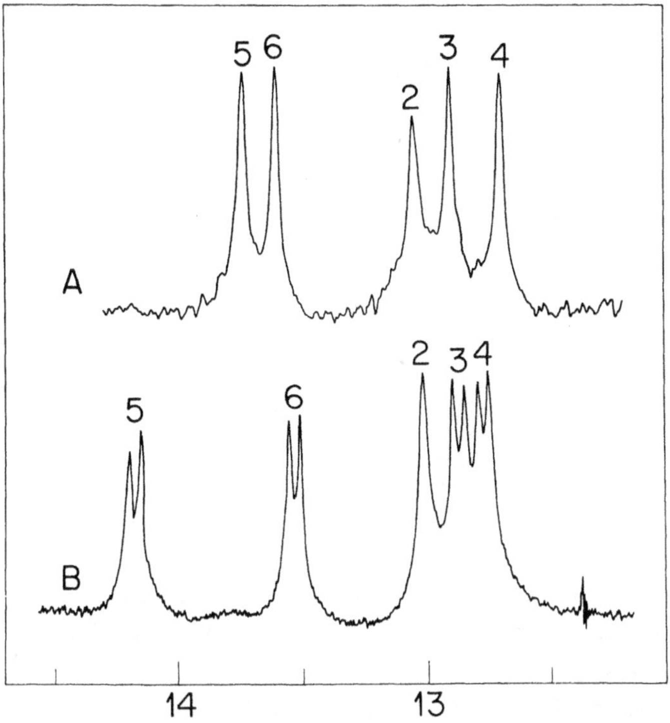

FIG. 11. The 498 MHz proton NMR spectra of the (A) AATT 12-mer and (B) the 1 netropsin per AATT 12-mer complex at 30°C. The spectrum of the duplex was recorded in 0.1M phosphate, 2.5 mM EDTA, 4:1 $H_2O:D_2O$, pH 6.95 with a line broadening of 5 Hz. The spectrum of the complex was recorded in 0.2M phosphate, 5 mM EDTA, 5:1 $H_2O:D_2O$, pH 6.86 with a line broadening of 1 Hz.

TABLE 7. *Chemical Shift Changes of the AATT 12-mer Thymidine H-3 and Adenosine H-2 Protons on Netropsin Complex Formation at 30° C*

	12-mer[1] δ, ppm	Complex[2] δ, ppm	Δδ, ppm[3]
Base Pair 5			
T(H-3)	13.73	14.19, 14.15	—0.44
A(H-2)	7.26	7.44, 7.37	—0.14
Base Pair 6			
T(H-3)	13.59	13.55, 13.51	+0.06
A(H-2)	7.61	8.09	—0.48

[1] Buffer: 0.1M phosphate, 2.5 mM EDTA, 4:1 $H_2O:D_2O$, pH 6.95.
[2] Buffer: 0.2M phosphate, 5 mM EDTA, 4:1 $H_2O:D_2O$, pH 6.9.
[3] Downfield and upfield complexation shifts are designated by minus and plus signs respectively.

TABLE 8. *Chemical Shift Changes of the Netropsin Pyrrole NCH_3, H-3 and H-5 Protons on AATT 12-mer Complex Formation at 30° C*

	Netropsin [1] δ, ppm	Complex [2] δ, ppm	Δδ, ppm [3]
Pyrrolle H-5	7.08	7.63	—0.55
	7.08	7.48	—0.40
Pyrrole NCH_3	3.78	3.89	—0.11
	3.79	4.02	—0.23
Pyrrole H-3	6.76	6.67	+0.09
	6.73	6.55	+0.18

[1] Netropsin chemical shifts in the dissociated antibiotic·d(GGAATTCC) complex in 0.1M NACl. 10 mM phosphate, 1 mM EDTA, D_2O at 90°C.
[2] Netropsin chemical shifts in the intact antibiotic·d(CGCGAATTCGCG) complex in 0.2M phosphate, 5 mM EDTA, D_2O, pH 6.9 at 30°C.
[3] Downfield and upfield complexation chemical shifts are designated by minus and plus signs respectively.

information may suggest that the pyrrole H-5 protons of netropsin are interacting with the DNA, the NOE data outlined below rule out this possibility.

Current models of the netropsin-DNA complex postulate antibiotic binding into the minor groove at dA·dT rich sites on DNA (Zimmer, 1975; Wartell, *et al.*, 1974; Patel and Canuel, 1977). We therefore searched for intermolecular NOEs between protons on the N-methyl pyrrole rings of the antibiotic and the adenosine H-2's of base pairs 5 and 6 located in the minor groove in the center of the AATT 12-mer duplex. The aromatic spectral region (6.5 to 8.5 ppm) of the netropsin · AATT 12-mer complex at 30°C is shown in Figure 12A. We observe large intermolecular NOEs at the superpositioned 8.09 ppm adenosine H-2 protons of dA · dT base pair 6 in the complex on saturation of the netropsin H-3 protons at 6.55 ppm (Figure 12B) and at 6.67 ppm (Figure 12C). The magnitude of these intermolecular NOEs as a function of the length of the saturation pulse are listed in Table IX and exhibit large values of -0.43 and -0.60 for a 1 second saturation pulse. Saturation of the 8.09 ppm adenosine H-2 proton of dA· ·dT base pair 6 results in intermolecular NOEs at the 6.55 ppm and 6.67 ppm netropsin H-3 protons of the complex (Figure 12D

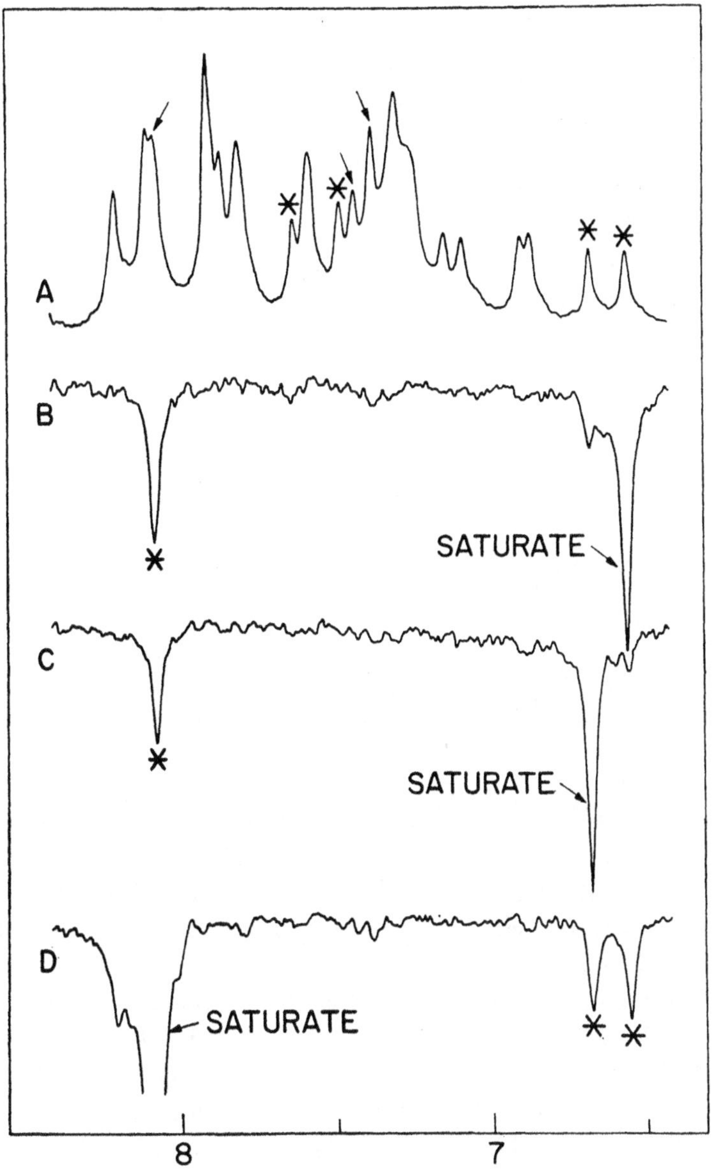

FIG. 12. (A) The 498 MHz proton NMR spectrum (6.5 to 8.5 ppm) of the 1 netropsin per d(CGCGAATTCGCG) duplex complex in 0.2M phosphate, 5 mM EDTA, D_2O, pH 6.9 at 30°C. The arrows designate the adenosine H-2 protons of dA·dT base pairs 6 (8.09 ppm) and 5 (7.44, 7.37 ppm) while the asterisks designate netropsin pyrrole H-5 (7.63, 7.48 ppm) and pyrrole H-3 (6.67, 6.55 ppm) protons. (B and C) A negative intermolecular NOE is observed at the 8.09 ppm adenosine H-2 proton of dA · dT base pair 6 on saturation of the 6.55 ppm netropsin pyrrole H-3 proton in difference spectrum B and on saturation of the 6.67 ppm netropsin pyrrole H-3 proton in difference spectrum C of the complex. (D) Negative intermolecular NOE effects are observed at the pyrrole H-3 protons at 6.67 and 6.55 ppm on saturation of the 8.09 ppm adenosine H-2 proton of dA · dT base pair 6 in the complex. The saturation pulse was on for 1000 msec during each accumulation.

TABLE 9. *Intermolecular NOE Effects Between the AATT 12-mer Adenosine H-2 Protons of Base Pair 6 and the Pyrrole Ring H-3 Protons of Netropsin in the Complex at 30° C*

Saturation at	NOE Observed at		
	Net H-3 [1]	Net H-3 [2]	DNA A(H-2) [3]
100 msec			
Net H-3 [1]			—0.12
Net H-3 [2]			—0.17
DNA A(H-2) [3]	—0.09	—0.13	
250 msec			
Net H-3 [1]			—0.29
Net H-3 [2]			—0.37
DNA A(H-2) [3]	—0.21	—0.22	
1000 msec			
Net H-3 [1]			—0.43
Net H-3 [2]			—0.60
DNA A(H-2) [3]	—0.34	—0.37	

[1] Netropsin H-3 resonance at 6.67 ppm in complex at 30°C.
[2] Netropsin H-3 resonance at 6.55 ppm in complex at 30°C.
[3] Adenosine H-2 resonance of dA·dT base pair at 8.09 ppm in complex at 30°C.

and Table IX) confirming the conclusions reached above. These results establish two points of contact between the antibiotic and the DNA with each N-methylpyrrole H-3 proton of netropsin in close proximity with each adenosine H-2 proton located in the center of the duplex (Scheme VII).

We do not observe intermolecular NOEs between the adenosine H-2's of the dA · dT base pairs and either the pyrrole H-5 or NCH_3 protons of the netropsin. We conclude that it is the netropsin pyrrole H-3 protons that are directed towards the DNA and the pyrrole H-5 and NCH_3 protons that face the exterior in the antibiotic-DNA complex.

We have also observed a weaker intermolecular NOE between the adenosine H-2 proton of dA·dT base pair 5 at 7.44 ppm and a proton resonance(s) at 4.12 ppm in the complex. The latter resonance is tentatively assigned to the guanidine methylene protons of netropsin and establishes a third point of interaction in the complex (Scheme VII). The remaining adenosine H-2 proton of dA·dT base pair 5 at 7.37 ppm is not well resolved and attempts to search for an intermolecular NOE to the amidinopropyl CH_2—CH_2 groups of netropsin have not yet been successful (Scheme VII).

We have generated the complex between netropsin and the AATT 12-mer subject to the constraints from the intermolecular NOEs between the antibiotic and the DNA and a CPK model of the complex is presented in Plate I. The three amide peptide groups of netropsin can hydrogen bond with the adenosine N3 and thymidine O2 acceptor groups in the minor groove of the dA·dT base pairs. The antibiotic spans four base pairs in the proposed model of the complex.

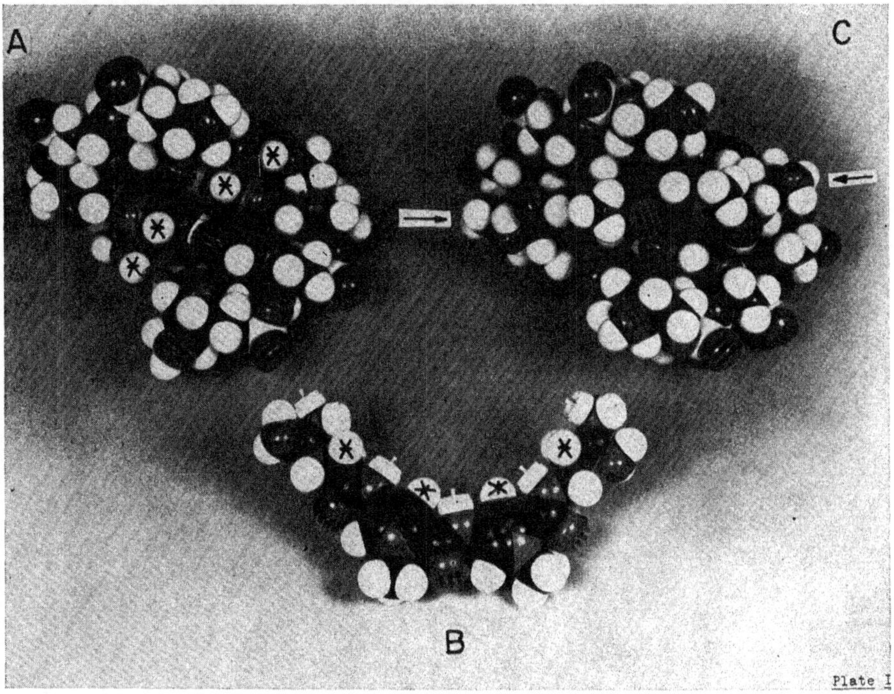

PLATE I: (A) A view of the minor groove in the CPK model of the d(AATT) tetranucleotide duplex segment. The adenosine H-2 protons on the four dA·dT base pairs are designated by asterisks. The DNA helix axis is inclined relative to the vertical of the page. (B) A CPK model of netropsin in a planar configuration. The peptide amide and pyrrole H-3 protons are on the concave face while the peptide carbonyl, pyrrole H-5 and NCH_3 groups are on the convex face. (C) A view of the CPK model of the netropsin·d(AATT) complex. The concave face of the antibiotic is directed into the minor groove of the nucleic acid. The ends of the netropsin molecule are designated by arrows. The DNA helix axis in the complex is inclined relative to the vertical of the page.

The netropsin interacts directly with the DNA and hence complex formation must involve displacement of the tightly bound water molecules that are observed in the minor groove dA · dT rich core of the AATT 12-mer in the crystalline state. The intermolecular NOE measurements readily explain why netropsin does not bind to dG · dC rich sites on duplex DNA. The magnitude of the intermolecular NOEs suggests that the pyrrole H-3 protons are proximal to the adenosine H-2 protons in the complex so that there would be no space to replace the adenosine H-2 proton by a bulky guanosine 2-amino group and still maintain the interaction.

11. *Kinetic Stabilization at and adjacent to the Antibiotic Binding Site*

We have compared the hydrogen exchange kinetics of the imino protons of base pairs at and adjacent to the netropsin binding site on the d(AATT) segment of the d(CGCGAATTCGCG) duplex (Pardi, *et al.*, 1983). Large differences are observed in the imino proton magnetization recovery rates between the AATT 12-mer duplex and its netropsin complex as monitored at dG · dC base pair 4 (Figure 13A), dA · dT base pair 5 (Figure 13B) and dA · dT base pair 6 (Figure 13C) in phosphate buffer at 52.5°C.

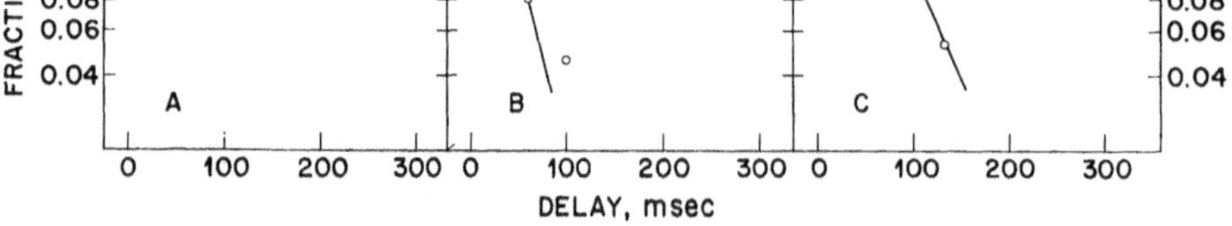

FIG. 13. Plots of the fractional saturation of the imino protons of (A) dG · dC base pair 4, (B) dA · dT base pair 5 and (C) dA · dT base pair 6 in the AATT 12-mer duplex in 0.1M phosphate, 4:1 $H_2O:D_2O$, pH 6.95, 52.5°C (O) and the complex of the AATT 12-mer with 1 equivalent of netropsin in 0.2M phosphate, 5:1 $H_2O:D_2O$, pH 6.86, 52.5°C (●). These experimental data were collected at 498 MHz.

The observed imino proton lifetimes are a measure of hydrogen exchange at this temperature and the experimental values for the dodecanucleotide and its antibiotic complex at 52.5° and 60°C are summarized in Table X.

TABLE 10. *Temperature dependence of the 498 MHz saturation recovery lifetimes (msec) of the imino protons in the AATT 12-mer* [1] *and the complex of the AATT 12-mer with 1 equivalent of netropsin* [2]

Temp. °C	Imino proton lifetimes in msec			
	G·C #3	G·C #4	A·T #5	AT #6
52.5°C				
12-mer	38	145	30	50
Complex	38	249	176	162
60.0°C				
12-mer	15	45	16	23
Complex	26	116	94	87

[1] Buffer: 0.1M phosphate, 2.5 mM EDTA, 4:1 $H_2O:D_2O$, pH 6.95.
[2] Buffer: 0.2M phosphate, 5 mM EDTA, 4:1 $H_2O:D_2O$, pH 6.86.

Netropsin binding at the dA · dT rich core of the AATT 12-mer duplex results in a kinetic stabilization of a factor of 3.5 at dA · dT base pair 6 and a factor of 6 at dA · dT base pair 5 at the complexation site (Table X). We also note a kinetic stabilization of ~2 at dG·dC base pair 4 which is adjacent to the netropsin binding site.

Arrhenius plots of the saturation recovery lifetimes of the AATT 12-mer and its netropsin complex at dG · dC base pair 3 (Figure 14A), dG · dC base pair 4 (Figure 14B), dA · dT base pair 5 (Figure 14C) and dA · dT base pair 6 (Figure 14D) indicate that the kinetic stabilization on antibiotic binding is maintained over the entire temperature range where exchange makes the predominant contribution. Further, similar activation barriers are observed for imino proton exchange in the absence and presence of netropsin (Figure 14) indicative of the same base pair opening process operating for the dodecanucleotide and its antibiotic complex (Pardi, *et al.*, 1983).

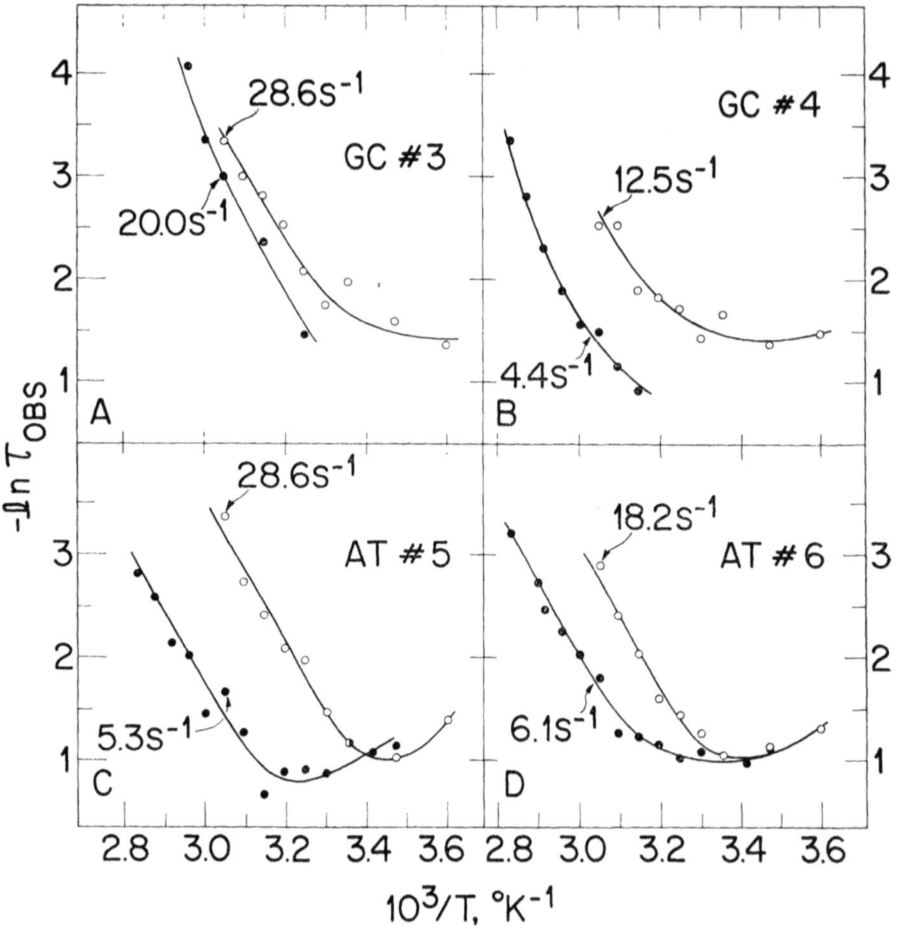

FIG. 14. Arrhenius plots for the inverse of the observed imino proton lifetimes of (A) dG·dC base pair 3, (B) dG·dC base pair 4, (C) dA·dT base pair 5 and (D) dA·dT base pair 6 in the AATT 12-mer duplex at pH 6.8 (O) and the complex of AATT 12-mer with 1 equivalent of netropsin at pH 7.0 (●) in 0.1M phosphate, 2.5 mM EDTA, 4:1 H_2O:D_2O. These experimental data were collected at 360 MHz.

SUMMARY

This review summarizes our recent NMR contributions on the role of sequence on the structure and base pair opening kinetics of DNA fragments and their antibiotic complexes in solution. The experimental techniques are outlined with examples on the d(CGATTATAATCG) duplex which contains the consensus TATAAT promotor sequence.

The well resolved imino and nonexchangeable base protons are assigned from one-dimensional NOE experiments and their temperature-dependent chemical shifts are used to monitor the premelting and melting transitions. We have monitored the distance-dependent NOEs between minor groove proton markers on adjacent base pairs to demonstrate that the dA · dT base pairs are propeller-twisted with a defined handedness which results in a sequence dependence to base pair stacking in solution.

We have monitored hydrogen exchange kinetics at the individual base pair level from saturation recovery measurements on the resolved and assigned imino protons in the d(CGATTATAATCG) duplex in 0.1M phosphate solution. A comparison of this exchange data with related parameters in the d(CGCGTATACGCG) and d(CGCGAATTCGCG) duplexes demonstrates a sequence dependence to base pair opening rates in DNA helices. Specifically, dA·dT rich sequences such as TATA and TATAAT which are common features of promotor sites exhibit transient opening kinetics which are a factor of 2 to 3 faster than the related AATT sequence.

The binding site of netropsin on the d(CGCGAATTCGCG) duplex has been elucidated from intermolecular NOE measurements between the antibiotic and the nucleic acid. These experiments demonstrate that netropsin binds in the minor groove at the AATT central region of the dodecanucleotide. A comparison of the hydrogen exchange kinetics of the imino protons in the dodecanucleotide and its antibiotic complex indicates that netropsin kinetically stabilizes the dA·dT base pairs at its binding site as well as flanking dG·dC base pair regions on the d(CGCGAATTCGCG) duplex.

The review has emphasized one dimensional NMR measurement of DNA fragments and their antibiotic complexes to estimate structural features of nucleic acids and their ligand complexation sites. Recent advances in two dimensional NMR techniques can readily be applied to nucleic acids resulting in the complete assignment of all the base and sugar protons in DNA fragments up to ~2 turns of helix. (Feigon, *et al.*, 1983; Haasnoot and Hilbert, 1983; Scheek, *et al.*, 1983; Hare, *et al.*, 1983; Weiss, *et al.*, 1984). Several laboratories are currently attempting to couple the distances derived from NOE measurements between pairs of proton spins with distance-geometry algorithms to derive detailed nucleic acid structures by NMR in solution.

Acknowledgement

The 498 MHz proton NMR spectra were recorded at the Francis Bitter National Magnet Laboratory, Massachusetts Institute of Technology (funded by National Institutes of Health Grant RR0095 and National Science Foundation Grant C-670).

REFERENCES

Alma, N.C.M., Harmsen, B.J.M., Hull, W.E., van der Marel, G., van Boom, J.H., and Hilbers, C.W., (1981), *Biochemistry*, **20**, 4419-4428. Double resonance experiments on gene-5 protein and its complex with octadeoxyriboadenylic acid.

Alma, N.C.M., Harmsen, B.J.M., van Boom, J.H., van der Marel, G., and Hilbers, C.W.

(1983), *Biochemistry*, **22**, 2104-2115. NMR study of the structure and structural alterations of gene-5 protein-oligo (deoxyadenylic acid) complexes.

Altona, C., (1982), *Recl. Trav. Chim. Pays-Bas.*, **101**, 413-433. Conformational analysis of nucleic acids. Determination of backbone geometry of single-helical RNA and DNA in aqueous solution.

Arnott, S., and Hukins, D.W.L., (1972), *Biochem. Biophys. Res. Commun.*, **47**, 1504-1509. Optimized parameters for A-DNA and B-DNA.

Arnott, S., Chandrasekaran, R., Hall, I.H., Puigjaner, L.C., and Walker, J.K., (1983), *Cold Spring Harbor Symposium Quantitative Biology*, **47**, 53-65. DNA secondary structures: Helices, Wrinkles and Junctions.

Berman, H. M. Neidle, S., Zimmer, C., and Thrum, H., (1979), *Biochem. Biophys. Acta*,. **561**, 124-131. Netropsin, A DNA-binding oligopeptide: Structural and binding studies.

Bothner-By, A.A., (1979), in *Biological Applications of Magnetic Resonance*, ed., Shulman, R.G., Academic Press, pp. 177-219. Nuclear Overhauser effects on protons and their use in the investigations of structures of biomolecules.

Broido, M.S., and Kearns, D.R., (1982), *J. Am. Chem. Soc.*, **104**, 5207-5216. Proton NMR evidence for a left-handed helical structure of poly(ribocytidylic acid) in neutral solution.

Calladine, C.R., (1982), *J. Mol. Biol.*, **161**, 343-362. Mechanics of Sequence-dependent stacking of bases in B-DNA.

Chou, S.H., Hare D.R., Wemmer, D.E., and Reid, B.R., (1983), *Biochemistry*, **22**, 3037-3041. Chemical synthesis and NMR assignment of the imino protons of λO_R3 operator deoxyribonucleic acid.

Dattagupta, N., Hogan, M., and Crothers, D.M., (1980), *Biochemistry*, **19**, 5998-6005. Interaction of netropsin and distamycin with DNA: Electric dichroism study.

Dickerson, R.E., Drew, H.R., Conner, B.N., Kopka, M.L., and Pjura, P.E., (1983), *Cold Spring Harbor Symposium on Quantitative Biology*, **XLVII**, pp. 13-24. Helix geometry and hydration in A-DNA, B-DNA and Z-DNA.

Dickerson, R.E., (1983), *J. Mol. Biol.*, **166**, 419-441. Base sequence and helix structure variations in B-DNA and A-DNA.

Early, T.A. Kearns, D.R., Hillen, W., and Wells, R.D., (1980), *Nucleic Acids Research*, **8**, 5795-5812. NMR study of a 12-base pair restriction fragment: Investigation of structure by relaxation measurements.

Early, T.A., Kearns, D.R., Hillen, W., and Wells, R.D., (1981a), *Biochemistry*, **20**, 3756-3764. NMR investigations of a 12-base pair DNA restriction fragment: Relaxation behavior of the low field resonances in water.

Early, T.A., Kearns, D.R., Hillen, W., and Wells, R.D., (1981b), *Biochemistry*, **20**, 3764-3769. Proton NMR investigations of DNA Restriction Fragments: Dynamic Properties.

Feigon, J., Wright, J.M., Denny, W.A., Leupin, W., and Kearns, D.R., (1983), *Cold Spring Harbor Symposia on Quantitative Biology*, **XLVII**, 207-217. Application of multiple-pulse H-NMR techniques to the study of two synthetic DNA decamers.

Haasnoot, C.A.G., and Hilbers, C.W., (1983), *Biopolymers* **22**, 1259-1266. Effective water resonance suppression in 1D- and 2D-FT proton NMR spectroscopy of biopolymers in solution.

Hahn, F.E., (1975) in *Antibiotics III*, eds. Corcoran, J.W., and Hahn, F.E., Springer-Verlag, pp. 79-100. Distamycin and Netropsin.

Hare, D.R., and Reid, B.R., (1982), *Biochemistry*, **21**, 1835-1842. Direct assignment of the Dihydrouridine-Helix imino proton resonances in transfer RNA NMR spectra by means of the NOE.

Hare, D.R., Wemmer, D.E., Chou, S.H., Drobny, G., and Reid, B.R., (1983), *J. Mol. Biol.*, *171*, 319-336. Assignment of the nonexchangeable proton resonances of d(CGCGAATT-CGCG) using two-dimensional NMR methods.

Hayashi, Y., Nakanishi, N., Tsuboi, M., Tazawa, I., and Inoue, Y., (1981), *Biochemica et. Biophysica Acta*, **656**, 120-122. Hydrogen-exchange kinetics in double-helical polynucleotide with the adenine-uracil base pair.

Heerschap, A., Haasnoot, C.A.G., and Hilbers, C.W., (1982), *Nucleic Acids Research,* **10**, 6981-7000. NMR studies on yeast $tRNA^{phe}$. Assignment of the imino proton resonances of the acceptor and D stem by means of nuclear Overhauser effect measurements.

Hilbers, C.W., (1979), in *Biological Applications of Magnetic Resonance*, ed., Shulman, R.G., Academic Press, pp. 1-43. Hydrogen-bonded proton exchange and its effect on NMR spectra of nucleic acids.

Hurd, R.E., and Reid, B.R., (1980), *J. Mol. Biol.*, **142**, 181-193. Helix-coil dynamics in RNA: The amino acid acceptor helix of E. coli $tRNA^{phe}$.

Johnston, P.D. and Redfield, A.G., (1977), *Nucleic Acids Research,* **4**, 3599-3315. An NMR study of the exchange rates for protons involved in the secondary and tertiary structure of yeast $tRNA^{phe}$.

Johnston, P.D., and Redfield, A.G., (1981), *Biochemistry,* **20**, 1147-1156. NMR and nuclear Overhauser effect study of yeast $tRNA^{phe}$ imino protons.

Kan, L.S., Cheng, D.M., Jayaraman, K., Leutzinger, E.E., Miller, P.S., and Ts'o, P.O.P., (1982), *Biochemistry,* **21**, 6723-6732. Proton NMR study of a self-complementary decadeoxyribonucleotide d(CCAAGCTTGG).

Kime, M.J., and Moore, P.B., (1983a), *Biochemistry,* **22**, 2622-2629. Nuclear Overhauser effect on the downfield proton spectra of 5S ribonucleic acid and its complex with ribosomal protein L25.

Kime, M.J., and Moore, P.B., (1983b), *Biochemistry,* **22**, 2615-2622. Nuclear Overhauser experiments on the downfield proton spectrum of a ribonuclease resistant fragment of 5S ribonucleic acid.

Krey, A.K., (1980), in *Progress in Molecular and Subcellular Biology,* 7, eds., Hahn, F.E., Kersten, H., Kersten, W., and Szybalski, W., Springer-Verlag, pp. 43-87. Non-intercalative binding to DNA.

Mandal, C., Kallenbach, N.R., and Englander, S.W., (1979), *J. Mol. Biol.*, **135**, 391-411. Base-pair opening and closing reactions in the double helix.

Mandal, C., Englander, S.W., and Kallenbach, N.R., (1980), *Biochemistry,* **19**, 5819-5825. Hydrogen-deuterium exchange analysis of ligand-macromolecule interactions: Ethidium-deoxyribonucleic acid system.

Neidle, S., and Berman, H., (1983), *Prog. Biophys. Molec. Biol.*, **41**, 43-66. X-Ray crystallographic studies of nucleic acids and nucleic acid-drug complexes.

Nick, H., Arndt, K., Boschelli, F., Jarema, M.A.C., Lillis, M., Sadler, J., Caruthers, M., and Lu, P., (1982), *Proc. Natl. Acad. Scs.* USA, **79**, 218-222. *Lac* repressor — *lac* operator interactions: NMR observations.

Pardi, A., and Tinoco, I., Jr., (1982), *Biochemistry,* **21**, 4686-4693. Kinetics for exchange of imino protons in deoxyribonucleic acid, ribonucleic acid and hybrid oligonucleotide helices.

Pardi, A., Morden, K.M., Patel, D.J., and Tinoco, I., Jr., (1982), *Biochemistry,* **21**, 6567-6574 Kinetics for exchange of imino protons in the d(CGCGAATTCGCG) double helix and in two similar helices that contain a G·T base pair and an extra adenine.

Pardi, A., Morden, K.M., Patel, D.J., and Tinoco, Jr., I., (1983), *Biochemistry,* **22**, 1107-1113. Kinetics for exchange of the imino protons of the d(CGCGAATTCGCG) double helix in complexes with the antibiotics netropsin and/or actinomycin.

Patel, D.J., and Hilbers, C.W., (1975), *Biochemistry,* **14**, 2651-2656. Proton NMR investigations of fraying in double-stranded d(ATGCAT) in H_2O solution.

Patel, D.J., and Canuel, L.L., (1977), *Proc. Natl. Acad. Scs.* USA., **74**, 5207-5211. Netropsin··poly(dA—dT) complex in solution: Structure and dynamics of antibiotic-free base pair regions and those centered on bound netropsin.

Patel, D.J., (1979), *Eur, J. Biochem.*, **99**, 369-378. Netropsin·d(GGAATTCC) complex. Antibiotic binding at adenine·thymine base pairs in the minor groove of the self-complementary oligonucleotide duplex.

Patel, D.J., and Canuel, L.L., (1979), *Eur. J. Biochem.*, **96**, 267-276. Helix-coil transition of the self-complementary d(GGAATTCC) duplex.

Patel, D.J., Kozlowski, S.A., Rice, J.A., Broka, C., and Itakura, K., (1981), *Proc. Natl. Acad. Scs.* USA., **78**, 7281-7284. Mutual interaction between adjacent dG·dC actinomycin binding sites and dA·dT netropsin binding sites on the self-complementary d(CGCGAA-TTCGCG) duplex in solution.

Patel, D.J., (1982), *Proc. Natl. Acad. Scs.* USA., **79**, 6424-6428. Antibiotic-DNA interactions: Intermolecular nuclear Overhauser effects in the netropsin—d(CGCGAATTCGCG) complex in solution.

Patel, D.J., Pardi, A., and Itakura, K., (1982a), *Science*, **216**, 581-590. DNA Conformation, Dynamics and Interactions in solution.

Patel, D.J., Kozlowski, S.A., Nordheim, A., and Rich, A., (1982b), *Proc. Natl. Acad. Scs.* USA., **79**, 1413-1417. Right-handed and left-handed DNA: Studies of B- and Z-DNA by using proton nuclear Overhauser effect and P NMR.

Patel, D.J., Kozlowski, S.A., Marky, L.A., Broka, C., Rice, J.A., Itakura, K., and Breslauer, K.J., (1982c), *Biochemistry*, **21**, 428-436. Premelting and melting transitions in the d(CGCGAATTCGCG) self-complementary duplex in solution.

Patel, D.J. Kozlowski, S.A., Ikuta, S., Itakura K., Bhatt, R., and Hare, D.R., (1983a), *Cold Spring Harbor Symposia on Quantitative Biology*, **XLVII**, 197-206. NMR Studies of DNA Conformation and Dynamics in Solution.

Patel, D.J., Ikuta, S., Kozlowski, S., and Itakura, K., (1983b), *Proc. Natl. Acad. Scs.* USA., **80**, 2184-2188. Sequence dependence of hydrogen exchange kinetics in DNA duplexes at the individual base pair level in solution.

Patel, D.J., Kozlowski, S.A., and Bhatt, R., (1983c), *Proc. Natl. Acad. Scs.* USA, **80**, 3908-3912. Sequence dependence of base-pair stacking in right-handed DNA in solution: Proton nuclear Overhauser effect NMR measurements.

Redfield, A.G., Roy, S., Sanchez, V., Tropp, J., and Figueroa, N., (1981), in *Second* SUNYA *Conversation in the Discipline Biomolecular Stereodynamics*, *I*, ed., Sarma, R.H., Adenine Press, N.Y., pp. 195-208. Nuclear Overhauser effect studies of tRNA: A progress report.

Reid, D.G. Salisbury, S.A., Bellard, S., Shakked, Z., and Williams, D.H., (1983a), *Biochemistry*, **22**, 2019-2025. Proton nuclear Overhauser effect study of the structure of a deoxyoligonucleotide duplex in aqueous solution.

Reid, D.G. Salisbury, S.A., and Williams, D.H., (1983b), *Biochemistry*, **22**, 1377-1385. Proton nuclear Overhauser effect study of the structure of an actinomycin D complex with a self-complementary tetranucleoside triphosphate.

Roy, S., and Redfield, A.G., (1981), *Nucleic Acids Research*, **9**, 7073-7083. Nuclear Overhauser Study and assignment of D-stem and reverse-Hoogsteen base pair proton resonances in yeast $tRNA^{Asp}$.

Sanderson, M.R., Mellema, J.R., van der Marel, G.A., Wille, G., van Boom, J.H., and Altona, C., (1983), *Nucleic Acids Research*, **11**, 3333-3346. Nonexchangeable base and H1′ resonance assignments of d(G—G—C*—C*—G—G—C—C) using the nuclear Overhauser effect.

Scheek, R.M. Zuiderweg, E.R.P., Klappe, K.J.M., van Boom, J.H., Kaptein, R., Ruterjans, H., and Beyreuther, K., (1983a), *Biochemistry*, **22**, 228-235. *Lac* repressor headpiece binds specifically to half of the *lac* operator: A proton NMR study.

Scheek, R.M., Russo, N., Boelens, R., and Kaptein, R., (1983b), *J. Am. Chem. Soc.*, **105**, 2914-2916. Sequential resonance assignments in DNA proton NMR spectra by two-dimensional NOE spectroscopy.

Schimmel, P.R., and Redfield, A.G., (1980), *Ann. Res. Biophys. Bioeng.*, **9**, 181-221. Transfer RNA in solution: Selected Topics.

Teitelbaum, H., and Englander, S.W., (1975a), *J. Mol. Biol.*, **92**, 57-78. Hydrogen exchange study of adenine-containing double helices.

Teitelbaum, H. and Englander, S.W., (1975b), *J. Mol. Biol.*, **92**, 79-92. Hydrogen exchange studies of cytosine-containing double helices.

Tran-Dinh, S., Neumann, J.M., Huynh-Dinh, T., Genissel, B., Igolen, J., and Simonnot, G., (1982), *Eur. J. Biochem.*, **124**, 415-425. DNA fragment conformations: d(ACATGT).

Tropp, J., and Redfield, A.G., (1981), *Biochemistry*, **20**, 2133-2140. Environment of ribothymidine in transfer ribonucleic acid studied by means of nuclear Overhauser effect.

Viswamitra, M.A., (1983), *Cold Spring Harbor Symposia on Quantitative Biology*, **XLVII**, pp. 25-31. Structural Diversity in DNA: From monomer structures to oligonucleotides.

Wang, A.H.J., Fujii, S., van Boom, J.H., and Rich, A., (1983), *Cold Spring Harbor Symposium on Quantitative Biology*, **XLVII**, pp. 33-44. Right-handed and left-handed double helical DNA: Structural Studies.

Wartell, R.M., Larson, J.E., and Wells, R.D., (1974), *J. Biol, Chem.*, **249**, 6719-6731. Netropsin. Specific probe for A-T regions of duplex DNA.

Weiss, M., Patel, D.J., Sauer, R.T., and Karplus, M., (1984), *Proc. Natl. Acad. Scs.* USA., **81**, 130-134. Two dimensional ^{1}H-NMR study of the lambda operator: A sequential assignment strategy.

Zimmer, C.H., (1975), *Prog. Nucleic Acids Res. Mol. Biol.* **15**, 285-318. Effects of the Antibiotics Netropsin and Distamycin A on the structure and function of nucleic acids.

ELEMENTS OF SPECIFIC RECOGNITION OF NON-INTERCALATING LIGANDS IN THE INTERACTION WITH DNA

CHRISTOPH ZIMMER AND GERHARD LUCK

Akademie der Wissenschaften der DDR
Zentralinstitut für Mikrobiologie und Experimentelle Therapie Jena
Abteilung Molekulare Biochemie
DDR 69 Jena, Beutenbergstr. 11
German Democratic Republic

ABSTRACT

Model studies on interactions of non-intercalating ligands with DNA have provided important information for the specific binding of proteins and DNA. The elements involved in their specific binding with DNA double helix may be used in molecular interactions of biological systems. Two classes of DNA-binding drugs, netropsin-like oligopeptides and antitumour-active compounds having quarternary ammonium structures, are discussed as representative non-intercalators.

It is documented that base pair specificity, groove binding and geometrical requirements of their interaction are related to the properties inherent in the structure and dynamics of both the interacting ligand and nucleic acid.

INTRODUCTION

Nucleic acid protein interactions play a central role in all biological processes of gene transcription, replication, recombination, repair or transposition. Efforts have been made for many years to understand the mechanism of recognition between proteins and DNA at the molecular level (for compilation see von

Abbreviations used: Nt, netropsin; Dst-3, distamycin A; Dst-2, Dst-4, Dst-5, analogs of Dst-3; bis-Nt, bifunctional Nt-like ligand containing two monomeric Nt-like units; NSC-101327 name for NSC number (National Cancer Institute, USA); SN, screening number for compounds as documented elsewhere (Baguley, 1982).

Hippel, 1979; Wells *et al.*, 1980 or Helene and Lancelot, 1982). For this reason research has been directed to different conformations and to the dynamics of nucleic acids as well as to the main factors, which determine the reactivity and complementarity between DNA and proteins (Gursky *et al.*, 1977; Helene and Lancelot, 1982). Elucidation of the binding of regulatory proteins to DNA is more complex. Small DNA-binding ligands were found to be more simple external reactants and provided important information on the nature of the reactivity and specificity in their interaction with DNA. Most of them bind non-covalently and reversibly to DNA. Two major classes of DNA-binding ligands as model systems are represented by intercalating and non-intercalating drugs. In addition to that many of these drugs became important tools in the biochemical and biological research of the genetic material. Many biologically active small ligands cause their genetic, biochemical and pharmacological effects as consequences of an interaction with DNA (for compilation see Waring, 1981). As intercalators actinomycin D or acridines are most extensively investigated and widely used in DNA biology (Neidle, 1979; Waring, 1981). In this case chromophoric ring systems are placed between base pairs. On the other hand the oligopeptide antibiotics netropsin (Nt) and distamycin A (Dst-3) represent one prototype of non-intercalating ligands exhibiting site-specific groove binding properties (Hahn, 1975; Krey, 1980; Zimmer, 1975, Gursky *et al.*, 1977).

This article outlines our efforts to elucidate essential elements involved in the specific interaction of non-intercalative DNA-binding drugs. The subject will be based mainly on the groove binding ligands of netropsin-type oligopeptides and of bisquarternary ammonium heterocyclic compounds (reviewed in previous and recent papers: Hahn, 1975; Zimmer, 1975; Grunicke *et al.*, 1976; Krey, 1980; Zimmer, 1983; Baguley, 1982). A large number of other substances interacting to the exterior of the DNA helical structure will not be considered herein. Some of these important types of non-intercalators interacting non-covalently with DNA are described in recent compilations (Waring, 1981; Baguley, 1982; Pullman *et al.*, 1983).

1. Structures of non-intercalators

As naturally-occurring compounds Nt and Dst-3 represent unusual basic oligopeptide structures consisting of 1-methyl-pyrrole-2-carboxamide residues linked to guanidinoacetamido and propionamidino groups in the first case and a formyl and amidinopropyl group in the second one (Fig. 1, A). They were found to be DNA-binding agents with antibacterial, antiviral and antineoplastic properties (Finlay *et al.*, 1951; Zimmer, 1975; Hahn, 1975; Grunicke, 1976). To modify and possibly to improve their biological effects as well as the DNA binding specificity, analogs were used and their synthesis was designed in several directions; e.g. modification of the chain length of Dst-3 (Fig. 1B) 1-methyl-pyrrolecarboxamide units from n = 1 to n = 5 (Arcamone et *al.*, 1968; Grokhovsky *et al.*, 1975, Grokhovsky *et al.*, 1978) and substitution of the methyl group by propyl or isoamyl residues in Dst-2 and replacement of two amide bonds by methylamide ones (Grokhovsky *et al.*, 1978; Grokhovsky *et al.*, 1982). More recently a new type of highly sequence specific DNA-binding ligands have been constructed containing two netropsin-like molecules (Khorlin, *et al.* 1980; Khorlin, *et al.* 1982; Fig. 1, C). The basic structures of Nt and Dst-3 carry several potential reaction sites for binding to DNA: positively charged terminal residues, amide groups as hydrogen donor sites as well as CH_2 groups for possible intermolecular contacts. The capacity of hydrogen donors and acceptors of the peptidic linkage increases with the chain length of the oligopeptides from n = 1 to n = 5 (Fig. 1, A).

Other DNA-binding ligands with similarities to Nt with respect to structural features and potential reactive centers are a series of antitumour compounds composed of bisquarternary ammonium structures containing benzene and heterocyclic ring systems (Baguley, 1982). The DNA-binding capacity of three members of this group compared with that of Nt are NSC-101327, SN-6999 and SN-18071 (Fig. 1, D, E, F).

The stereochemistry of netropsin and of the distamycin analog Dst-2 (Fig. 1, A) has been elucidated by crystal structure analysis

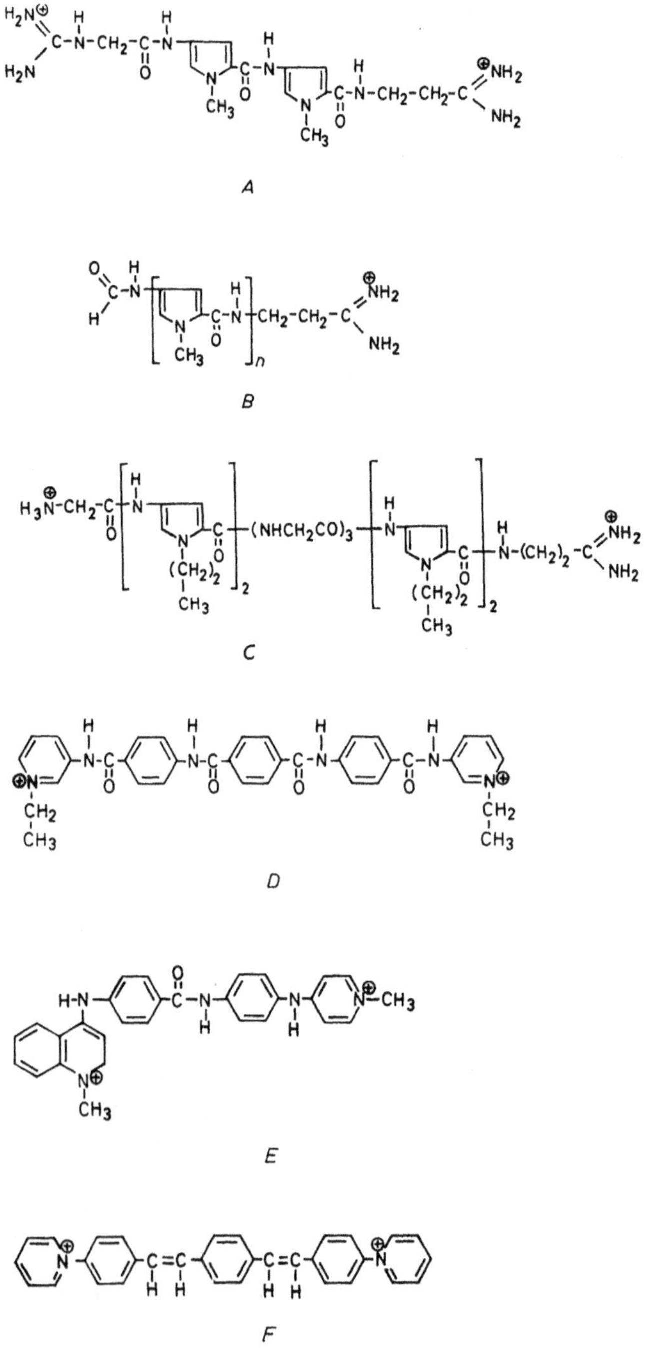

FIG. 1. Chemical structures of nonintercalating DNA-binding drugs. Netropsin-like structures. A: netropsin (Nt); B: distamycin A ($n = 3$: Dst-3) and analogs ($n = 1, 2, 4$, and 5); C: bis-netropsin-like fragments (bis-Nt (→→→), according to Gursky *et al.*, 1983. Bisquarternary ammonium heterocycles (Baguley, 1982). D: NSC-101327; E: SN-6999 F: SN-18071.

(Berman *et al.*, 1979; Gurskaya et al., 1979). The data suggest important information on the geometry of reactive sites which are responsible for the binding to DNA. Essential features of the conformation of netropsin sulfate in the crystal structure are shown in Fig. 2. There is a concave side persisting of amide groups (A in Fig. 2) which are themselves hydrogen bonded to water

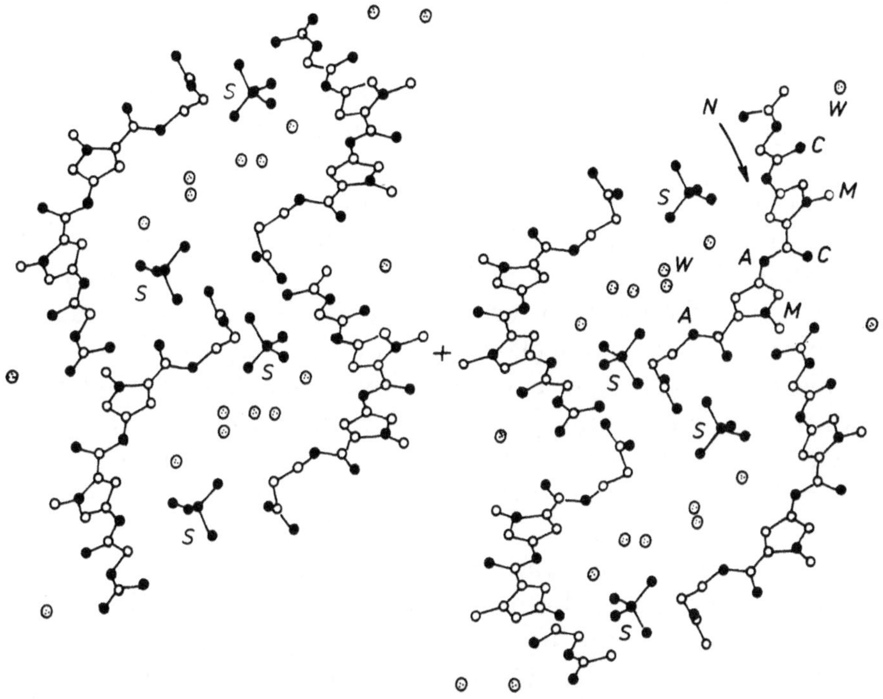

FIG. 2. Crystal structure of netropsin sulfate according to Berman *et al.*, 1979. The structure contains alternating hydrophilic and hydrophobic sections. Hydrophilic regions are represented by netropsin cations with concave sides (N) facing water molecules (spots: W) and sulfate anions (S) sandwiched between positively charged terminal residues. The convex sides of netropsin conformation contain carbonyl (C) and methyl groups (M); A. amide groups.

molecules (spots, W). By this way an oligopyrrole backbone is formed. The convex sides of the netropsin molecules bearing the carbonyl (C) and methyl residues (M) form hydrophobic sections while the terminal cations together with their concave sides form hydrophilic sections. Orientation and bonding of potential reaction centers in the crystal, especially NH groups in the middle of the concave side (N) of the molecule show their hydrogen donor function and mirror the ability of the flexible Nt molecule to adopt a bowed backbone-like structure. This has some relevance for an isohelical orientation of Nt in a binding model with DNA (below).

2. Interaction of netropsin-type ligands

a) *Base pair specificity*

The binding of Nt and Dst-3 to DNA was demonstrated by using optical and hydrodynamical methods (Zimmer, 1975). Based on numerous results of melting temperature increase, UV spectral changes, induced circular dichroism and viscosity changes, a high specificity for dA · dT rich regions was shown to be important in the interaction of both oligopeptides with DNA.

As one of the most powerful spectroscopic binding increments, the extrinsic Cotton effects caused in the CD spectra of DNA complexes with Nt and Dst-3 directly reflect bound drug molecules in solution (Zimmer *et al.*, 1972; Luck *et al.*, 1974; Zasedatelev *et al.*, 1974). Typical CD spectra of the Nt complex formed with poly(dA) · poly(dT) are shown in Fig. 3. Nt free in solution is optically inactive. The induced CD around 315nm shows binding of Nt with increasing total ligand concentration. Thus under comparable conditions the CD amplitude of this extrinsic band demonstrates the binding efficiency of the drugs with different DNA's (Table 1). Greater CD signals of dA · dT containing duplex DNA suggest an extremely high affinity of Nt and Dst-3 to dA · dT base pairs whereas weak or no binding was detected for duplexes with 100% dG · dC pairs. Binding of Nt and Dst-3 is also absent for RNA and for single-stranded DNA's (Table 1). The specificity and tight binding to dA · dT base pairs in clusters is reflected by several lines of evidence. The oligopeptides are attached to DNA in the minor groove without intercalation (Zimmer, 1975) as evidenced by the findings that bulky groups in the major groove do not lower their binding efficiency (Zasedatelev *et al.*, 1974; Wartell *et al.*, 1974). Binding of Nt as well as Dst-3 to DNA's is increasingly sensitive to ionic

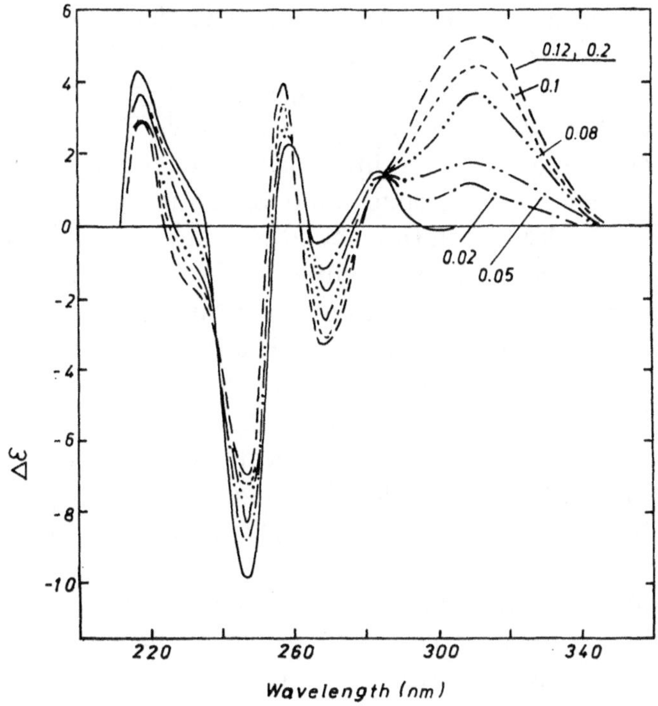

FIG. 3. CD spectra of the Nt complex with poly(dA) · poly (dT) in 0.1M NaCl, attached numbers designate r' of Nt.

TABLE 1. *Binding of Nt and Dst-3 detected by circular dichroism at 0.02 M NaCl*

Polynucleotide	Δ ε (M^{-1} cm^{-1})	
	Nt	Dst-3
Poly (dA) · Poly (dT)	4.9	10
Poly (dA-dT) · Poly (dA-dT)	4.0	17
Poly (dF-dC) · Poly (dI-dC)	3.0	11.0
DNA (58 mole-% A+T)	3.2	7.0
Poly (dA-dC) · Poly (dG-dT)	2.0	7.0
DNA (28 mole% AT)	1.9	5.0
Poly (dG-dC) · poly (dG-dC)	no binding	0.3
Poly (dG) · Poly (dC)	no binding	0.3 (2.0)
ds RNA (f2phage)	no binding	no binding
ss DNA (øX174)	no binding	no binding
Poly (dT), Poly (dA)	no binding	no binding

Δ ε was measured at r' = 0.1; at 315 - 320 nm for Nt and at 325 - 328 nm for Dst-3.

strength as the dA · dT content decreases (Luck *et al.*, 1974; Zasedatelev *et al.*, 1974; Wartell *et al.*, 1974). The highest stability of a Nt complex is formed with poly(dA) · poly(dT). As indicated by the CD signal at 315nm in Table 2 even at 5M LiCl no dissociation occurs. A first marked decrease in $\Delta\varepsilon$ is caused in the presence of hydrogen bond breaking agents, such as urea or guanidine hydrochloride, but dissociation is still not complete. This demonstrates a weakening of the Nt binding and suggests a dominating role of hydrogen bonding involved in the specific complex formation with double-stranded DNA. The high degree of binding specificity to dA · dT base pairs in clusters of DNA is also supported by comprehensive viscosity data (Reinert, 1971; *et al.*, 1979). Raman spectroscopic analysis provided further physical evidence for hydrogen bonding (Martin *et al.*, 1979). Specific binding of Nt-type oligopeptides is predominantly achieved by hydrogen bonds between NH groups of the drugs and 02 atoms of thymidine residues and/or N3 atoms of adenine bases. The role of amide groups as primary hydrogen donor sites is indirectly evidenced by the finding, that the distamycin analog Dst-2 (Fig. 2, B) containing two of the three amide groups replaced by methylamide ones, failed to interact with poly(dA) · · poly(dT) (Grokhovsky *et al.*, 1982, Krylov *et al.*, 1979). Some essential binding constants, reviewed in Table 3 are in accordance with a dA · dT specific binding of Nt and Dst-3 as well as their analogs. The collected literature data, however, deserve a comment. Binding constants exhibit a relatively broad scatter for DNA complexes which is due to different conditions and methods used.

Direct comparable data are based on a method by using fluorescent labeled Dst analogs (Krylov *et al.*, 1979; Gursky *et al.*, 1983). The decrease of the binding constant of Dst-3 corresponds two orders of magnitude in going from poly(dA-dT) · · poly(dA-dT) to the dG · dC containing duplexes (Table 3). In close agreement with an dA · dT specific interaction of Nt, Gursky *et al.*, 1983 estimated from their binding data the free-energy changes for the interaction of a single amide group of the distamycin molecule, which is $\Delta G = -2.1$ for thymine, $\Delta G = -1.1$ for adenine, $\Delta G = -0.9$ for cytosine and $\Delta G =$

= +0.9 kcal · M^{-1} for guanine. This nicely supports the conclusions on the highest reactivity of Nt-NH groups with 02 of thymidine residues. The free energy contribution of a pyrrolecarboxamide unit is about 2 kcal per mole for its interaction with an dA · dT pair and 0.95 kcal/mole for a dG · dC pair (Krylov *et al.*, 1979). The presence of dG · dC pairs weakens the bindding affinity of Nt as well as of Dst-3 because cytosine 02 atoms are involved in hydrogen bonding with guanine and hence they represent very weak acceptors. In duplexes containing dI · dC pairs, however, the 02 position of cytosine in the minor groove has an increased acceptor function due to unshielding of N3 of inosine. The binding of Nt and Dst-3 to this duplex polymer is nearly as effective as in the case of dA · dT duplexes (Table 2).

TABLE 2. *Effect of salt, urea and guanidine hydrochloride (GHCl) on the binding of Nt to poly(dA) · poly(dT) detected by CD.*

Conditions	Δ ε (315 nm)	
	r′ = 0.1	r′ = 0.2
0.02 M	4.9	4.9
0.1 M NaCl	4.8	4.8
4 M LiCl	3.8	3.9
5 M CsCl	2.4	2.5
2 M LiCl +		
6 M urea	1.7	2.9
2 M LiCl +	1.8	2.6
3 M GHCl		

Recently Marky *et al.*, 1983 reported the calorimetrically-determined binding enthalpy of —9.2 kcal · M^{-1} for the Nt complex with poly(dA-dT) · poly(dA-dT). If the main contribution results from hydrogen bonding of amide groups one obtains -3 kcal per hydrogen bond, a reasonable value, which may explain the specific binding to three dA · dT pairs due to the presence of three NH donor groups. A fourth hydrogen bond seems also possible through one of the terminal residues. The positively charged terminal groups of Nt and Dst-3 (Fig. 1) are, however, potential sites for strong electrostatic interactions with phosphate sites of the DNA backbone (Zimmer, 1975). This has no marked effect on the specificity of drug binding, e.g. removal of the guanidine as well as the propionamidin side chain from Nt (Fig. 1) does not eliminate the specific binding of this Nt derivative to poly(dA) · poly(dT) (Wähnert *et al.*, 1975; Berman *et al.*, 1979). In principle the analogs Dst-4 or Dst-5 (Fig. 1) containing 5 and 6 NH donor sites more efficiently bind to dA · dT duplex polymers. On the other hand with chain elongation up to that of Dst-5 the potential increases to form more or less stable complexes with dG · dC containing duplexes (Zimmer *et al.*, 1983 b), which can be attributed to hydrogen bond formation with dG · dC pairs. This context is considered below in section (B). It appears therefore, that Nt and Dst-2 (Fig. 1) can be classified as highly dA · dT specific ligands (Zimmer *et al.*, 1983 b). The binding specificity to dA · dT pairs has been considerably improved by the synthesis of bis-Nt molecules (Gursky *et al.*, 1983). The highest binding efficiency to poly(dA) · poly(dT) is observed for bis-Nt which contains the two fragments oriented in parallel sequence (Fig. 1 C). The CD binding data (Gursky *et al.*, 1983) suggested a much greater specificity of bis-Nt as compared to that of the monomeric molecule. Summarizing the data on base pair specificity of Nt-like DNA-binding oligopeptides this class of nonintercalating ligands specifically recognize dA · dT pairs in the minor groove.

b) *Sequence specificity*

Analysis of DNA complexes with Nt, Dst-3 and their congeners revealed a selective interaction with dA · dT base pairs in clusters (Zimmer, 1975; Wartell *et al.*, 1974; Reinert *et al.*, 1979). A site size of four to five base pairs was obtained for Nt binding in studies with dA · dT duplex polymers (Luck *et al.*, 1974; Zasedatelev *et al.*, 1974; McGhee, 1976). At least two or three consecutive dA · dT pairs are required for a strong Nt binding. Recent NMR data on imino proton redistribution suggest that the Nt molecule interacts with the dA · dT tetranucleotide core of a selfcomplementary dodecamer causing considerable stabilization (Patel *et al.*, 1981; Patel *et al.*, 1982). The sequence specificity of the Nt binding is indirectly reflected by our CD and enzymatic studies

(Zimmer *et al.*, 1979; Zimmer *et al.*, 1980 a). Fig. 4 A shows CD titration curves obtained for binding of Nt to poly(dA) · poly(dT) and poly(dA-dC) · poly(dG-dT). The saturation level of Nt to the dG · dC containing sequential polymer is markedly lower than that of poly(dA) · poly(dT) indicating a greater binding specificity to the latter. A sharp dissociation by increasing salt concentration of Nt from the complex of poly(dA-dC) · poly(dG-dT) further verifies a much lower binding strength for this sequence, whereas the dA · dT containing duplex polymers are highly stable (Fig. 4 B). This means that placement of one dG · dC pair between two dA · dT pairs strongly reduces the

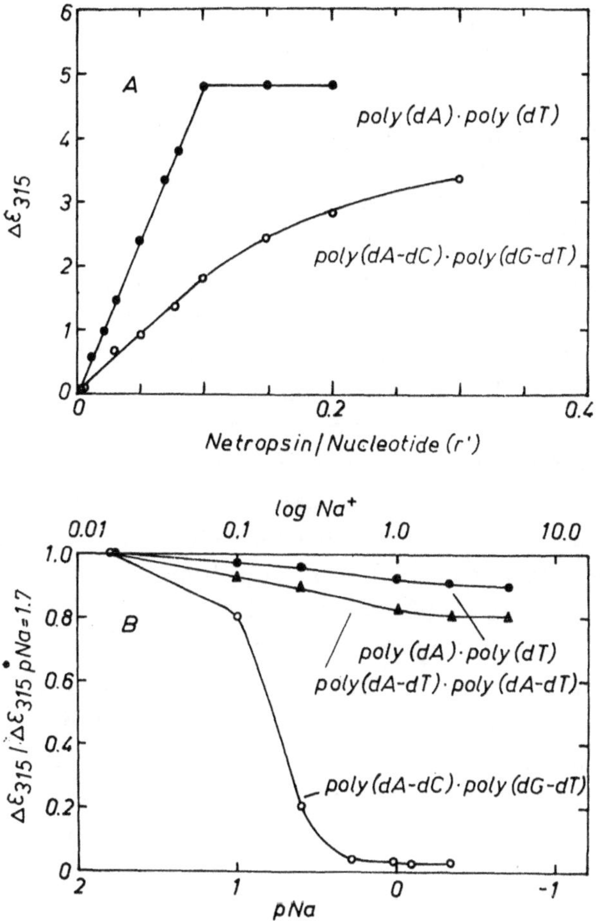

FIG. 4. A: CD titration at 315 nm of duplex polymers with Nt in 0.15M NaCl. B: Variation of the stability of Nt complexes of duplex polymers with increasing salt concentration at r' = 0.1. $\Delta\varepsilon$ at 315 nm (ordinate) was normalized by taking its intensity at pNa = 1.7 (0.02 M NaCl) as 1.0.

specificity of the Nt interaction due to the absence of neighbouring of 02 atoms of thymines as primary acceptor sites and the presence of additional steric hindrance of the adjacent guanine amino group (Wartell *et al.*, 1974, Zimmer, 1983; Gursky, 1983). A selective protection effect of DNA duplex polymers against DNase I cleavage caused by Nt interaction also mirrors the higher binding strength to pure dA · dT containing polymers as a consequence of a greater binding specificity. As shown in Table 4 the degree

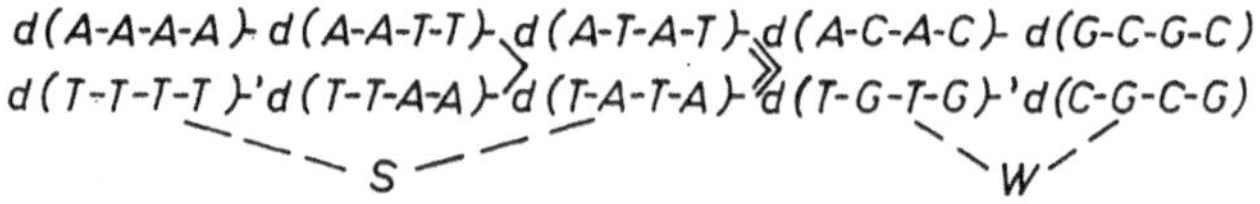

FIG. 5. Order of sequence preferred binding of Nt to its site size of a tetramer duplex: s, strong binding sites; w, weak binding sites.

of the Nt protection from the endonucleolytic DNA degradation is 100% for poly(dA) · poly(dT), about 65% for poly(dA-dT) · · poly(dA-dT) and is completely absent for the two dG · dC containing duplexes. The order of a decreasing binding specificity of Nt to different tetramer-duplex sequences can be suggested as given in Fig. 5. In extracting most relevant binding data (Table 3) one may consider two different binding constants for Nt around $K_m \sim 10^8 M^{-1}$ and $K_m \sim 10^5$ to $10^6 M^{-1}$ which roughly could account for strong and weak types of binding regions. In agreement with some of the collected binding data in Table 3 one may classify pure dA · dT clusters as strong sites and alternating dG · dC containing sequences as weak affinity sites for Nt. The reduced affinity of Nt for the alternating dA · dT copolymer relative to that of the homopolymeric sequence of poly(dA) · poly(dT) (Fig. 4 B, Table 4) can be well interpreted by the finding that the structure of the former has a lower minimum of the electrostatic potential in the minor groove if it exists in an alternating B form (Pullman and Pullman, 1981). A direct technique for determining binding sites of Nt and Dst-r by drug-protection in a cleavage pattern caused by methidiumpropyl-EDTA · Fe (II) complex elegantly showed a minimal protected segment of four base pairs (Van Dyke *et al.*, 1982). The strong binding sites of Dst-3 to a

TABLE 3. *Binding constants (K_a) of netropsin and distamycin complexes with DNA's*

Oligopeptide	DNA	K_a (M^{-1})	References
Nt	Calf thymus	$2.9 \cdot 10^5$	a
Nt	Various bacterial DNA's	$9 \cdot 10^4$-$2.7 \cdot 10^5$	b
Nt	Poly (dA) · Poly (dT)	$4.9 \cdot 10^5$	b
Nt	Poly (dI) · Poly (dC)	$4.8 \cdot 10^5$	b
Nt	Poly (dA-dT) · Poly (dA-dT)	$4.0 \cdot 10^5$	b
Nt	Poly (dA-dT) · Poly (dA-dT)	$5.0 \cdot 10^8$	c
Bis-Nt	Poly (dA) · Poly (dT)	$\sim 10^{12}$	d
Dst-3	Calf thymus DNA	$1.2 \cdot 10^6$	a
Dst-3	SPP1 phage DNA	$2.4 \cdot 10^9$	a
Dns-Gly-Dst-2	Poly (dA-dT) · Poly (dA-dT)	$(18 \pm 3) \cdot 10^5$	d
Dns-Gly-Dst-2	Poly (dA-dC) · Poly (dG-dT)	$(10 \pm 5) \cdot 10^3$	d
Dns-Gly-Dst-2	Poly (dG-dC) · Poly (dG-dC)	$1 \cdot 10^3$	d
Dns-Gly-Dst-3	Poly (dA-dT) · Poly (dA-dT)	$(18 \pm 3) \cdot 10^6$	d
Dns-Gly-Dst-3	Poly (dA-dC) · Poly (dG-dT)	$(2 \pm 1) \cdot 10^4$	d
Dns-Gly-Dst-3	Poly (dG) · Poly (dC)	$(5 \pm 3) \cdot 10^4$	d

[a] Luck *et al.*, 1974;
[b] Wartell *et al.*, 1974;
[c] Mc Ghee, 1976;
[d] Gursky *et al.*, 1983.

TABLE 4. *Protection of duplex DNA's against DNase I cleavage by specific interaction with Nt, Dst-3 and Dst-A.*

Duplex DNA	Degree of protection from enzymatic cleavage (%) Nt	Dst-3	Dst-5
Poly (dA) · Poly (dT)	100	—	—
Poly (dA-dT) · Poly (dA-dT)	65	43	47
Poly (dA-dC) · Poly (dG-dT)	0	12	37
Poly (dG-dC) · Poly (dG-dC)	0	—	—

Decree of protection is given by the percentage of the inhibition of DNase I cleavage of the free polymer caused by the ligand added at $r' = 0.1$; kinetics of the reaction was measured by hyperchromicity caused by DNase I (10 ug/ml) at saturation after 30 min as given by Zimmer *et al.*, 1980.

fragment of pBR322 DNA at low ligand concentration are composed in almost all cases of at least two adjacent thymine residues of the protected dA · dT tetranucleotide region. This again testifies the highly specific binding to dA · dT clusters with consecutive thymines. Some deviation in the sequence binding affinity was recently reported for the elongated chain, such as for

Dst-5 (Zimmer *et al.*, 1983 b). In contrast to Nt or Dst-2, analogs having three to five methylpyrrolecarboxamide units tend to form complexes with dG · dC containing sequential duplex polymers. Such a Dst-5 complex is even not completely dissociated at 1 M or at 4 M NaCl (Fig. 6). Also there is a protection by Dst-3 and

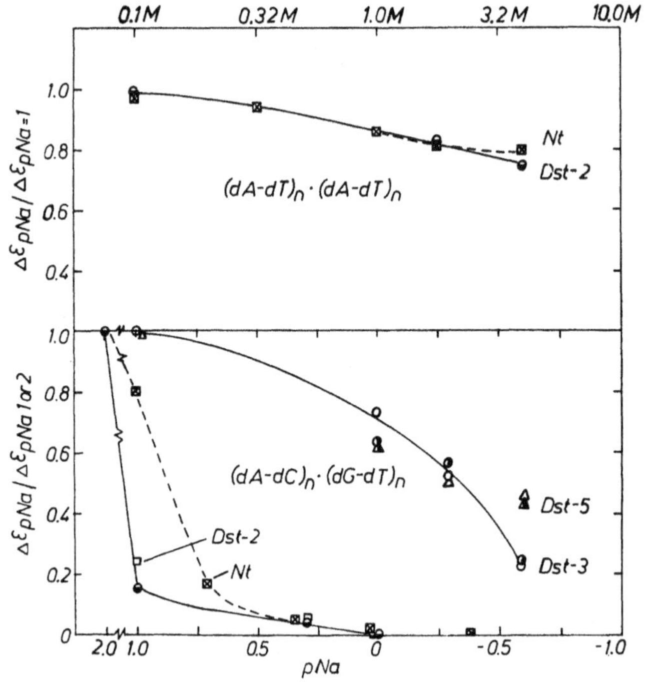

FIG. 6. Influence of increasing salt concentration (NaCl) on the complex formation of synthetic duplex DNA polymers at the total ratio r′, drug per nucleotide, as designated by symbols ◒ 0.1 Dst-2, □ 0.2 Dst-2; ◑ 0.1 Dst-3; ○ 0.2 Dst-3; ◧ 0.1 Dst-5, △ 0.2 Dst-5; ■ 0.1 Nt. Δε was normalized by taking its intensity at pNa = 1 (or pNa = 2) as 1.0; Δε at the CD maxima is used for Nt at 315 nm, Dst-2 at 320 nm, Dst-3 at 330 nm, Dst-5 at 338 nm (according to Zimmer *et al.*, 1983).

Dst-5, e.g. of poly(dA-dC) · poly(dG · dT) in the endonucleolytic cleavage of DNase I (Table 4). These data seem to suggest, that Nt and Dst-2 possess the higher sequence specificity, which means they do not show binding affinity to dG-dC pairs. Interestingly the binding behaviour of the Nt-type ligands exhibit a relationship to their reactive sites to the oligopeptide structure (Table 5). Obviously the interaction with dG·dC pairs is favoured when the keto-acceptor sites or amide-donor sites exceed n = 3 (Table 5). The possible function of the peptide group of Dst-5 as an acceptor for guanine in the minor groove has been suggested (Zimmer *et al.*, 1983 b). This type of interaction is probably weaker than hydrogen donation of the amide group. In agreement with the high hydrogen-bond donating capacity of Nt in the crystal structure (Fig. 2) three amide nitrogen atoms interacting with 02 atoms of neighboured thymines can explain this sequence specificity. In case of bis-Nt (Fig. 1C, Table 3) the two Nt-like units react independently and the binding specificity is strongly increased (Gursky *et al.*, 1983).

TABLE 5. *Relationship between oligopeptide binding with dG · dC containing duplexes and the number of reactive groups of non-intercalating oligopeptides.*

Oligo-peptide	Δ ε (AT)	Δ ε (AC/GT)	Δ ε (GC)	Methylpyrrole-Carboxamide-	Number (n) of Interacting Sites Donor	Acceptor	Charged Groups N+
Dst-2		0.3	0	2	3	3	1
Nt	15	1.8	0	2	3 (4)	3	2
Dst-D	—	3.8	0	3	3	3	2
Dst-3	18	6.6	0.9	3	4	4	1
Dst-4	26	11.3	2.0	4	5	5	1
Dst-5	40	26	3.0	5	6	6	1

Δ ε correspond to CD maxima of bound ligands from 315 nm to 330 nm at r′ = 0.1 in 0.1 M NaCl:
Δ ε (AT) poly (dA-dT) · poly (dA-dT), Δ ε (AC/GT) poly (dA-dC). poly (dG-dT), Δ ε (GC) poly (dG-dC) · poly (dG-dC) or homopolymer duplex.

c) *Geometry of the interaction*

Binding of Nt and Dst-3 to the double-helix structure requires the B conformation of DNA as evidenced by lacking of any binding signal for double-stranded RNA (Table 2). Poly(rA) · poly-(rU) and a natural DNA · RNA hybrid, which is believed to exist in the A conformation also failed to interact specifically with Nt (Wartell *et al.*, 1974; Zimmer, 1975). CD studies clearly demonstrate that Nt and Dst-3 binding to A-type DNA in solution cause a reversal to the B-type conformation (Ivanov *et al.*, 1974; Minchenkova and Zimmer, 1980). The CD spectral changes of the DNA-Dst-3 complex illustrate this behaviour (Fig. 7). The low-humidity A-type conformation of DNA (curve 0) characterized by the large CD signal at 270 nm is drastically decreased upon binding of Dst-3, which is directly monitored by the CD at 330 nm. The Ā-DNA existing at 82% ethanol (Ivanov *et al.*, 1974) is converted to B-DNA with increasing binding of the ligand (Fig. 7, below). The CD signal at 270 nm of Ā-DNA approaches that of the B-DNA at 65% ethanol on interaction with Dst-3.

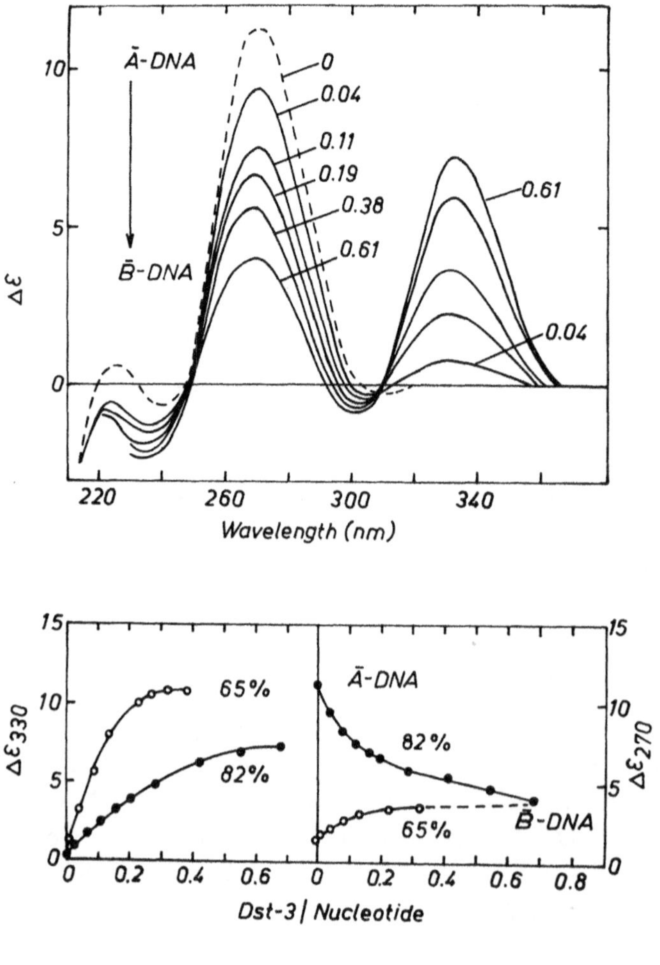

Fig. 7. Reversal of Ā to B̄ conformation of calf thymus DNA in 82% ethanol/water (v/v) induced by interaction with Dst-3 at $3 \cdot 10^{-4}$M NaCl. Below: Titration curves of Ā-DNA and B̄-DNA with Dst-3 at $3 \cdot 10^{-4}$M NaCl; 82% ethanol: Ā-DNA, 65% ethanol: B-DNA (according to Minchenkova and Zimmer, 1980).

This demonstrates that the Nt-type oligopeptides specifically interact with the B geometry and do not react in a similar way with RNA and Ā-DNA as well.

This conformational specificity of Nt has been used to investigate the presence or induction of B-like structures in certain DNA · RNA hybrid duplexes in solution (Zimmer *et al.*, 1982; Marck *et al.*, 1982). CD titration curves for Nt of A · T, A · U and I · C duplexes containing deoxyribose, ribose or 2′-deoxy-2′-fluororibose as sugar on either strand are compared in Fig. 8. The hybrids with dAfl, rA, dIfl and dCfl in one strand are pro-

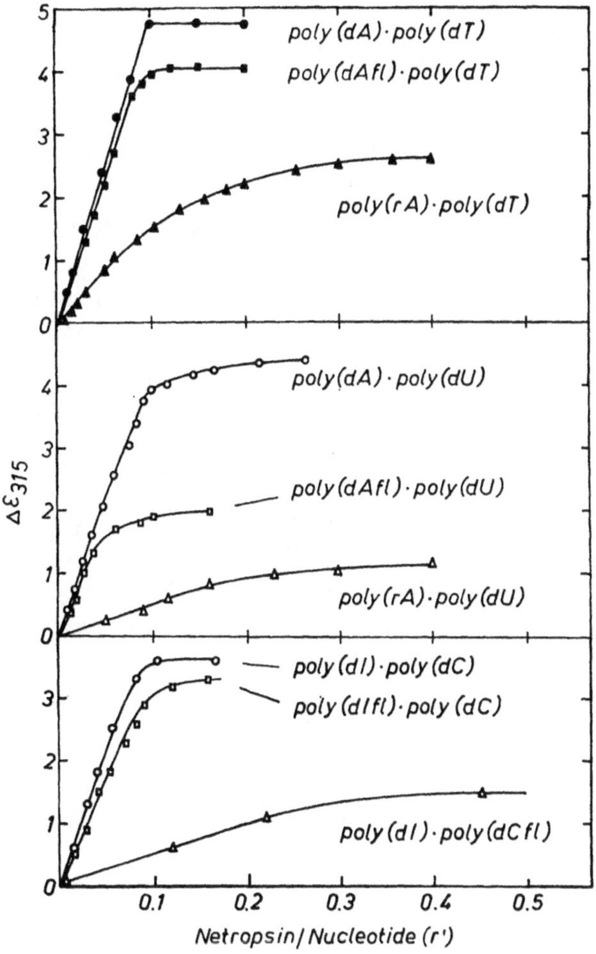

FIG. 8. CD titration at 315 nm (binding curves) of synthetic DNA duplexes and DNA·RNA hybrids with Nt (r′) at 0.1 M NaCl. A: thymidine, B: uridine and C: inosine-cytidine containing duplexes. dAfl, 2′-deoxy-2′-fluororibo-adenosine; dIfl, 2′-fluororiboinosine; dCfl, 2′-flouroribocytidine (according to Zimmer *et al.*, 1982; Marck *et al.*, 1982).

bably A-type duplexes (Zimmer *et al.*, 1982; Marck *et al.*, 1982). Nt binds to all six hybrids but shows lower CD signals over the whole titration range in comparison to that of the three B-type duplexes of poly(dA) · poly(dT), poly(dA) · poly(dU) and poly(dI) · poly(dC). This has been considered as an evidence that the hybrid duplexes may adopt a B-like conformation upon binding with Nt, which takes place to a different degree. Dst-3 and analogs show similar behaviour. The importance of the sugar conformation in the modulation of the geometry under the influence of Nt is obvious from the slightly lower binding signals of the duplexes containing fluorine in the purine nucleoside. The poly(rA) duplexes show much lower binding tendency for Nt in both series with poly(dT) and poly(dU) Fig. 8. Also the weaker binding signals found for hybrids with poly(dU) suggests that the absence of the methyl group favours the A-form more readily than the B-like structure. The data on this subject nicely suggest that Nt-type oligopeptides act as stabilizers of the B geometry and show at the same time some elements playing an essential role in the geometry of the Nt interaction.

Besides the B and A conformations the recently discovered left-handed Z-DNA has received great importance (Wang *et al.*, 1979; Drew *et al.*, 1980; Sage and Leng, 1980; for compilations see Wells *et al.*, 1989; Zimmermann, 1982). It was of interest to test the binding affinity to Z-DNA of the Nt-type oligopeptides, since left-handed Z-conformations seem to be of importance in natural DNA and in biological systems (Brahms *et al.*, 1982; Nordheim, *et al.*, 1981). One of the most investigated Z-DNA structure is that of poly(dG-dC) · poly(dG-dC) which shows a B to Z transition at high salt solution or in the presence of ethanol at low salt concentration (reviewed in Wells *et al.*, 1980; Zimmermann, 1982). As dA · dT specific ligand Nt does not interact with poly(dG-dC) · poly(dG-dC) under moderate and high salt concentration (Table 1). Binding occurs, however, at very low ionic strength due to electrostatic interactions. As demonstrated in Fig. 9 A the Nt binding to the B form of poly(dG-dC) · poly(dG-dC) at very low salt requires higher ligand concentration. No interaction occurs with the non-reversible Z-conformation of

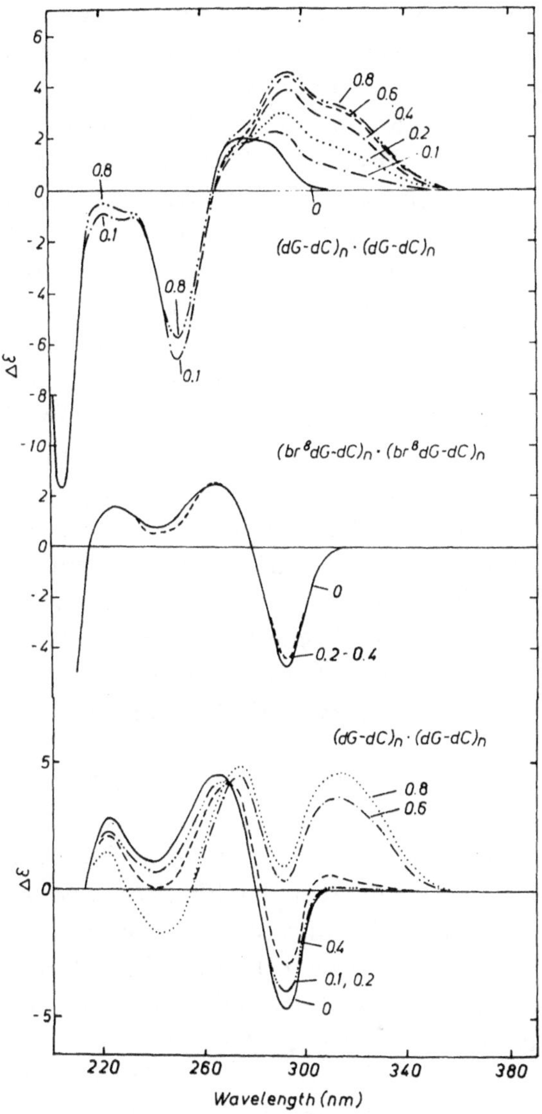

FIG. 9. CD spectra of poly(dG-dC) · poly(dG-dC) with increasing amounts of Nt. A: B-type conformation in $2 \cdot 10^{-3}$M NaCl; B: Z-type conformation of poly (br^{8}dG-dC) · poly(br^{8}dG-dC) in 10^{-3}M NaCl; C: Z-type conformation in 1.5×10^{-3} M CsCl, 60% ethanol (data from Zimmer *et al.*, 1983).

poly(br^{8}dG-dC) · poly(br^{8}dG-dC) (Fig. 9 B) whereas the reversible Z-form existing in 60% ethanolic solution is converted to the B-type conformation (Fig. 9 C). The ability of Nt to induce the Z to B transition was demonstrated in our recent work (Zimmer *et al.*, 1983 a). Therein the possible factors responsible for a conversion of non-B-DNA structures induced by netropsin have been related to differences in geometries and replacement of water shells in the hydration spine in the minor grooves.

Essential contribution to the explanation for the high affinity of Nt-type oligopeptides to the B-DNA geometry has been provided by theoretical works of Pullman and his coworkers (Pullman and Pullman, 1981; Pullman *et al.*, 1983). Interestingly the distribution of the electrostatic potential of the surface envelopes in the minor groove suggests a relation to the specific binding of Nt and Dst-3 in B-DNA. The minimum electrostatic potentials of the minor groove account very well for a high dA · dT affinity of the drugs in B DNA of poly(dA) · poly(dT) and of other DNA's, but it clearly disfavours their binding in the A conformation (Lavery and Pullman, 1981). In good agreement with our data, interaction of Nt and Dst-3 with the left-handed Z conformation is disfavoured (Pullman and Pullman, 1981). The calculated potentials of poly(dG-dC) · poly(dG-dC) also explain the restricted binding of Nt and Dst-3 to this B geometry while the surface potential of the minor groove for the dI · dC containing B form accounts for a favoured binding (Lavery and Pullman, 1981) as it is found experimentally (Table 1).

The specific interaction of Nt and Dst-3 with the B-DNA conformation involves an isohelical orientation of the oligopeptide backbone within the groove (Zimmer, 1975; Zasedatelev *et al.*, 1976, 1978; Berman *et al.*, 1979) suggesting a binding model in which the methyl groups of the pyrrole rings project away from the groove and the amide groups form hydrogen bonds with 02 atoms of thymine residues (Fig. 10). This is also implicated in the crystal structure analysis of Nt (Berman *et al.*, 1979; see Fig. 2) and of Dst-2 (Gurskaya *et al.*, 1979). Evidence was provided by Raman spectroscopy (Martin *et al.*, 1978) and by the fact that substitution of the methyl groups on the pyrroles by propyl (Krylow *et al.*, 1979) or isoamyl residues (Grokhovsky *et al.*, 1982) did not affect binding of distamycin analogs. In principle, the groove binding model of Nt (Fig. 10) has been supported by very recent crystal data analysis of Dickerson's group (unpublished data of Kopka *et al.*, 1983). The orientation of the Nt molecule to the

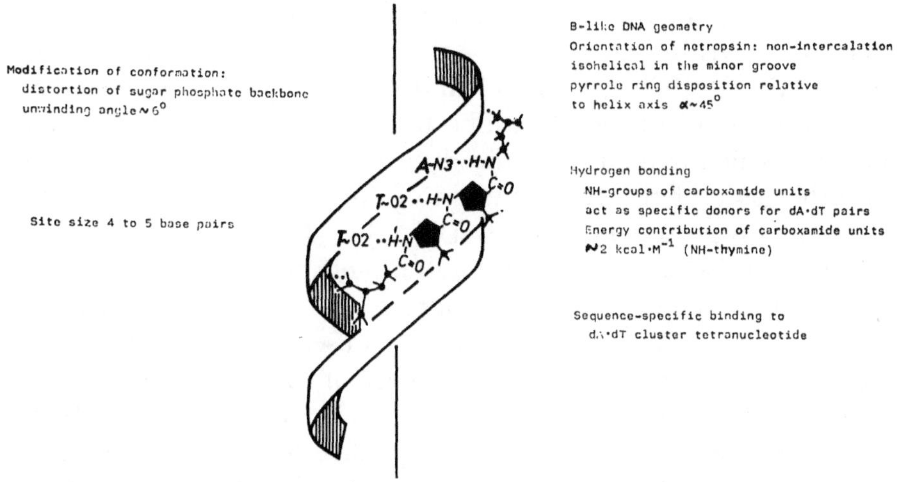

FIG. 10. Schematic model of the coordination of netropsin to the AT-rich tetranucleotide sequence of the minor groove of B-DNA. (Data according to Zimmer, 1975; Luck *et al.*, 1974; Zasedatelev *et al.*, 1976, 1978; Wartell *et al.*, 1974; Patel, 1982; Snounou and Malcolm, 1983.)

dA · dT tetranucleotide in the middle of a dodecamer co-crystal structure shows, however, interesting variations.

According to their results it appeared that Nt is directly centered within the groove by forming most probably bifurcated hydrogen bonds, and positively charged terminal groups are located also in the groove. It seems that Nt may replace the ordered hydration spine of dA · dT rich segments in the minor groove of B-DNA (Drew and Dickerson, 1981). This would also explain the stabilizing effect of Nt and Dst-3 on B-DNA, since the regular, ordered spine of water molecules in the minor groove is an essential element in the B-helix stability relative to A or Z geometry.

The specific groove binding of Nt is associated with a local conformational change (Reinert, 1971; Zimmer *et al.*, 1980 b). Interaction with superhelical DNA cause a slight unwinding not higher than 6° per bound oligopeptide (Luck *et al.*, 1974). This proves non-intercalation between base pairs. Interestingly, the recent finding of Snounou and Malcolm, 1983 demonstrated that Nt binding increases the linking number of DNA and may favour the creation of positively supercoiled molecules. This could reflect a change in the backbone geometry resulting in an increase of helical twisting due to Nt groove binding. Previous NMR results (Patel, 1979 a, b; Patel and Canual, 1977) suggested Nt-induced changes in the glycosidic torsion angles of a DNA duplex polymer which extends into the drug non-bound base pair regions on either side of the Nt-bound site. An alteration of the conformational properties in natural DNA by Dst-3 binding has been related to a transmission of an allosteric effect (Hogan *et al.*, 1979). Careful viscosimetric studies on natural DNA's have revealed that conformational changes occur as a consequence of Nt binding causing an increase in contour length of DNA and a stiffening in the bound region (Reinert *et al.*, 1979; Reinert *et al.*, 1980). Groove binding of Nt is associated with different binding modes and can be ascribed to various forms of dA · dT clusters having different secondary structure due to differing nucleotide sequences (Reinert *et al.*, 1980). The hydrodynamic data showed that bound distamycin molecules modify locally certain DNA helix parameters. A distinct deviation was observed in the DNA interaction studies between Nt and Dst-3. In contrast to Nt, Dst-3 and analogs cause a local helix bending without essential elongation at low degree of ligand binding (Reinert, 1981). At high degree of binding Dst-3 leads to an increase in contour length. Thus, it appears that the stereochemistry of the bound oligopeptide differs between Nt and Dst-3. Findings on the binding of distamycin analogs (Luck *et al.*, 1977; Zimmer *et al.*, 1983 b) point to this notion. The conclusion needs, however, further experiments.

Stereochemical models have been developed for the molecular recognition of dA · dT base pairs by Nt and Dst-3. Suggestions involve an isohelical interaction in the minor groove of the B-type double helix by forming specific hydrogen bonds with the bases to the same DNA strand (Zasedatelev *et al.*, 1978; Berman *et al.*, 1979) whereas for Nt, also groove binding with both strands was considered (Wartell *et al.*, 1974). Our CPK model studies imply both possibilities (Zimmer *et al.*, 1979). Fig. 11 illustrates that the concave side of the Nt molecule is faced into the minor groove to a tetramer of an dA · dT base pair segment of B-DNA. One charged terminal group is placed towards oxygens of phos-

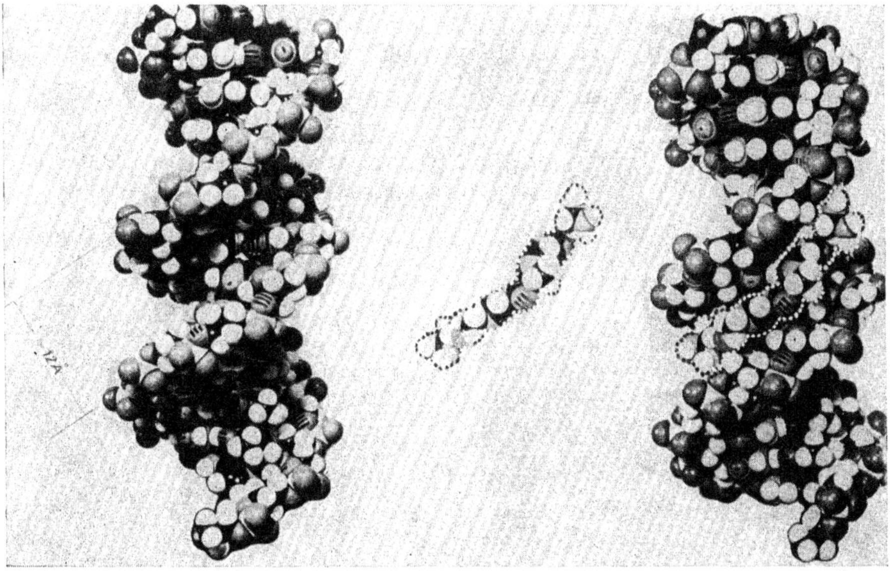

FIG. 11. CPK model of B-DNA interacting with netropsin within the minor groove (According to Zimmer *et al.*, 1979).

phate sites while its bulky end chain mainly occupies the space of the groove. All essential elements involved in the specific binding of Nt-type oligopeptides are collected in Fig. 10. Location of the ligands in the minor groove also agrees with their protection of sites in this groove against methylation with dimethylsulfate (Kolchinsky *et al.*, 1975). Specificity, hydrogen bonding and geometry including energetical features concern the considered topics above. Recent intermolecular nuclear Overhauser effects of a Nt complex with a dodecamer further strongly support the contacts of Nt in the minor groove of the bound d(A-A-TT) tetranucleotide region (Patel, 1982).

3. INTERACTION OF BISQUARTERNARY AMMONIUM HETEROCYCLES

Design of synthesis and recent DNA binding studies have been devoted to several bisamidine and bisquarternary ammonium heterocyclic drugs (Braithwaite and Baguley, 1980; Baguley, 1981). This group of compounds contains interesting structural modifications related to the Nt-type oligopeptides (Fig. 1 D to F).

The majority of these agents show cytotoxic activity, but some of them were described as antitrypanosomal and antileukaemic drugs (Baguley, 1981).

Binding to DNA of bisquarternary ammonium drugs was shown by some spectrophotometric changes and studies on ethidium bromide displacement from DNA (Braithwaite and Baguley, 1980). In a recent comprehensive spectroscopic analysis we were able to obtain more information on the binding properties of one type of these compounds, NSC-101327 (Fig. 1 D). Very similar as found for the DNA-Nt complex (Fig. 3) the drug NSC-101327 exhibits an induced CD upon binding to double-stranded DNA (Fig. 12).

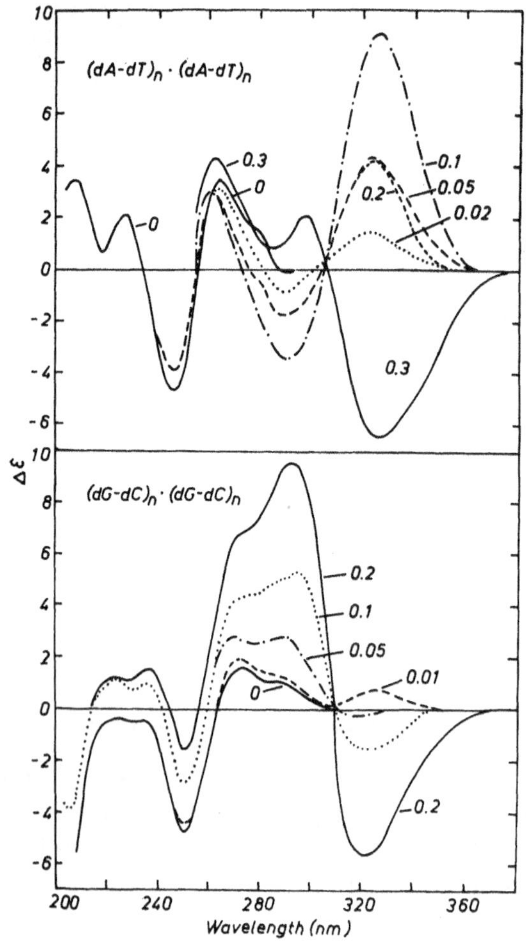

FIG. 12. Binding of NSC-101327 to DNA as reflected by CD spectra of poly(dA-dT) · poly(dA-dT) and poly(dG-dC) · poly(dG-dC) complexed with increasing amounts of the ligand at 0.1M NaCl attached numbers designate total ligand ratio, r′ (from Luck *et al.*, 1984).

Bisquarternary ammonium heterocyclic structures shown in Fig. 1 D to F, are optically inactive when existing free in solution. Thus, the extrinsic CD band around 328 nm is taken as indicator of bound ligand molecules. As indicated by the relative large positive CD signal at 328 nm (Fig. 12) interaction of NSC-101327 to poly(dA-dT) · poly(dA-dT) becomes inverted to a negative one between ligand per nucleotide ratios $r' = 0.1$ and $r' = 0.3$ which is attributable to a two step binding mechanism (Luck *et al.*, 1984). Interestingly binding of NSC-101327 also occurs to poly(dG-dC) · poly(dG-dC) at 0.1 M NaCl associated with a negative CD signal within the entire titration range (Fig. 12). This result differs from Nt-type oligopeptide.

The spectroscopic binding properties of the drug NSC-101327 are summarized in Table 6. The analysis of its complex formation with nucleic acids has revealed several striking features which differ from the specific DNA binding of Nt. First, NSC-101327 interacts with dA · dT and dG · dC containing duplex polymers although its affinity to the former is higher than to the latter (Table 6). Second, the drug binds to some extent to RNA and

TABLE 6. *Spectroscopic data on interaction of NSC-101327 with DNA's and RNA's in 0.1 M NaCl, measured at r' = 0.1 ligand per nucleotide*

Polymer	$\Delta\varepsilon$ ($M^{-1} cm^{-1}$)	$\varepsilon_{max_1} \cdot 10^3$ ($M^{-1} cm^{-1}$)	$\varepsilon_{max_2} \cdot 10^3$ ($M^{-1} cm^{-1}$)	Δ Tm (°C)
Poly (dA) · Poly (dT)	10.2	22.5	− 16.2	+ 16.5
Poly (dA-dT) · Poly (dA-dT)	9.3	20.3	− 4.6	+ 15.4
Poly (dI-dC) · Poly (dI-dC)	5.6	4.8[a]	− 29.9[a]	—
Calf thymus DNA	1.0	9.0	− 13.6	+ 2.0
Poly (dA-dC) · Poly (dG-dT)	− 2.8	9.3	− 8.5	+ 0.3
Poly (dG-dC) · Poly (dG-dC)	− 1.5	6.5	− 8,0	—
Poly (br^8 dG-dC) · Poly (br^8 dG-dC)	0.6	3.9[a]	− 33.8[a]	—
Poly (rA) · Poly (rU)	21.6	28.9	− 14.1	+ 15.0[b]
ds RNA (f2 phage)	6.3	22.4	− 13.3	+ 0.3
Poly (dT)	12.5	4.8	− 19.3	no[c]

$\Delta\varepsilon$ corresponds to the CD maxima around 323 to 327 nm; ε_{max_1} and ε_{max_2} are maxima and minima of UV absorption difference spectra (NSC-101327 complex versus free NSC 101327) around 330 to 340 nm and 300 to 305 nm, respectively; Δ Tm is the melting temperature increase caused by interaction with NSC-101327 in 0.1 M NaCl; [a] at $r' = 0.3$, [b] Tm increase of a fraction relative to melting of free duplex; [c] slight hyperchromic change with broad melting.

to the left-handed Z conformation of poly(br[8]dG-dC) · poly-(br[8]dG-dC) (Table 7) and hence is less selective for the DNA B geometry in comparison to Nt and Dst-3. NSC-101327 also exhibits a binding tendency to single-stranded polynucleotides, e.g. to poly(dT) (Table 7). With respect to the first binding step, interaction of NSC-101327 with double-stranded DNA occurs in the minor groove. Like Nt, the bisquarternary ammonium heterocyclic drug most probably forms hydrogen bonds between amide groups of peptide linkages and base pair edges as well as ionic contacts of the terminal groups and DNA phosphate sites. A competitive binding experiment with Nt (Fig. 13) suggests that

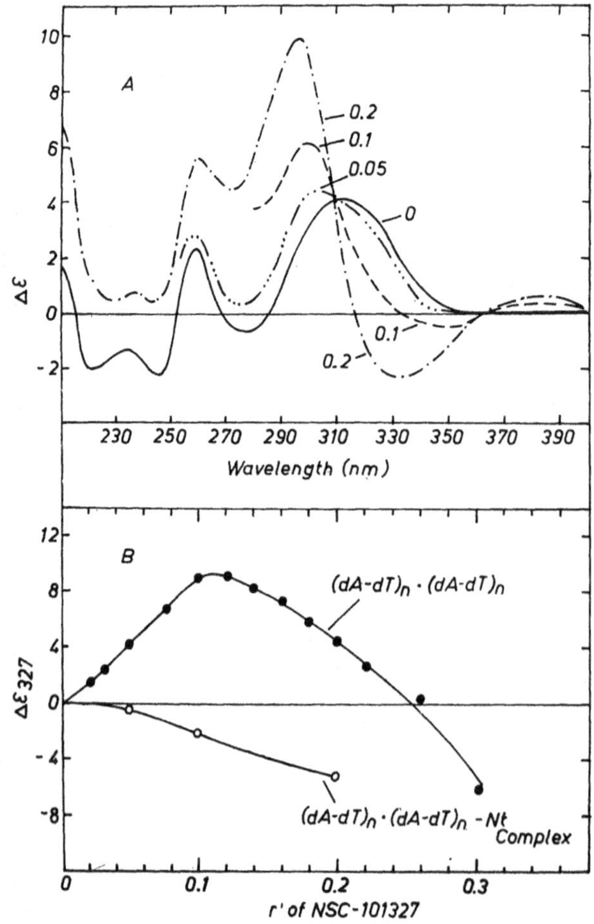

FIG. 13. A: Binding of NSC-101327 to poly-(dA-dT) · poly(dA-dT) complexed with Nt (at r' = 0.1, curve 0) B: CD titration of poly(dA-dT) · poly (dA-dT) and its Nt-complex with NSC-101327 (from Luck *et al.*, 1984).

NSC-101327 binds to the minor groove by non-intercalative means. Nt, which exclusively interacts with this helical groove inhibits both binding steps of NSC-101327 (Luck *et al.*, 1984). Fig. 13 illustrates the blocking of the first binding step of NSC-101327 to poly(dA-dT) · poly(dA-dT) by pre-bound Nt. Its non-intercalative binding is also based on some viscosity data of this class of ligands (Braithwaite and Baguley, 1980). In principle, the structural composition of NSC-101327 consists of subunits which possess, similar as in case of Nt-type oligopeptides, hydrogen donating capacity and electrostatic functional sites. Other structural differences seen in the chemical structure of NSC-101327 account for the variation of its nucleic acid binding properties indicated by a lower degree of base pair specificity and conformational selectivity. The predominant contribution of positively charged side chains to the binding affinity is also suggested by our CD binding results of SN-18071, which has no amide groups as specific hydrogen donor sites (Fig. 1 F). Binding of SN-18071 to DNA duplexes also does not reflect a clear-cut base pair specificity (Table 7).

TABLE 7. *Binding of non-intercalating bisquarternary ammonium compounds to DNA's as measured by circular dichroism.*

Polinucleotide	$\Delta\varepsilon$ (M^{-1} cm^{-1}) NSC-101327	SN-6999	SN-18071
Poly (dA-dT) · Poly (dA-dT)	9.3	14.8	4.0
DNA (58 mole-% A + T)	1.0	5.5	2.6
Poly (dG-dC) · Poly (dG-dC)	— 1.5	no binding	0.5
Poly (br^8 dG-dC) · Poly (br^8 dG-dC)	0.6	0.2[a]	—
ds RNA (f2 phage)	22.4	no binding	—
Poly (rA) · Poly (rU)	28.9	—	0.9

Measurements were made in 0.1 M NaCl at $r' = 0.1$, for NSC-101327 at $\lambda_{max} = 323$ to 327 nm, SN 6999 at $\lambda_{max} = 385$ nm, SN 18071 at $\lambda_{max} = 405$ nm; [a] at 10^{-3} M NaCl.

On the other hand we found that SN-6999 more specifically interacts with dA · dT base pairs in duplex DNA and failed to react with RNA (Table 7). This contrasts the behaviour of NSC-101327, but is comparable to that of the DNA-specific Nt binding.

We also observed a direct competition in the DNA binding between SN-6999 and Nt (unpublished results of the authors). We propose that SN-6999 interacts in the minor groove like Nt by formation of hydrogen bonds and ionic contacts.

One further element that emerges from the differential binding of other non-intercalators with DNA is the influence of the ligand structure on base pair specificity. As in case of a weaker, but significant binding affinity of Dst-5 to dG · dC pairs (Luck *et al.*, 1977; Zimmer *et al.*, 1983 b), the drug NSC-101327 is less sequence-specific and less selective for the B-DNA conformation in comparison to Nt and SN-6999.

It has been shown that other non-intercalators such as the fluorochrome Hoechst 33258 (Houssier and Fredericq, 1975) or the novel antitumour antibiotic CC-1065 (Chidester *et al.*, 1981; Li *et al.*, 1982) are potential dA · dT specific groove binding ligands. These compounds consist also of structural elements having a hydrogen donating capacity and therefore they may react similarly as Nt by isogeometric attachment in the minor groove of B-DNA.

4. Conclusions and biological implications

The interaction of small DNA-binding ligands with the DNA double-helix involves a number of essential elements which influence or determine the reactivity of the DNA. These molecular features are certainly important for the understanding of the recognition between DNA and proteins. First the majority of non-intercalative groove binding ligands specifically react with dA · dT base pairs from the minor groove. On the other hand intercalators, such as actinomycin D recognize dG · dC pairs in the minor groove of B-DNA (Gellert *et al.*, 1965; Gursky *et al.*, 1983). The primary role of hydrogen bonding in the specific recognition of dA · dT pairs in the minor groove is reflected by non-intercalative binding of several drugs. Possible acceptor and donor sites at the edges of base pairs, proposed by Seeman *et al.*, 1976 (e.g. 02 atoms of thymine and N3 of adenine bases) are experimentally verified.

Second, sequence-specificity and conformational dependency of the interaction of various non-intercalative ligands with DNA demonstrate the stereochemical requirements and complementarity. For Nt-type oligopeptides the best molecular correspondance is achieved for a pure dA · dT containing sequence of the DNA B-geometry. This depends on the nature of the interacting ligand, e.g. different from Nt the drug NSC-101327 shows binding affinity to the B- and to the A-conformation. The sequence-specificity exhibited by Nt or Dst-3 is considerably enhanced on increasing the specific reaction centers in bis-Nt (Gursky *et al.*, 1983). Sequence-dependence and stereochemistry of the oligopeptide-DNA complex illustrate a model which may be used in the molecular correspondence between proteins and DNA. Interesting proposals have been made for the complementary and a possible recognition code. These are based on the β-sheet structures of the polypeptide interacting along the minor groove through hydrogen bonding of its peptide groups (Gursky *et al.*, 1975; Gursky *et al.*, 1977).

A third important factor is involved in the geometrical variability of the DNA structure under the action of Nt-type oligopeptides. Non-B DNA conformation can be modified to B-like conformation upon binding of Nt or Dst-3. Specific binding of Nt-type oligopeptides induces some conformational changes in B-DNA (Patel and Canual, 1977; Reinert, 1971, Zimmer *et al.*, 1980; Patel, 1982), which may be associated with an allosteric effect (Hogan *et al.*, 1978). Although Nt-binding causes no drastic conformational effects of B-DNA, it distorts the sugar phosphate backbone (Patel and Canual, 1977; Snounou and Malcolm, 1983). These effects may be of significance in the molecular process of DNA enzyme recognition.

The sequence-specific binding properties and conformational requirements of non-intercalative drugs imply models for recognition between regulatory proteins and DNA, e.g. the repressor-operator or promotor-RNA polymerase interactions (Von Hippel, 1979; Wells *et al.*, 1980). In the light of improved binding specificity of DNA-groove binding ligands modes of binding of regulatory proteins, such as of the Cro protein and lac repressor, were ascribed to β-associated oligopeptide interaction with DNA (Gursky 1983).

As sequence-specific binding ligands, Nt-type oligopeptides and other non-intercalators represent inhibitors of the template-activity of the DNA-dependent DNA- and RNA-polymerase systems in vitro (Zimmer, 1975, Gursky *et al.*, 1983). These antibiotics also interfere with cleavage sites of restriction endonucleases and specifically block the DNase I cleavage in segments of dA · dT base pair clusters (Zimmer *et al.*, 1980, Nosikov and Sain, 1977). These examples along with numerous other data (for compilation see Zimmer, 1983) show that non-intercalative DNA binding ligands are useful tools in DNA biochemistry and in biology.

ACKNOWLEDGEMENTS

We are grateful to Mrs. Christel Radtke for drawing of the figures and Mrs. Ingrid Tittel for typing the manuscript. The authors are also indebted to Professor R.E. Dickerson and Mrs. Mary Kopka for communicating their results before publication.

REFERENCES

Arcamone, F., Nicolella, V., Penco, S. and Redaelli, S., 1969, Gazz. chim. Italiana *99*, 632.
Baguley, B.C., 1982, Mol. Cellul. Biochem. *43*, 167.
Berman, H., Neidle, S., Zimmer, Ch. and Thrum, H., 1979, Biochim. Biophys. Acta *561*, 124.
Brahms, S., Vergne, J., Brahms, J.G., Di Capna, E., Bucher, Ph. and Koller, Th., 1982, J. Mol. Biol. *162*, 473.
Braithwaite, A.W. and Baguley, B.C., 1980, Biochemistry *19*, 1101.
Chidester, C.G., Krueger, W.C., Mizsak, S.A., Duchamp, D.J. and Martin, D.C., 1981, J. Amer. chem. Soc. *103*, 7629.
Drew, H.R. and Dickerson, R.E., 1981, J. Mol. Biol. *151*, 535.
Drew, H., Tanaka, S., Itakura, K. and Dickerson, R.E., 1980, Nature *286*, 567.
Finlay, A.C., Hochstein, F.A., Sobin, B.A. and Murphy, F.X., 1951, J. Amer. chem. Soc. *73*, 341.
Gellert, M., Smith, C.E., Neville, D. and Felsenfeld, G., 1965, J. Mol. Biol. *11*, 445.
Grokhovsky, S.L., Zhuze, A.L. and Gottikh, B.P., 1975, Bioorg. Chem. (USSR) *1*, 1616.
Grokhovsky, S.L., Zhuze, A.L. and Gottikh, B.P., 1978, Bioorg. Chem. (USSR) *4*, 1078.
Grokhovsky, S.L., Zhuze, A.L. and Gottikh, B.P., 1982, Bioorg. Chem. (USSR) *8*, 1070.
Grunicke, H., Puschendorf, B. and Werchau, H., 1976, Rev. Physiol. Biochem. Pharmacol. *75*, 69.
Gurskaya, G.V., Grokhovsky, S.L., Zhuze, A.L. and Gottikh, B.P., 1979, Biochim. Biophys. Acta *563*, 336.
Gursky, G.V., Tumanyan, V.G., Zasedatelev, A.S., Zhuze, A.L., Grokhovsky, S.L. and Gottikh, B.P. 1975, Mol. Biol. (USSR) *9*, 635.
Gursky, G.V., Tumanyan, V.G., Zasedatelev, A.S., Zhuze, A.L., Grokhovsky, S.L. and Gottikh, B.P., 1977, Nucleic Acid-Protein Recognition (Ed. H.J. Vogel), Academic Press p. 189.

Gursky, G.V., Zasedatelev, A.S., Zhuze, A.L., Khorlin, A.A., Grokhowsky, S.L., Streltsov, S.A., Surovaya, A.N., Nikitin, S.M., Krylov, A.S., Retchinsky, V.O., Mikhoilov, M.K., Beabealashvili, R.S. and Gottikh, B.P., 1983, Cold Spring Harbor Symp. Quantit. Biol., *47*, 367.

Hahn, F.E., 1975, Antibiotics: Mechanism of action of antimicrobial and antitumor agents (Eds. J.W. Corcoran and F.E. Hahn) Springer, Berlin-Heidelberg-New York, 3, p. 79.

Helene, C. and Lancelot, G., 1982, Prog. Biophys. molec. Biol. *39*, 1.

Hogan, M., Dattagupta, N. and Crothers, D.M., 1979, Nature *278*, 521.

Houssier, J. and Fredericq, C., 1975, Nucl. Acids Res. *2*, 971.

Ivanov, K.J., Minchenkova, L.E., Minyat, E.E., Frank-Kamenetskii, M.D. and Schyolkina, A.K., 1974, J. Mol. Biol. *87*, 817.

Khorlin, A.A., Krylov, A.S., Grokhovsky, S.L., Zhuze, A.L., Zasedatelev, A.S., Gursky, G.V. and Gottikh, B.P., 1980, FEBS Lett. *118*, 311.

Khorlin, A.A., Grokhovsky, S.L., Zhuze, A.L. and Gottikh, B.P., 1982, Bioorg. Chem. (USSR), *8*, 1982.

Kolchinsky, A.M., Mirzabekov, A.D., Zasedatelev, A.S., Gursky, G.V., Grokhovsky, S.LL., Zhuze, A.L. and Gottikh, B.P., 1975, Mol. Biol. (USSR) *9*, 14.

Krey, A.K., 1980, Prog. Molec. Subcell. Biol. *7*, 43.

Krylov, A.S., Grokhovsky, S.L., Zasedatelev, A.S., Zhuze, A.L., Gursky, G.V. and Gottikh, 1979, Nucl. Acids Res. *6*, 289.

Lavery, R., and Pullman, B., 1981, Nucl. Acids Res. *18*, 4677.

Li, L.H., Swenson, D.H., Schpok, S.L.F., Kuentzel, S.L., Dayton, B.D. and Krueger, W.C., 1982, Cancer Res. *42*, 999.

Luck, G., Triebel, H., Waring, M. and Zimmer, Ch. 1974, Nucl. Acids Res. *1*, 503.

Luck, G., Zimmer, Ch., Reinert, K.-E. and Arcamone, F., 1977, Nucl. Acids Res. *4*, 2655.

Luck, G. Zimmer, Ch. and Baguley, B., 1984, Biochim. Biophys. Acta, in press.

Marck, Ch., Kakiuchi, N. and Guschlbauer, W., 1982, Nucl. Acids Res. *10*, 6147.

Marky, L.A., Blumenfeld, K.S. and Breslauer, K.J., 1983, Nucl. Acids Res. *11*, 2857.

Martin, J.C., Wartell, R.M. and O'Shea, D.C., 1978, Proc. Natl. Acad. Sci. USA *75*, 5483.

McGhee, J.D., 1976, Biopolymers *15*, 1345.

Minchenkova, L.E. and Zimmer, Ch., 1980, Biopolymers *19*, 823.

Neidle, S., 1979, Prog. Medicinal Chemistry (eds. G.P. Ellis and G.B. West), Elsevier, North-Holland, vol. *16*, 151.

Nordheim, A., Pardue, M.L., Lafer, E.M., Möller, A., Stollar, B.D. and Rich, A., 1981, Nature *294*, 417.

Nosikov, V.V. and Sain, B., 1977, Nucl. Acids Res. *4*, 2263.

Patel, D.J., 1979a, Eur. J. Biochem. *99*, 369.

Patel, D.J., 1982, Proc. Natl. Acad. Sci. USA *79*, 6424.

Patel, D.J. and Canuel, L.L., 1977, Proc. Natl. Acad. Sci. USA *74*, 5207.

Patel, D.J., 1979b, Accounts Chem. Res. *12*, 118.

Patel, D.J., Kozlowski, S.A., Rice, J.A., Broka, C. and Itakura, K., 1981, Proc. Natl. Acad. Sci. USA *78*, 7281.

Pullman, B. and Pullman, A., 1981, studia biophysica *86*, 95.

Pullman, B., Pullman, A. and Lavery, R., 1983. Structure, Dynamics, Interactions Evol. Biol. Macromolec. (Ed. C. Helene) D. Reidel Publish, Comp. p. 23.

Reinert, K.-E., 1971, J. Mol. Biol. *72*, 593.

Reinert, K.-E., 1981, Biophys. Chem. *13*, 1.

Reinert, K.-E., Stutter, E. and Schweiss, H., 1979, Nucl. Acids Res. *7*, 1375.

Reinert, K.-E., Thrum, H. and Sarfert, E., 1980, Nucl. Acids Res. *8*, 5519.

Sage, E. and Leng, M., 1981, Nucl. Acids Res. *9*, 1241.

Snounou, G. and Malcolm A.D.B., 1983, J. Mol. Biol. *167*, 211.

Van Dyke, M.W., Hertzberg, R.P. and Dervan, P.B., 1982, Proc. Natl. Acad. Sci. USA *79*, 5470.

Von Hippel, P.H., 1979, Biol. Regulation and Development (Ed. R.F. Goldberger), Plenum Publish. Corp. 1979, vol. *1*, 279.

Wahnert, U., Zimmer, Ch., Luck, G. and Pitra, Ch., 1975. Nucl. Acids Res. *2*, 391.

Wang, A.H.-J., Quigley, G.J., Kolpak, F.J., Crawford, J.L., Van Boom, J.H., Van der Marel, G. and Rich, A., 1979, Nature *282*, 680.

Waring, M., 1981, Mol. Basis Antibiotic Action (Eds. E.F. Gale, E. Coundliffe, P.E. Reynolds, M.H. Richmond, M.J. Waring, Wiley, London, p. 258.

Wartell, R.M., Larson, J.E. and Well, R.D., 1974, J. Biol. Chem. *249*, 6719.

Wells, R.D., Goodman, T.G., Hillen, W., Horn, G.T., Klein, R., Larson, D., Muller, U.R., Neuendorf, S.K., Panayotatos, N. and Stirdivant, S.M., 1980, Prog. Nucleic Acid Res. Mol. Biol. *24*, 167.

Zasedatelev, A.S., Gursky, G.V., Zimmer, Ch. and Thrum, H., 1974, Mol. Biol. Rep. *1*, 337.

Zasedatelev, A.S., Zhuze, A.L., Zimmer, Ch., Grokhovsky, S.L., Tumanyan, V.G., Gursky, G.V. and Gottikh, B.P., 1976, Dokl. Akad. Nauk SSSR *231*, 1006.

Zasedatelev, A.S., Zhuze, A.L., Zimmer, Ch., Grokhovsky, S.L., Tumanyan, V.G., Gursky, G.V. and Gottikh, B.P., 1978, studia biophysica *67*, 47.

Zimmer, Ch., 1975, Prog. Nucleic Acid Res. Mol. Biol. *15*, 285.

Zimmer, Ch., 1983, Comments Mol. Cell. Biophys. *1*, 399.

Zimmer, Ch., Luck, G. Thrum, H. and Pitra, Ch., 1972, Eur. J. Biochem. *26*, 81.

Zimmer, Ch., Marck, Ch. Schneider, Ch. and Guschlbauer, W., 1971, Nucl. Acids Res. *6*, 2831.

Zimmer, Ch., Luck, G., Lang, H. and Burckhardt, G. 1979, Proc. 12th FEBS meeting (Eds. S. Rosenthal, H. Bielka, C. Coutelle and Ch. Zimmer) Pergamon, Oxford, *51*, 83.

Zimmer, Ch., Luck, G. and Nuske, R., 1980a, Nucl. Acids Res. *8*, 2999.

Zimmer, Ch., Marck, Ch., Schneider, Ch., Thiele, D., Luck, G., and Guschlbauer, W., 1980b. Biochim. Biophys. Acta *607*, 232.

Zimmer, Ch., Kakiuchi, N. and Guschlbauer, W., 1982, Nucl. Acids Res. *10*, 1721.

Zimmer, Ch., Marck, Ch. and Guschlbauer, W., 1983a, FEBS lett. *154*, 156.

Zimmer, Ch., Luck, G., Birch-Hirschfeld, E., Weiss, R., Arcamone, F. and Guschlbauer, W. 1983b, Biochim. Biophys. Acta, in press.

Zimmermann, S.B., 1982, Ann. Rev. Biochem. *51*, 395.

INTERACTIONS OF TRANSFER RNAS WITH THEIR BIOLOGICAL PARTNERS

J.P. EBEL, R. GIEGE, D. MORAS and P. REMY

Institut de Biologie Moléculaire et Cellulaire du CNRS

15, rue René Descartes - 67084 Strasbourg - France

ABSTRACT

The three-dimensional structure of two elongator tRNAs, $tRNA^{Phe}$ and $tRNA^{Asp}$ from yeast, has been solved at high resolution. Their secondary cloverleaf structures are folded in a tertiary L-shaped conformation with a more open structure for $tRNA^{Asp}$. Both tRNAs present different molecular dynamics as revealed by temperature factors of the different parts of the molecules. These features are related to different conformational states of the tRNAs: $tRNA^{Phe}$ would possess the structure of a free tRNA and $tRNA^{Asp}$ that of a tRNA on the ribosome. The solution structure of both tRNAs was probed with chemical reagents and compared with the crystal structures. Structural similarities as well as differences were detected reflecting the versatility of tRNA structures.

Comparing the accessibility of tRNAs, in their free state and complexed to aminoacyl-tRNA synthetases or elongation factor EF-Tu, to chemical or enzymatic probes, it was possible to determine the regions of tRNAs in close contact with their macromolecular partners. Upon complex formation with aminoacyl-tRNA synthetases a multistep adaptation of both macromolecules takes place, analogous to an induced fit. This process determines the specificity of the tRNA aminoacylation reaction. In the particular case of the yeast phenylalanine system, the wybutine residue next to the anticodon of $tRNA^{Phe}$ has been identified as an important element in this mechanism. More precise structural information will arise in future from studies on crystallized complexes. Such a complex, that between $tRNA^{Asp}$ and aspartyl-tRNA synthetase, has been crystallized and is under investigation.

I. Introduction

Transfer RNAs (tRNAs) represent a fascinating family of nucleic acids, because they participate in numerous biological functions. The function which has been the most extensively studied is their role in the mRNA mediated protein biosynthesis (for a review see Schimmel *et al*, (1979) (Table IA)). However this is not the only cellular role played by these molecules. Transfer RNAs are also involved in other important functions. For example they participate in the mRNA independent transfer of amino acids into different cellular components, such as peptidoglycans in the bacterial cell wall biosynthesis, complex lipids and even the terminal NH_2 groups of synthesized proteins (Table IB). All these functions involve numerous interactions with different biological partners. Other interactions are related to tRNA processing and modification (Table IC). Finally there is more and more evidence for the involvement of tRNAs in functions where they interact with genomic components. Several tRNA species are

TABLE I. *Interactions between tRNAs and several ligands related to different tRNA functions*

A) *mRNA mediated protein biosynthesis* Interaction of tRNA with: — aminoacyl-tRNA synthetases — initiation and elongation factors — mRNA — ribosome	B) *mRNA independent transfer by several tRNAs of amino acids into*: — peptidoglycans — complex lipids — NH_2 terminus of synthesized proteins.
C) *tRNA processing and modification* Interaction of tRNAs or of their precursors with: — processing enzymes — nucleoside modifying enzymes - in situ modifications of nucleosides (methylated nucleosides, dihydrouridine, pseudouridine, hypermodified nucleosides, etc.) - insertion of Q base	D) *Interaction of tRNA with genomic components* — involvement of several tRNA species as primers in reverse transcription of viral RNAs (retroviruses) — regulatory functions of tRNA (regulation of biosynthesis of amino acids, aminoacyl-tRNA synthetases, tRNA, rRNA).

involved as primers of reverse transcriptase and participate in regulatory processes (Table ID) (for a review see La Rossa and Söll, 1980).

Among these various functions, the best understood role of tRNAs is their participation in protein synthesis. In that process their function is to carry amino acids to the ribosomes, to decode the messenger RNAs and to incorporate the correct amino acid into the protein sequence. These functions lead tRNAs to many interactions with different proteins and nucleic acids. With aminoacyl-tRNA synthetases, enzymes which attach the correct amino acid to the 3′-end of their cognate tRNAs, the molecular recognition must be highly specific. The same is true for the decoding of the genetic code at the messenger RNA level. However, with the elongation factor, which carries the aminoacylated tRNAs to the ribosome and also with the ribosomal components which are involved in peptide bond synthesis, the common partners imply common features. In this study, we will concentrate on two interactions which are examples of these two situations and which have been extensively studied in our laboratory: the highly specific interaction of tRNAs with aminoacyl-tRNA synthetases and the less specific interaction of aminoacyl-tRNAs with the bacterial elongation factor EF-Tu.

II. Structure of tRNA

The three-dimensional structures of two tRNAs, yeast $tRNA^{Phe}$ and $tRNA^{Asp}$, have been solved at high resolution. Both are elongator tRNAs with a short extra-loop. In this section, we will focus on the structure of $tRNA^{Asp}$ with reference to that of $tRNA^{Phe}$.

1. *Primary structure*

The nucleotide sequence of yeast $tRNA^{Asp}$, shown on Figure 1 together with that of $tRNA^{Phe}$, presents some characteristic features (Gangloff *et al.*, 1971, RajBhandary and Chang, 1968). It contains a high number of G-C base pairs, except in the D-stem where two G-U base pairs are present. The variable loop is made of four nucleotides versus five in $tRNA^{Phe}$ (for convenience of comparisons we kept the same numbering, assuming a deletion at positions 47). The D-loop has the same length as that of $tRNA^{Phe}$ but the two conserved Gs, which are crucial for D and T-loops tertiary interactions, are at positions

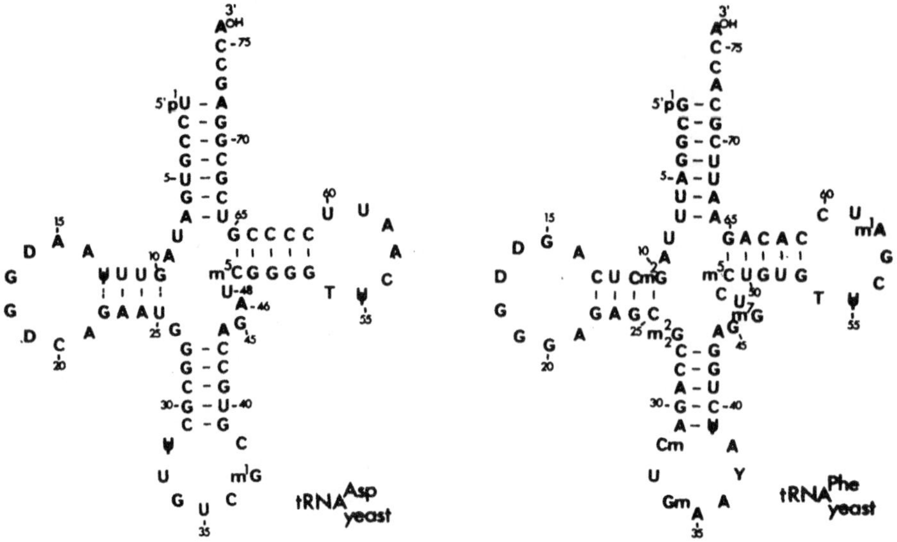

FIG. 1. The nucleotide sequence of yeast $tRNA^{Asp}$ and $tRNA^{Phe}$. For convenience the numbering system of the nucleotides is that of yeast $tRNA^{Phe}$; in the 75 nucleotide- long $tRNA^{Asp}$ position 47 in the variable loop has been omitted. Non-classical Watson-Crick base pairs are indicated by broken lines.

17 and 18 instead of 18 and 19, thus making α and β regions of the D-loop quite symmetrical. Last but not least, the anticodon GUC presents the peculiarity to be self-complementary, with a slight mismatch at the uridine position. This feature was first noted by Grosjean *et al.* (1978) who showed the existence of a significant interaction in solution and suggested it to be a tempting model to study tRNA-mRNA recognition.

2. *Crystal structure*

Structure determination: The molecular structure of phenylalanine specific tRNA from yeast has been solved independently in different laboratories. The groups at MIT and Duke have reported their results on the orthorhombic crystal from an analysis at 2.5 Å resolution (Kim *et al.* 1974; Quigley *et al.*, 1975 a; Sussman *et al.*, 1978). The MRC group has reported his work on the monoclinic crystal form at 2.5 Å from MIR (multiple isomorphous replacement) maps and also a real-space refinement of the model at 2.5 Å resolution (Ladner *et al.*, 1975; Jack *et al.*, 1976). The same crystal form was solved independently at Madison by Stout *et al.* (1978).

For yeast $tRNA^{Asp}$ two structures have been solved from a multiple isomorphous replacement (MIR) X-ray analysis of two crystal forms (Moras *et al.*, 1980). The transition between forms A and B is temperature dependent but it can also be induced around 20° by pH changes or the addition of some heavy atoms derivatives (Huong *et al.*, 1984). The structure of one form, the lower temperature one, has been refined in reciprocal space using the restrained least-square method of Konnert and Hendrickson (1980) and in real-space with the graphic modelling program FRODO (Jones, 1978). Both programs were adapted for nucleic acid handling.

Boomerang versus L-shape: A view of the $tRNA^{Asp}$ molecule together with a similar view of $tRNA^{Phe}$ is shown on Figure 2. The topological organization of the cloverleaf sequence is similar to that first found for yeast $tRNA^{Phe}$. This gives the L-shaped

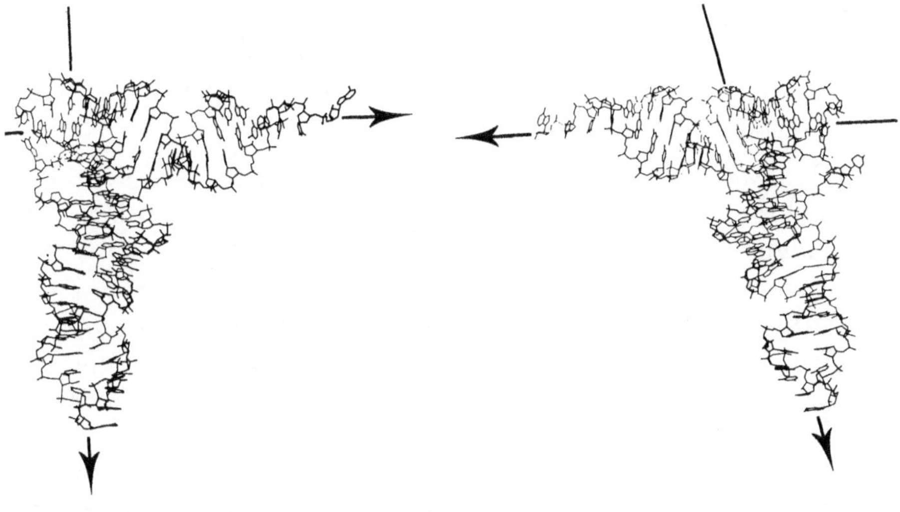

FIG. 2. Two views of the three-dimensional structures of yeast $tRNA^{Asp}$ (right) and $tRNA^{Phe}$ (left). The coordinates of $tRNA^{Asp}$ refer to the refined low temperature form; the R-factor for these data is presently 24.5% at 3 Å resolution. The CCA-end part however is not yet fully defined and was set in the standard helical conformation. The coordinates of yeast $tRNA^{Phe}$ are those of the orthorhombic crystal form (Quigley *et al.*, 1975 b).

structure formed by two units: vertically the anticodon and the D-stems, horizontally the T- and the amino acid accepting stems. However in the case of $tRNA^{Asp}$ the two branches forming the L are more open (by more than 10°) than in $tRNA^{Phe}$ conferring to this tRNA a boomerang-like shape. This results in a different positioning of the anticodon and of the T-stems and loops with respect to fixed acceptor and D-stems which superpose well to the corresponding part of $tRNA^{Phe}$ (Moras *et al.*, 1980).

Anticodon-anticodon interaction: In the orthorhombic crystal lattice (space group C 222_1) $tRNA^{Asp}$ molecules are associated through a two-fold symmetry axis parallel to the crystallographic *b* direction by anticodon-anticodon interactions. Figure 3 represents the local conformation (Westhof *et al.*, 1983). The GUC anticodon triplets of symmetrically related molecules form complementary hydrogen bonded base pairs, arranged in a normal helical conformation. This small helix is stabilized by stacking of the modified base m^1G37 on both sides. This packing confers a great stability to the dimeric structure and explains the good quality of the electron density map in the anticodon region. A contact between the anticodon loops of two tRNAs also exists in the orthorhombic form of yeast $tRNA^{Phe}$. In that case, however, the G_mAA anticodons cannot be base paired and they are arranged in a stacked conformation.

Dynamic aspects: Crystallographic refinement leads to the determination of the so-called Debye-Waller temperature factor. Although temperature factors contain various components, it has been shown (Frauenfelder *et al.*, 1979 and Artymiuk *et al.*, 1979) that their variation along a macromolecular backbone has some physical meaning. In Figure 4 are represented the temperature factors of each phosphate along the polynucleotide chain in $tRNA^{Asp}$ and in $tRNA^{Phe}$.

For $tRNA^{Asp}$, it is apparent that the stems are more rigid than the loops, except for the end of the amino acid acceptor stem. This variation is not so apparent for $tRNA^{Phe}$, although the effect is present. Also, in $tRNA^{Asp}$, the T-loop presents higher temperature factors than the anticodon loop. This is in marked a

b

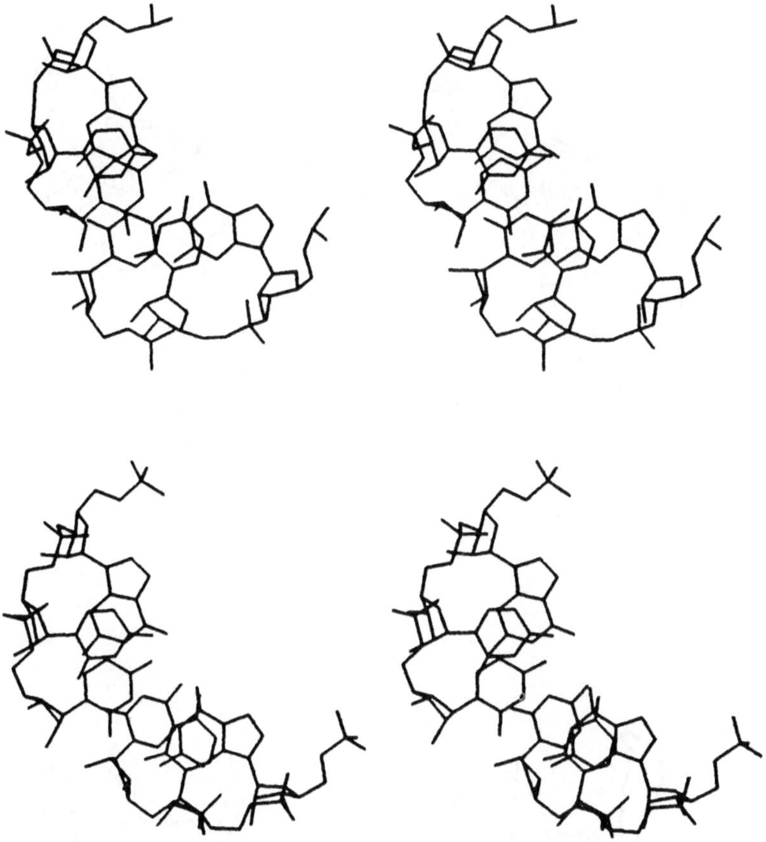

FIG. 3. Stereo-views of the (a) anticodon triplet (GUC) base pairing in yeast tRNA^{Asp} and (b) of the corresponding triplet in the RNA-11 conformation (Arnott *et al.*, 1972).

contrast to the situation in tRNA^{Phe}, where the anticodon loop presents the highest temperature factors and the T-loop the lowest ones. There is a similar but less pronounced reversal in the region of the P10-loop, where P10 is at a minimum in tRNA^{Asp} and at a maximum in tRNA^{Phe}. The origins of these differences are difficult to pinpoint. We think that the different behaviour of the " flexibility " of the tRNA molecule in the two crystals arises from the different packing of the molecules in crystals of tRNA^{Asp} and of tRNA^{Phe}, since there is anticodon-anticodon base-pairing in the former crystals and not in the latter ones.

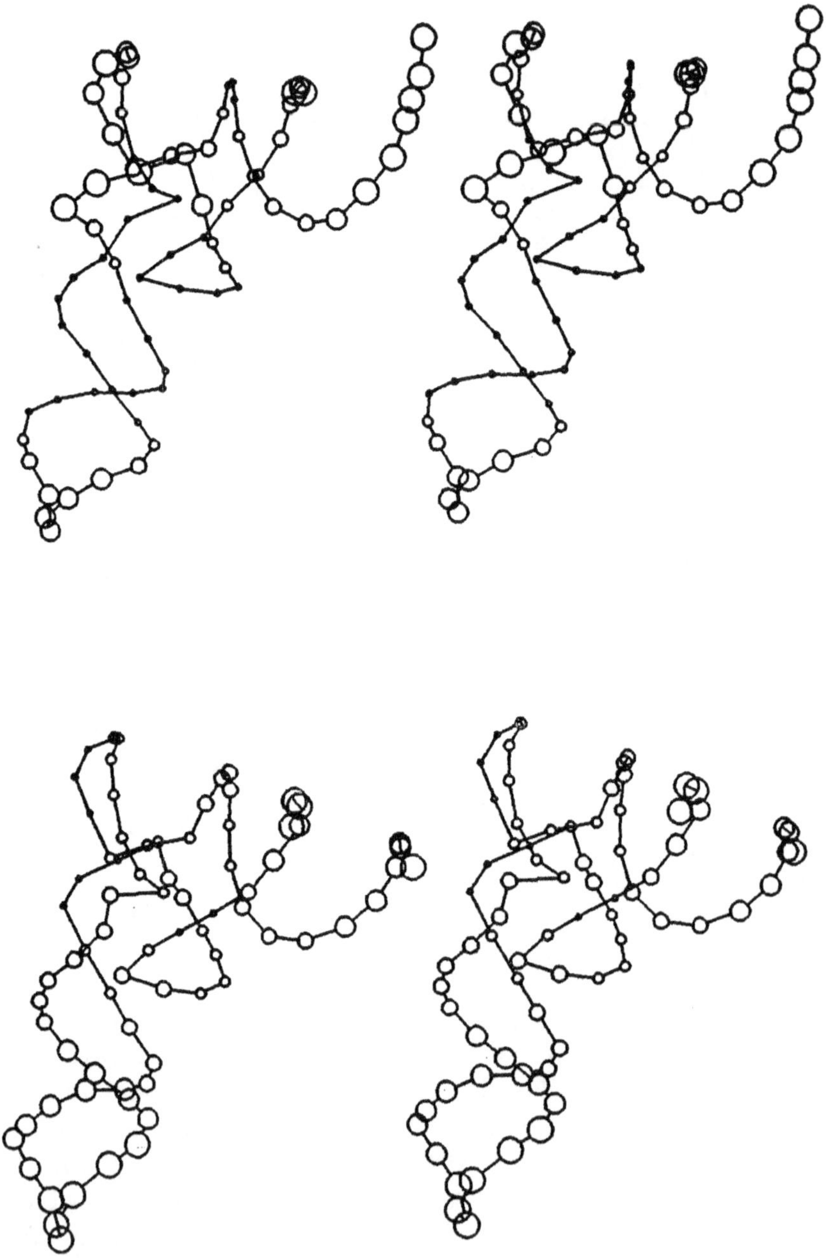

FIG. 4. Thermal vibration of the backbones. Two stereo-views of yeast $tRNA^{Asp}$ (up) and $tRNA^{Phe}$ (bottom) are shown. Each phosphate group is presented by a ball proportional to the value of the temperature factor.

To summarize, we can say that the structure of $tRNA^{Asp}$ shows the conformational state of a tRNA on the ribosome. In the crystals one GUC triplet from one tRNA molecule mimics a codon of mRNA interacting with the GUC anticodon of a second molecule. This interaction might act as a signal and trigger conformational changes elsewhere. The open structure of the tRNA might be a result of such a mechanism. Influence on the D and T-loop association is strongly suggested by thermal factors. The structure of $tRNA^{Phe}$ would be that of a free tRNA, with flexible anticodon stem and loop.

3. *Solution structure*

Experimental approaches: Solution structures of tRNAs have already been approached by a variety of physical and biochemical techniques (for a review see Schimmel and Redfield, 1980). Here we have used new biochemical methodologies which permit to probe the accessibility of the tRNA molecule to chemical reagents specific for different functional groups of the nucleic acid. Enzymatic probes can also be used (e.g. Wrede *et al.*, 1979; Favorova *et al.*, 1981; Boutorin *et al.*, 1982; Florentz *et al.*, 1982), but in this study we have given the preference to the chemical reagents, because of their small size compared to the bulky enzymes. This allows one to probe discrete conformational features in the tRNA. The principle of the different methods derives from the chemical sequencing methodologies of nucleic acids (Peattie, 1979; Maxam and Gilbert, 1980) and relies on their statistical and low yield modification at each potential target, in such a way that each tRNA chain undergoes less than one modification. Two experimental conditions are usually chosen; one which maintains the native structure of the tRNA and another one where the tRNA is denatured in such a way that all potential targets become accessible and can react with the probes. The tRNA molecules labelled at their 3′ or 5′ end with radioactive $[^{32}P]$ ATP are then specifically split at the modified positions and the resulting end-labelled oligonucleotides are analyzed by high voltage electrophoresis on sequencing gels followed by autoradiography. The assignment of the bands is done by comparing their migrations to ladders obtained after limited T_1 RNAse digestion or alkaline hydrolysis. In such a way it becomes possible to probe the entire tRNA molecule in one experiment.

Three chemical reagents, ethylnitrosourea (ENU), dimethylsulfate (DMS) and diethylpyrocarbonate (DEPC) were used. All three are potent carcinogenic agents which react at different sites of nucleic acids (Leonard *et al.*, 1971; Kusmierek and Singer, 1976) but with a pronounced preference for those indicated in Figure 5. The modification reactions labilize the ribophosphate backbone so that it can be split easily at the modified positions according to the scheme shown in Figure 5.

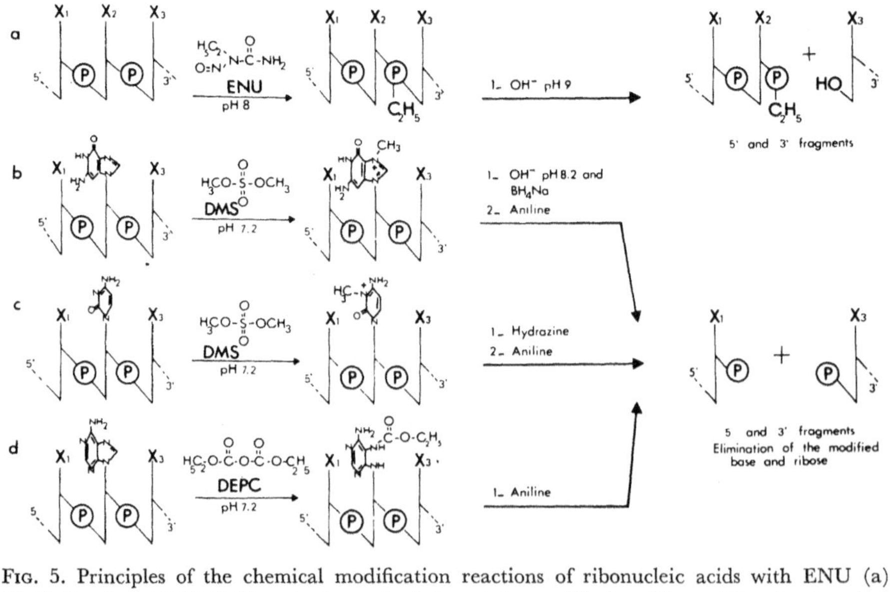

FIG. 5. Principles of the chemical modification reactions of ribonucleic acids with ENU (a), DMS (b, c) and DEPC (d) and of strand-scission at the modified positions. The ribophosphate backbone is schematized; (P) represents a phosphate group; X_n are the bases. Experimental details for the ENU reaction were described by Vlassov et al., 1980 and for base alkylations by Peattie and Gilbert, 1980.

ENU probes the accessibility of phosphates (Vlassov *et al.*, 1980). The reactivity per residue is particularly low: 0.1 to 0.3 ethyl group incorporated per tRNA molecule. The splitting of the ribophosphate chain at the modified positions is done at alkaline pH; 3′ and 5′ fragments are liberated, the latter one carrying the ethyl groups. The method was established with yeast $tRNA^{Phe}$ and it was shown that the results could be interpreted in terms of accessibilities to the reagent of the phosphates in the tertiary structure of the tRNA. This was best illustrated in a comparison between the chemical phosphate reactivities in native yeast $tRNA^{Phe}$ and the calculated accessible areas of the anionic oxygens of the phosphates for Na^+ ions and for water (Thiyagarayan and Ponnuswamy, 1979; Vlassov *et al.*, 1981; Lavery *et al.*, 1980 a). This comparison gave a quasi-perfect identity between the phosphodiester reactivity toward ENU and the steric accessibility of the phosphate groups (Figure 6).

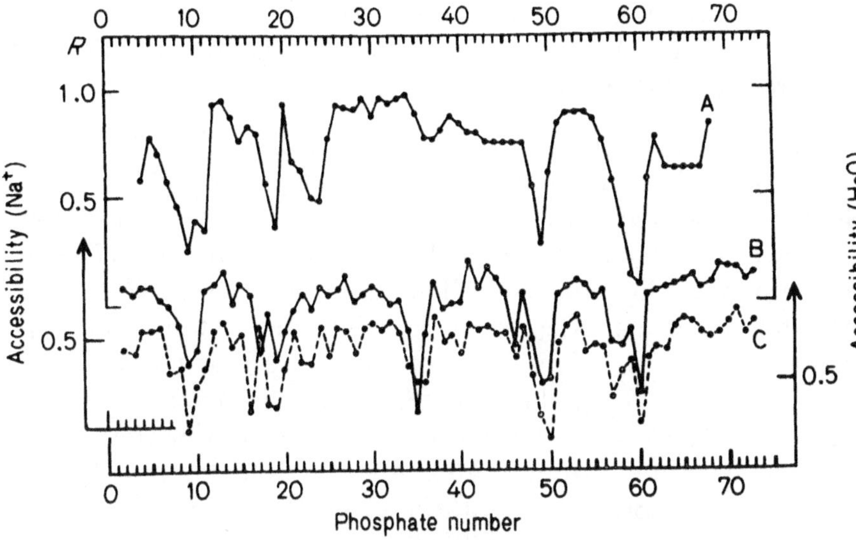

FIG. 6. A comparison between the chemical phosphate reactivites in the native yeast $tRNA^{Phe}$ (A) with calculated (Thiyaharayan and Ponnuswamy, 1979; Lavery et al, 1980 a) accessible areas of the anionic oxygens of the phosphates for Na^+ ions (B) and for water (C) in the crystal structure of the tRNA. In (A) the ratio R is defined as in Figure 8.

DMS and DEPC probe the accessible bases and were introduced by Peattie and Gilbert (1980) who established the method with yeast $tRNA^{Phe}$. The relevance of the approach was supported by accessibility and potential computations of the base environments (Lavery *et al.*, 1980 b). DMS and DEPC react at the N7 position of G and A; for A the modification is particularly sensitive to stacking. DMS also alkylates the N3 position of C; this position can be involved in Watson-Crick pairings thus making this modification a powerful probe for secondary or tertiary interactions. The extent of base alkylation is about 10 fold that of phosphate modification by ENU, a favourable fact from an experimental point of view (Barciszewski *et al.*, 1982). However, the presence of naturally occuring N7 methylated purines in tRNA is a disadvantage since it could be responsible for the splitting of all molecules at these positions. Therefore the conditions for chain scission are chosen so that the splitting is only partial.

The structural environment of the studied positions (either accessible or inaccessible to the reagents) was analyzed on a graphic interacting display system (MPS from Evans and Sutherland at EMBL, Heidelberg, Germany) using the graphic modelling program FRODO (Jones, 1978).

Accessible and buried residues: Chemical modification experiments were done in parallel on $tRNA^{Asp}$ and $tRNA^{Phe}$ under identical experimental conditions. Figure 7 gives an example of a DEPC experiment in which the accessibility of the N7 position of adenines has been probed. The appearance of a band indicates that this particular adenine is accessible under the given experimental conditions. This is for instance the case for the constant residue A21 in the D-loop of tRNA which reacts with DEPC when the tRNA is under native conditions. On the contrary, other residues, for instance A62, A64 in $tRNA^{Phe}$ and A44, A46 in $tRNA^{Asp}$ are protected in the native tRNAs. These examples illustrate how chemical modification experiments are interpreted. In the frame of this report we do not intend to present all accessibility results obtained with $tRNA^{Asp}$ and $tRNA,^{Phe}$ but we only want to focus on 3 particular positions: P60, A21 and G45.

Structural similarities in tRNAs — The conformational environment of phosphate 60: A comparison of the patterns of phosphate alkylation in tRNAs from yeast and *E.coli* (Vlassov *et al.*, 1981) revealed a striking similarity essentially in the T-arm and particularly at position 60 where a total absence of alkylation can be noticed.

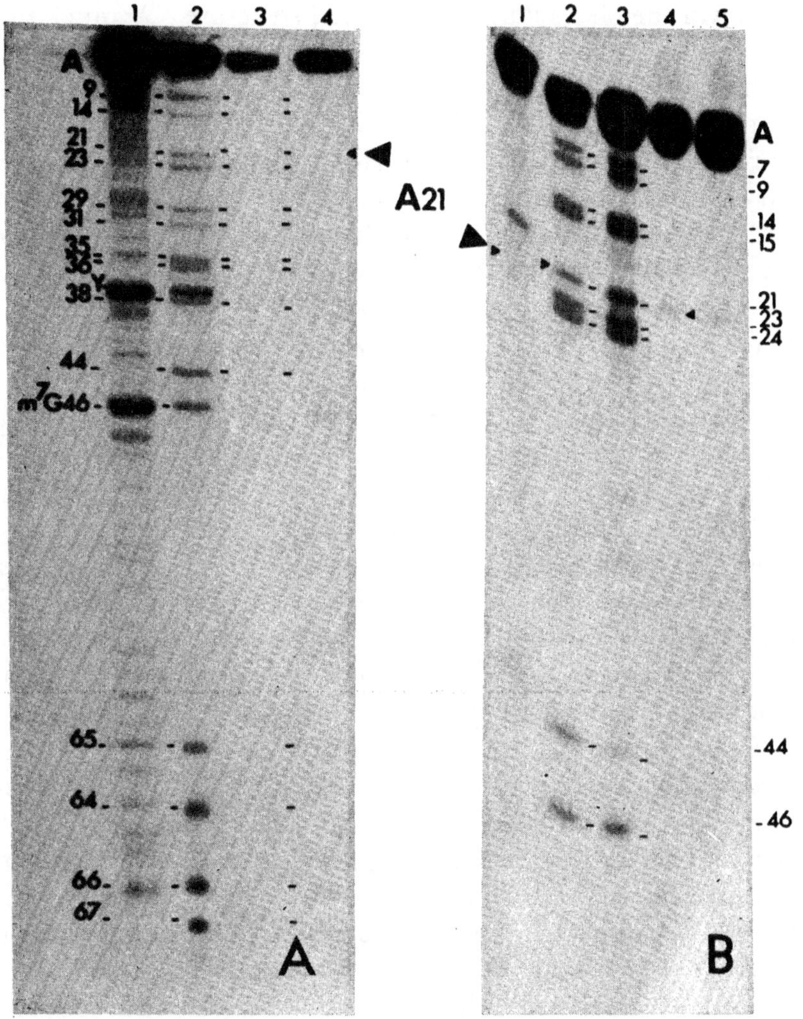

FIG. 7. Adenosine accessibility experiment. Autoradiogram of 15% acrylamide gels of an alkylation experiment with DEPC on tRNA^{Phe} (a) and tRNA^{Asp} (b). (1) Formamide ladder, (2) denatured conditions: 1 mM EDTA, 50 mM sodium cacodylate buffer pH 7.2, 90° C, (3) semi-denatured conditions: 1 mM EDTA, same buffer, 37° C, (4-5) native conditions: 10 mM $MgCl_2$, same buffer, 37° C. Adenosine positions are numbered.

In the three-dimensional structure of tRNA^{Phe} (Quigley *et al.*, 1975a; Jack *et al.*, 1976; Stout *et al.*, 1978; Sussman *et al.*, 1978), as well as in that of tRNA^{Asp} (Moras *et al.*, 1980), phosphate 60 is located in the hinge region where the two helical domains of tRNA join to form the characteristic L-shaped structure. A quick inspection of both molecular structures suggest a good explanation. Phosphate 60 appears to be buried in a pocket formed by A58, C60 and C61 on the one side and the deep groove of the T-stem on the other side. The resulting steric hindrance induced by the tertiary folding would thus account for the absence of alkylation. This interpretation is best illustrated in $tRNA^{Phe}$ by the computation of steric inaccessibility of phosphate 60 to sodium ions or water (Thiyagarayan and Ponnuswamy, 1979; Lavery *et al.*, 1980 a). In both $tRNA^{Asp}$ and $tRNA^{Phe}$ the existence of a hydrogen bond between phosphate 60 and the N4 amino group of C61 and the 2′-hydroxyl group of ribose 58 is another good explanation for the lack of reactivity of this phosphate. It includes the steric inaccessibility in the hydrogen bond direction and has the additional effect of rigidifying the local conformation.

The ENU reaction on the 3′half-molecule of $tRNA^{Asp}$ clearly shows the similar absence of reactivity of phosphate 60. Figure 8 gives the experimental data. This experiment suggests a similar folding of the end of the T-stem which helps to stabilize the structure of the T-loop in a similar conformation in all tRNAs.

Structural differences in tRNA^{Asp} *and tRNA*^{Phe} — *The conformational environment of A*21 *and G*45: DMS and DEPC probe N7 positions of purines and N3 positions of cytosines. For simple structural comparison one should concentrate only on those residues located at similar positions in $tRNA^{Asp}$ and $tRNA^{Phe}$. These positions are indicated in Figure 9. Some residues exhibit similar behaviours in both tRNAs. Other nucleosides display different chemical reactivities; here we will discuss two of them, A21 and G45.

In $tRNA^{Asp}$ adenosine 21 is accessible to modification at N7 by DEPC; it is protected in $tRNA^{Phe}$. The explanation is linked to a different extra-loop structure in both tRNAs. $tRNA^{Phe}$ possesses a 5 bases long extra-loop, whereas $tRNA^{Asp}$ has only 4 bases in this domain. In the three-dimensional structure of both tRNAs, A21 is stacked between positions 46 and 48. Position 47 is an uridine residue in $tRNA^{Phe}$ and bulges out of the structure; the corresponging residue is missing in $tRNA^{Asp}$. This confers a

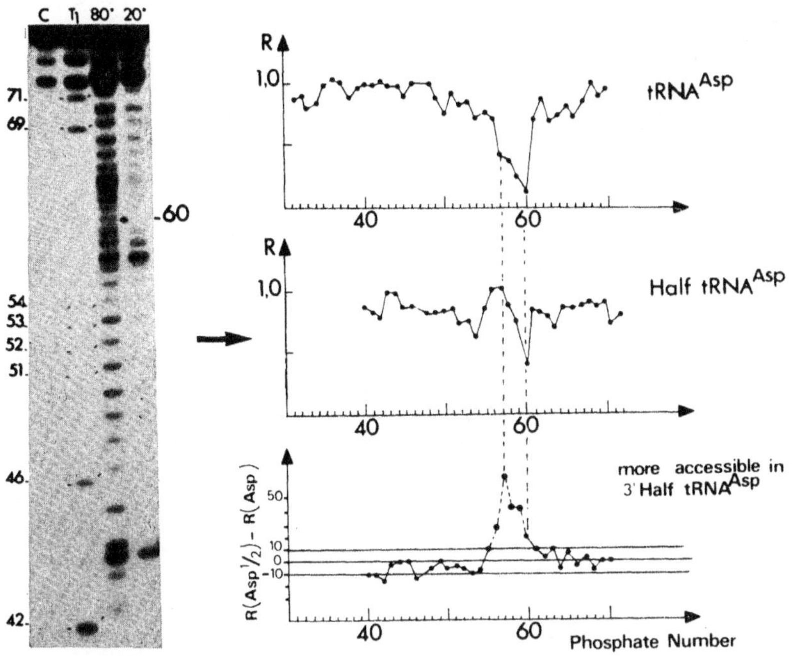

FIG. 8. Comparison of phosphate reactivities toward ENU between native and 3′half-molecule of $tRNA^{Asp}$. The figure shows an autoradiogram of a 15% acrylamide gel of an alkylation experiment on the 3′ half-molecule of $tRNA^{Asp}$ labelled at its 5′ end: (C) control incubation under conditions stabilizing the tRNA structure, (T_1) partial ribonuclease T_1 ladder, (80°, 20°) alkylations at 80° C or 20° C under conditions unfolding (80°) or stabilizing (20° C) the half-molecule. The absence of band at position 60 is indicated. The interpretation of data is given in plots of R versus phosphate number for $tRNA^{Asp}$ and half-$tRNA^{Asp}$. *R* values are the ratios between the intensities of the corresponding electrophoretic bands of the alkylated folded und unfolded molecules. A difference plot between half- and complete $tRNA^{Asp}$ is given showing the same shielding of P 60 in both molecules, but otherwise more accessible phosphates in the T-loop of the half- $tRNA^{Asp}$. This means that these phosphates are protected by folding effects in the complete tRNA; these effects are lost when the T-loop is isolated from its structural context.

slightly different positioning of residues 46 and 48 in the two tRNAs as can be seen in Figure 10. The conformations of the extra-loops permit a perfect stacking of A21 between bases 46 and 48 in $tRNA^{Phe}$ and a worse one in $tRNA^{Asp}$. As a consequence the N7 position of A21 is protected in $tRNA^{Phe}$ and accessible in $tRNA^{Asp}$.

As to G45, the situation is inversed. The N7 position of the guanine residue is accessible to alkylation by DMS in $tRNA^{Phe}$ and protected in $tRNA^{Asp}$. The reason for that is linked to a dif-

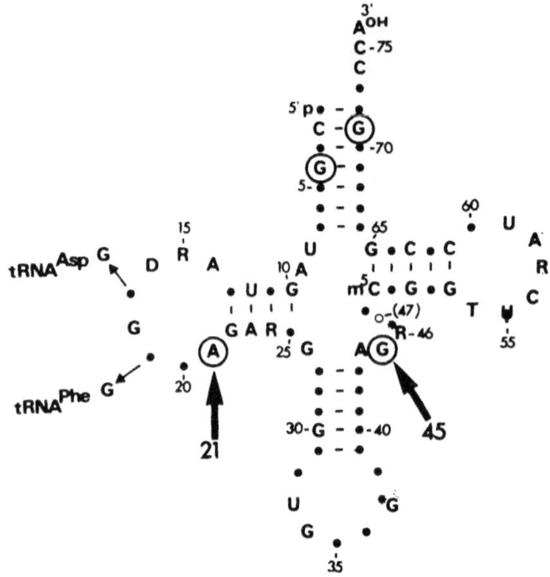

FIG. 9. Composite secondary structure of yeast $tRNA^{Asp}$ (Gangloff *et al.*, 1971) and $tRNA^{Phe}$ (RajBhandary and Chang, 1968) with nucleotides common to both tRNAs. R is for common purines and Y for common pyrimidines. Those residues which exhibit a different reactivity towards DMS or DEPC are encircled. Uridine residues are not tested.

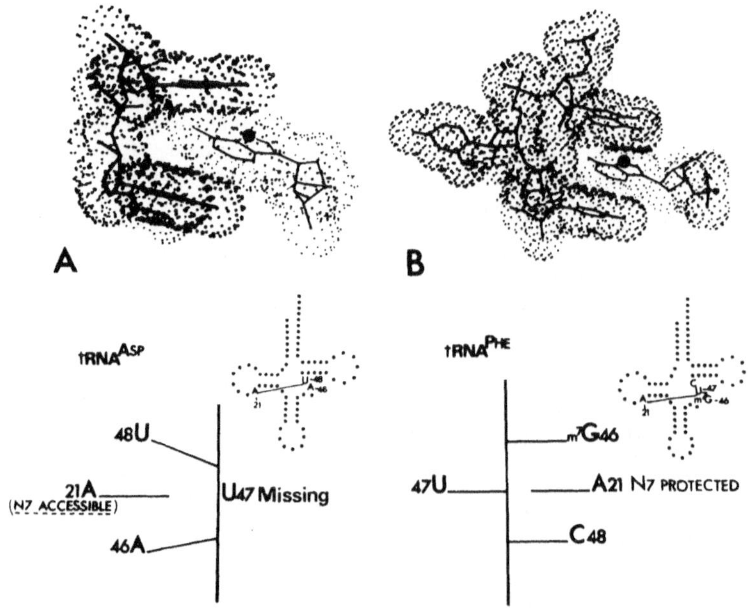

FIG. 10. Graphic modelling of the structural environment of adenosine 21 in $tRNA^{Asp}$ (A) and $tRNA^{Phe}$ (B) and interpretation of the display pictures. The Van der Waals spheres of the atoms are shown and the N7 position of A 21 is indicated by a dark dot.

ferent stacking of residue A9 between positions 45 and 46. This difference is due to the presence in $tRNA^{Phe}$ of m^7G in position 46 and of A46 in $tRNA^{Asp}$. The bulky methyl group of m^7G hinders A9 to be stacked correctly in $tRNA^{Phe}$ and as a consequence G45 becomes accessible to DMS. In contrast, in $tRNA^{Asp}$ A9 is perfectly stacked between residues 45 and 46 explaining the protection of G45.

III. Interaction between tRNA and aminoacyl-tRNA synthetase

1. *Crystallization of the $tRNA^{Asp}$ — Aspartyl-tRNA synthetase complex*

Both $tRNA^{Asp}$ and aspartyl-tRNA synthetase could be crystallized in various media including low salt conditions (Giegé *et al.*, 1977; Dietrich *et al.*, 1980). The best diffracting crystals, however, could only be grown in the presence of ammonium sulfate at 62% saturation for tRNA and at 54% for the synthetase. For the complex, crystallization was observed to occur only under high salt conditions, between 48 and 53% saturation of ammonium sulfate depending upon the concentration of macromolecular components and the temperature. The optimal crystallization conditions for the different components of the aspartic acid system are summarized in Table II. A full account of the

Table II. *Crystallization conditions of $tRNA^{Asp}$, aspartyl-tRNA synthetase and the complex formed between these two macromolecules*

Molecules	$tRNA^{Asp}$	AspRS	Complex [tRNA]/[AspRS] = 2
Molecular weight	24160	125000	175000
Precipitant			
— Nature	$(NH_4)_2SO_4$	$(NH_4)_2SO_4$	$(NH_4)_2SO_4$
— Concentration	62%	54%	50%
Macromolecular concentrations ($mg \cdot ml^{-1}$)			
—tRNA	3 - 5	—	1 - 5
— Enzyme	—	4	3 - 12
pH	6.8	6.7	7.0 - 8.0

crystallization procedure was reported by Giegé *et al.* (1980) and Lorber *et al.* (1983).

For the complex, one of the most important factors in crystal formation is the stoechiometry of the components. No crystals can be grown for a molecular ratio tRNA/enzyme inferior to 1.8 and the quality of the crystals degrades when that ratio increases to more than 2.2. These values only reflect the experimental uncertainty of the concentration measurements; the best crystals are obtained when the mother liquor contains two tRNA molecules for one dimeric enzyme.

The first problem to solve in the presence of putative complex crystals is to ascertain the coexistence, within the crystal, of both the nucleic acid and the protein. The presence of $tRNA^{Asp}$ and aspartyl-tRNA synthetase was tested by gel electrophoresis and by biochemical activity assays. All experiments were done with crystals carefully washed so that the possibility of contamination by non-crystallized material could be eliminated.

The presence of aspartyl-tRNA synthetase was demonstrated by the aminoacylation of exogeneous tRNA added to the medium. For the tRNA two different experiments allowed one to demonstrate its presence. With and without addition of exogeneous enzyme it was possible to recover equivalent amounts of trichloroacetic acid precipitable radioactivity, clearly establishing the aspartylation of a significant amount of tRNA.

Crystals belong to the cubic system with a unit cell parameter of 354 Å. The systematic absence of hkl for $h+k+l = 2n+1$ leads to the space group I432, with 48 asymmetric units unit cell.

Assuming one molecule of enzyme and two molecules of tRNA per asymmetric unit the value of Vm (crystal volume per unit of macromolecular weight) is 5.3$Å^3$/dalton, outside the standard range for proteins (1.68 to 3.53). For a partial specific volume of 0.7 $cm^3 . g^{-1}$ the solvent content of the crystal would then be 78%, a large value consistent with the particular softness of the crystals.

For the best crystals, diffracted intensities extend up to 7,5Å resolution but routinely the limit of resolution is 8,5Å.

2. *Solution studies: Static aspects*

One way to determine the part of two macromolecules interacting with each other is to compare the accessibility of these molecules to a chemical probe in the free and complexed states. This can easily be done on tRNA, a molecule for which many specific chemical reagents are available which can attack functional groups on the individual nucleotides. The regions interacting with the probe in the free tRNA but not in the complexed molecule can be considered in close contact with the interacting macromolecule, an aminoacyl-tRNA synthetase for example, and thus protected by it.

Here we present results obtained with ENU, the alkylating reagent which essentially ethylates the phosphate residues of nucleic acids. Methodological problems were discussed in the section " Solution structure of tRNA ". In the presence of aminoacyl-tRNA synthetases, experimental conditions must allow both chemical reactivity and good complex formation. Therefore samples are incubated for 3h at pH 8.0 and at 20°C in the presence of magnesium at a rather low ionic strength and with concentrations of tRNA (1.5 μM) and enzyme (5 μM) in the range where the complex formation is guaranteed. Detailed experimental procedures were published for experiments with $tRNA^{Phe}$ (Vlassov *et al.*, 1983).

Typical alkylation experiment of 5′-labelled $tRNA^{Asp}$ is shown on Figure 11. This experiment allows to probe the phosphate groups located between positions 6 to 55. Similar assays with 3′-labelled tRNA probing the phosphates from the 3′ to the 5′ side of the molecule were also carried on. In the presence of aspartyl-tRNA synthetase the splitting at some phosphate positions is nearly suppressed or is strongly reduced (for instance P27 to P33), suggesting that these groups are protected from alkylation by the enzyme and thus most likely are in close contact with the protein. The results are summarized on Figure 11 in which the protected phosphate groups are indicated by arrows on the cloverleaf structure of yeast $tRNA^{Asp}$. Among the parts of the molecule not tested it is clear that the CCA-end, or at least the

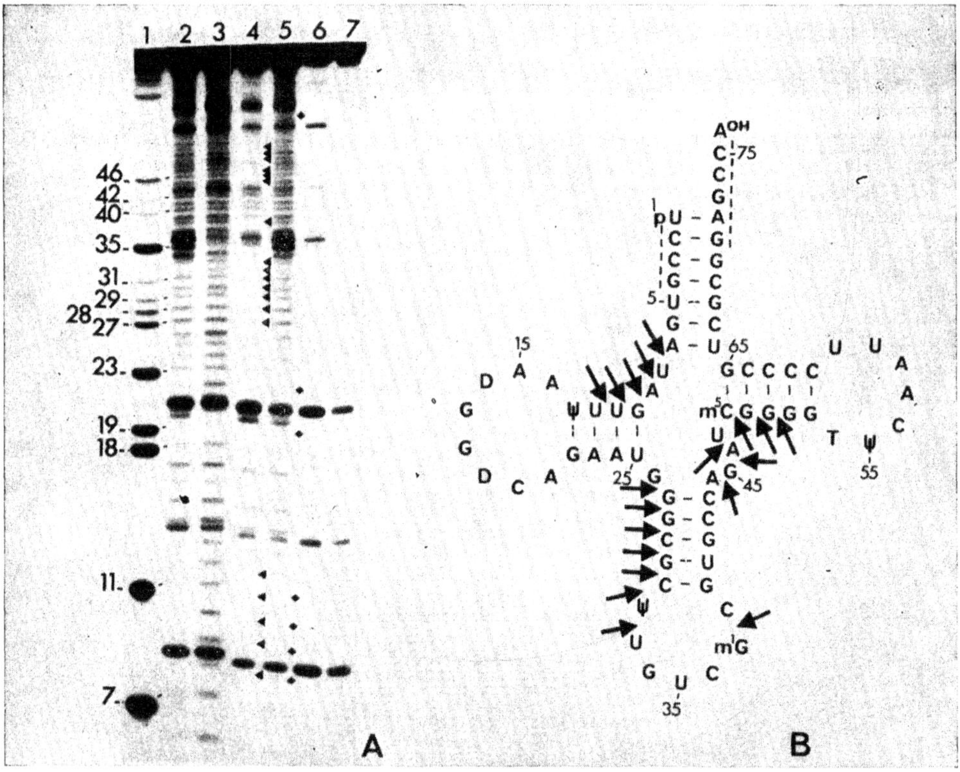

FIG. 11. *A*. Autoradiogram of 15% acrylamide gel of a phosphate alkylation experiment with ENU of 5′ end-labelled $tRNA^{Asp}$ from yeast in the presence of yeast aspartyl-tRNA synthetase. Lane 1, partial ribonuclease T_1 digest; lanes 2 and 5, alkylation at 20° C in the absence of enzyme; lane 3, alkylation at 80° C in the absence of enzyme; lane 4, alkylation at 20° C in the presence of aspartyl-tRNA synthetase; lanes 6 and 7, control incubations in the absence of reagent and in the presence of enzyme (6) and in the absence of reagent (7). The phosphates protected against alkylation in the native tRNA (lanes 2 and 5) and in the tRNA complexed with its synthetase (lane 4) are indicated by diamonds and arrows respectively. *B*. Cloverleaf structure of yeast $tRNA^{Asp}$ (Gangloff *et al.*, 1971) with positions of the phosphates strongly protected by yeast aspartyl-tRNA synthetase against ENU. Full arrows show the phosphates protected by the aspartyl-tRNA synthetase, regions not tested for technical reasons (Vlassov *et al.*, 1983) are indicated by dashed lines.

terminal adenosine, is also in contact with the enzyme for catalytic necessity.

In yeast $tRNA^{Phe}$ the cognate phenylalanyl-tRNA synthetase protects certain phosphates located in all four stems and in the anticodon and extra-loop of the tRNA (Vlassov *et al.*, 1983). Particularly strong protections occur on phosphate 34 in the anticodon loop and on phosphates 23, 27, 28, 41 and 46 in the D and anticodon stems (Figure 12).

In yeast $tRNA^{val}$ complexed with yeast valyl-tRNA synthetase the protected phosphates are essentially located in the corner between the amino acid-accepting and D-stems, in the D-loop, anticodon stem and variable region of the tRNA. Three guanosine residues, located in the D-stem, and another one in the 3′- part of the anticodon stem were also found protected by the synthetase against alkylation by 4-(N-2-chloroethyl-N-methylamino)-benzylamine (Vlassov *et al.*, 1983). In mammalian $tRNA^{val}$, complexed with the cognate but heterologous yeast valyl-tRNA synthetase, the protected phosphates lie in the anticodon stem, in the extra-loop and in the T-arm (Figure 12).

If one compares the protected groups in the various tRNAs tested so far by similar approaches (Vlassov *et al.*, 1983 and references therein), the striking feature which emerges is the large difference between observed contact areas from one tRNA to another. For the three following systems from yeast: phenylalanine, valine, and aspartic acid, the only common contact areas, besides the terminal adenosine, are the variable loop and the neighbourhood of P9. Not surprisingly, these regions are very close in space. Biochemical experiments have also emphasized the involvement of U8 in a contact interaction of tRNAs with their cognate synthetase (Starzyk *et al.*, 1982). The other contact areas of tRNAs differ markedly.

For $tRNA^{Asp}$ one side of the L-shaped molecule is clearly involved in protein-nucleic acid association. This face includes the variable loop, the 5′- end of the anticodon stem and part of the 3′- end of the amino acid arm. The surface involved is quite important and this observation is consistent with neutron diffraction results which lead to the existence of large contact areas between the protein and the nucleic acid (Moras *et al.*, 1983) For $tRNA^{Phe}$ the distribution of the protected phosphate groups not so well defined. At least, three different topological zones are involved in these contacts. For most of the phosphates, their localization differs significantly from the one derived from the model based on cross-linking experiments (Rich and Schimmel,

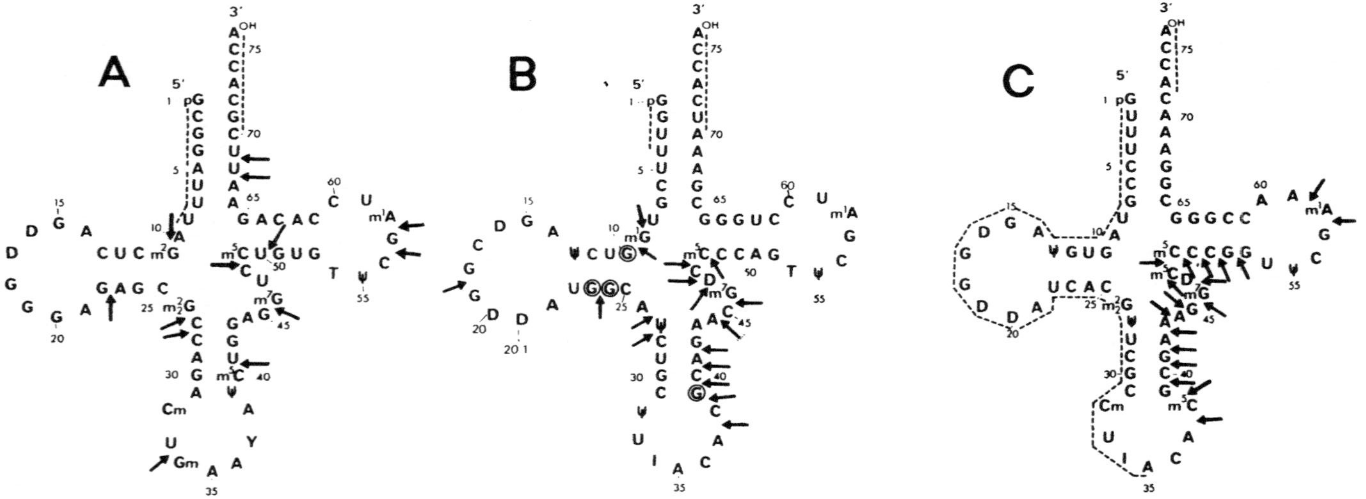

FIG. 12. Cloverleaf structure of yeast tRNA^{Phe} (A), tRNA^{Val} (B) and mammalian tRNA^{Val} (C) with positions of the residues strongly protected by yeast phenylalanyl-tRNA synthetase (A) and valyl-tRNA synthetases (B, C). Sequence data are according to RajBhandary and Chang (1968), Bonnet *et al.* (1974) and Jank *et al.* (1977) and the numbering system of nucleotides according to that of tRNA^{Phe} from yeast. Full arrows show the phosphates protected by the synthetases against alkylation by ENU; regions not tested are indicated by dashed lines Encircled guanosine in yeast tRNA^{Val} (B) are those residues protected against alkylation by 4-(N-2-chloroethyl-N-methylamino)-benzylamine in the complex with yeast valyl-tRNA synthetase (guanosines 1, 2 and 7 were not tested).

1977). If one assumes a similar folding for $tRNA^{Asp}$, another type of interaction between the enzyme and the tRNA must be postulated.

These observations underline the differences which are likely to exist in the recognition between tRNAs and their cognate synthetases. It is worthwhile noticing that the three aminoacyl-tRNA synthetases used in the investigation are quite different as already stated: aspartyl-tRNA synthetase is a dimer (α_2) of MW ≈ 125,000 daltons which can bind two tRNA molecules, whereas phenylalanyl-tRNA synthetase is an $\alpha_2\beta_2$ tetramer (MW ≈ 270,000) which also binds two $tRNA^{Phe}$ molecules and valyl-tRNA synthetase is a large monomer (MW ≈ 130,000) which binds one $tRNA^{Val}$ molecule.

3. *Solution studies: Dynamic aspects*

In the last few years, the elucidation of the tertiary structure of $tRNA^{Phe}$ (Kim *et al.*, 1974, Robertus *et al.*, 1974), as well as the identification of the areas of tRNAs in tight interaction with aminoacyl-tRNA synthetases (e.g. Söll and Schimmel, 1974; Wubbeler *et al.*, 1975; Ofengand, 1977, Rich and Schimmel, 1977; Ebel *et al.*, 1979; Renaud *et al.*, 1979 a; Rosa *et al.*, 1979; Renaud *et al.*, 1981 a; Vlassor *et al.*, 1983) afforded a rather accurate static picture of the recognition step. Very soon, it became evident that these static features could not account for the specificity of the aminoacylation since (i) numerous tRNAs can interact with a given aminoacyl-tRNA synthetase with substantial affinity (Bonnet and Ebel, 1975); the aminoacylation rate cannot be directly correlated to the strength of interaction. (Kern *et al.*, 1972; Ebel *et al.*, 1973; Roe *et al.*, 1973; Giegé *et al.*, 1974; Von der Haar and Cramer, 1978; Renaud *et al.*, 1979 b); (ii) the areas of tRNAs interacting with aminoacyl-tRNA synthetases were found very similar, both in the correct and incorrect systems (Renaud *et al.*, 1981 a); (iii) the formation of Schiff's base between the 3′-end of oxidized tRNAs and lysyl groups of the enzymes catalytic centres was observed in correct and incorrect systems (Giegé *et al.*, 1980). Since then, more attention has been paid to the dynamics of the interaction process: conformational changes were detected at the level of both macromolecules (Krauss *et al.*, 1976; Rigler *et al.*, 1976; Willick and Kay, 1976; Zaccai *et al.*, 1979; Favre *et al.*, 1979; Ehrlich *et al.*, 1980; Lefevre *et al.*, 1980; Lefevre *et al.*, 1981; for a recent review see Rigler and Wintermeyer, 1983). Simultaneously, it was observed that the cores of tRNAs are able to trigger conformational changes of aminoacyl-tRNA synthetases leading to important modifications of the catalytic properties of the enzymes (Renaud *et al.*, 1981 b).

It will be shown that upon complex formation between tRNAs and aminoacyl-tRNA synthetases a multistep adaptation of both macromolecules takes place (analogous to an induced fit) which may be a determining step in the specificity of aminoacylation. In the particular case of the yeast phenylalanine system, the wybutine residue in the anticodon loop of $tRNA^{Phe}$ has been identified as an important element in this induced fit mechanism.

Conformational changes in the tRNA during complex formation: The study was conducted in the phenylalanine system from yeast, taking advantage of the presence of a built-in fluorescent probe (the wybutine residue adjacent to the anticodon) or using a modified $tRNA^{Phe}$ in which the terminal 3′-adenosine was replaced by a fluorescent structural analogue: the formycin. Previous results (Ehrlich *et al.*, 1980) showed that conformational changes occurred in the anticodon-loop of $tRNA^{Phe}$ upon binding to phenylalanyl-tRNA synthetase. This is illustrated in Figure 13 A, which shows that the fluorescence intensity of the wybutine residue

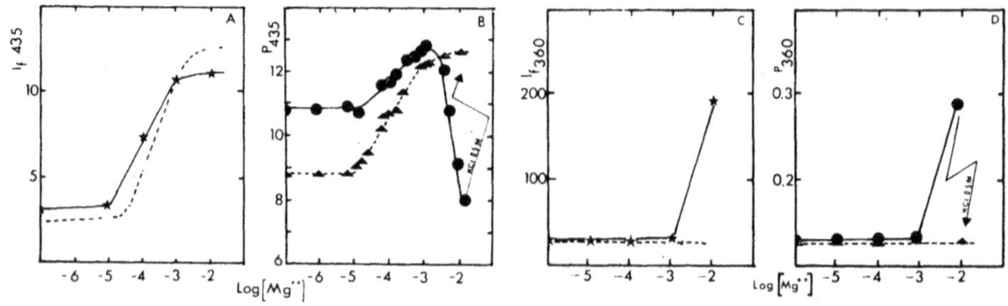

FIG. 13. Variation of the fluorescence intensity of wybutine (A) and 3′-formycin (C) as a function of Mg^{++} in the isolated (○—○) and complexed (★-★) $tRNA^{Phe}$. Variation of the polarization degree of the fluorescence of wybutine (B) and 3′-formycin (D) in the isolated (▲—▲) and complexed (●—●) $tRNA^{Phe}$.

in the isolated or complexed tRNA increases with the Mg^{++} concentration. This stimulation results from a better stacking of the wybutine between adjacent bases, shielding the fluorophore against dynamic quenching by water molecules. This interpretation is confirmed by the values observed for the lifetime of the excited state. Phenylalanyl-tRNA synthetase appears to enhance the fluorescence of wybutine at low [Mg^{++}] and to quench it at high [Mg^{++}], the crossing point being located at 1 mM [Mg^{++}]. Figure 13 B shows the evolution of the polarization degree of wybutine luminescence as a function of [Mg^{++}] in the free and complexed $tRNA^{Phe}$. In the unbound tRNA, the polarization degree increases monotonically, reflecting a decrease of the rotational tumbling of wybutine due to the better stacking. In the bound tRNA, the polarization degree first increases and then abruptly decreases above 1 mM [Mg^{++}], showing an expulsion of the wybutine residue from a stacked region. This increased mobility of wybutine is confirmed by the calculation of the mean rotational correlation time of wybutine in the free and complexed tRNA (Table III). This conformational change in the anticodon

TABLE III. *Mean rotational correlation time of wybutine and formycin in the free and complexed* $tRNA^{Phe}$

Fluorescent probe	System	(ns)
Wybutine	Isolated $tRNA^{Phe}$	23
	Complex $tRNA^{Phe}$–PheRS	9
	Isolated $tRNA_F^{Phe}$	5
3'-Formycin	Complex $tRNA_F^{Phe}$–PheRS	175
	Complex $tRNA_F^{Phe}$ minus wybutine–PheRS	33

loop is specific for the complex formation, since upon addition of high salt concentrations which destabilize the complex, a normal polarization degree is restored for wybutine (Figure 13). Furthermore, in the presence of a non cognate aminoacyl-tRNA synthetase like aspartyl-tRNA synthetase, no expulsion of the wybutine takes place (data not shown).

By measuring the accessibility of the fluorophore to deuterated water as well as the circular anisotropy of wybutine luminescence, we could show that, once expelled, the wybutine interacted tightly with the enzyme (Ehrlich *et al.*, 1980). This had already been suggested by the observation that specific irradiation of the tRNA-enzyme complex in the wybutine absorption band resulted in the covalent cross-linking of both macromolecules (Baltzinger *et al.*, 1979). It was of course of interest to look for possible conformational changes elsewhere in the tRNA molecule. The search for a transconformation of the 3′-accepting end was undertaken using a modified $tRNA^{Phe}$ in which the 3′-adenosine was replaced by the fluorescent formycin. Figure 13 C shows the evolution of formycin luminescence as a function of $[Mg^{++}]$ in the free and complexed tRNA. Whereas the fluorescence intensity remains constant in the unbound tRNA, it is markedly increased in the complexed tRNA, above 1 mM $[Mg^{++}]$. This increase of fluorescence results from a combination of three different factors (Lefevre *et al.*, 1981): (i) a prototropic effect in the active centre of the enzyme, mimicking an alkaline environment (possibly due to the lysyl group responsible for the Schiff's base formation with oxidized $tRNA^{Phe}$); (ii) a shift of the tautomeric equilibrium of the formycin fluorophore; (iii) a partial destacking of the 3′-formycin from the adjacent C residue, resulting in a release of static quenching. Figure 13 D shows that the fluorescence stimulation is accompanied by a marked increase of the polarization degree due to a decrease of the fluorophore mobility. This is confirmed by the important increase of the mean rotational correlation time, as shown in Table III. That the immobilization of formycin occurs in the enzyme active centre is shown by the fact that $tRNA^{Phe}_{Formycin}$ can be aminoacylated although the rate of aminoacylation is 20 to 50 fold lower, due to a conformation of formycin intermediate between the *syn* and *anti* configurations. This was confirmed by the observation that an enzyme, the active site of which had been blocked by affinity labelling (Lefevre *et al.*, 1981), failed to induce the modifications of formycin luminescence.

Conformational changes in the enzyme during complex formation:

It was shown (Ehrlich *et al.*, 1980) that the binding of tRNA also led to structure modifications as shown by the change in both the tryptophan fluorescence and circular dichroism of the protein. The variation of the ellipticity at 222 nm revealed an important loss of α-helical regions. It is worth noticing that conformational changes, similar to the one observed in the phenylalanine system from yeast, have been reported in other systems (Willick and Kay, 1976; Favre *et al.*, 1979).

Correlation between conformational changes and biological activity: Several lines of evidence support the idea that the conformational changes observed are correlated to the expression of the biological activity. Indeed on the one hand, we could recently show (Lefevre *et al.*, 1981) that the rate of transfer of the phenylalanine moiety from the isolated enzyme: adenylate complex to the tRNA was close to zero at very low [Mg^{++}]. It increases slowly with [Mg^{++}] and bounces off around 1 mM [Mg^{++}] (Figure 14) con-

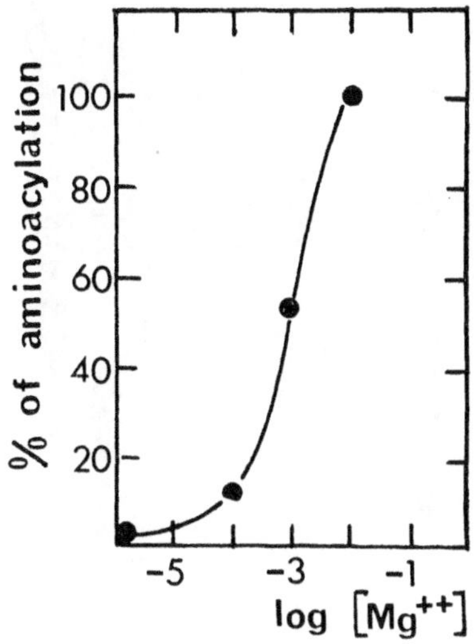

FIG. 14. Rateof transfer of the phenylalanine moiety from the adenylate: enzyme complex to $tRNA^{Phe}$ as a function of [Mg^{++}], expressed as % of the maximum rate.

centration which was precisely observed to trigger the conformational changes of complexed $tRNA^{Phe}$ (see above).

On the other hand, the existence of an important conformational change in phenylalanyl-tRNA synthetase upon binding of tRNA prompted us to search for a possible alteration of the catalytic properties. This was carried out by studying the acylation activity of aminoacyl-tRNA synthetases toward simple substrates like isolated adenosine or CCA-oligonucleotide (Renaud *et al.*, 1981 b):

In the phenylalanine system, as shown in Figure 15 A (lane 5), phenylalanyl-tRNA synthetase alone is unable to catalyse the aminoacylation of isolated adenosine or CCA-trinucleotide. But upon addition of a $tRNA^{Phe}$ lacking 1, 2 or 3 terminal 3′-nucleotides, an efficient acylation of adenosine or CCA takes place

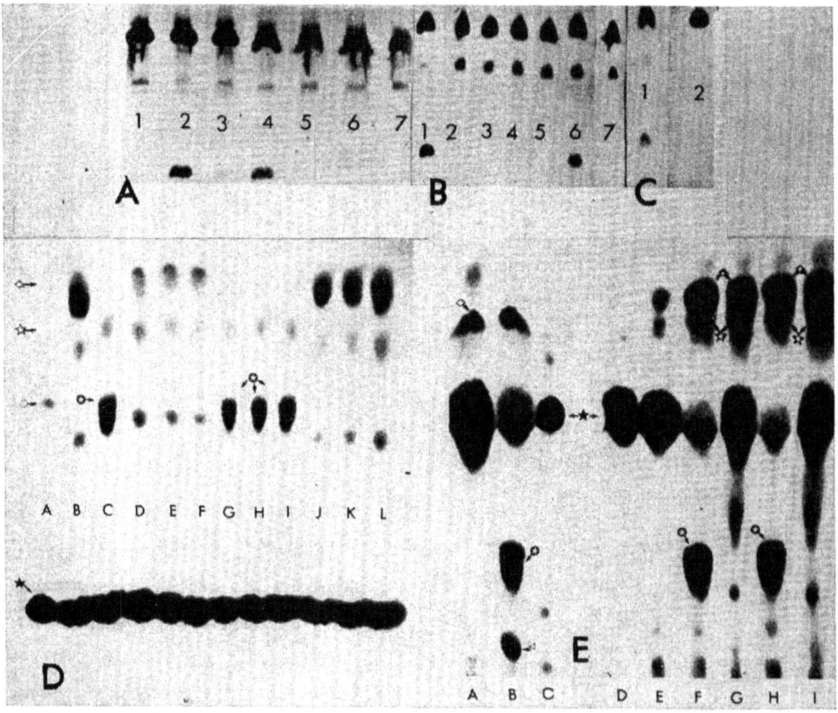

FIG. 15. Paper electrophoretic characterization of aminoacyl-adenosine arising from catalytic site activation: the reaction mixture contains 1μM aminoacyl-tRNA synthetase, 10μM of modified tRNA lacking all or part of the CCA-end, 5mM free adenosine or CCA-trinucleotide. In the case of CCA aminoacylation a pancreatic RNAse treatment was performed prior to electrophoresis.

15 A. Phenylalanine system:

Lane 1, control reaction mixture containing only $tRNA^{Phe}$ minus A
Lane 2, reaction mixture containing $tRNA^{Phe}$ minus A, plus free adenosine
Lane 3, reaction mixture containing $tRNA^{Phe}$ minus CCA, plus free adenosine

Lane 4, authentic sample of phenylalanyl-adenosine
Lane 5, control reaction mixture containing only free adenosine
Lane 6, reaction mixture containing $tRNA^{Phe}$ minus CCA, plus isolated CCA-trinucleotide
Lane 7, control reaction mixture containing only $tRNA^{Phe}$ minus CCA.

15 B. Valine system:

Lane 1, authentic sample of valyl-adenosine
Lane 2, control reaction mixture containing only isolated CCA-trinucleotide
Lane 3, control reaction mixture containing only free adenosine
Lane 4, control reaction mixture containing only $tRNA^{Val}$ minus CCA
Lane 5, control reaction mixture containing only $tRNA^{Val}$ minus A
Lane 6, reaction mixture containing $tRNA^{Val}$ minus CCA, plus isolated CCA-trinucleotide
Lane 7, reaction mixture containing $tRNA^{Val}$ minus A, plus free adenosine.

15 C. Effect of wybutine excision on catalytic site activation in the phenylalanine system:

Lane 1, mixture containing $tRNA^{Phe}$ minus A, plus free adenosine
Lane 2, mixture containing $tRNA^{Phe}$ minus A deprived of wybutine plus free adenosine.

15 D. Aspartic acid system:

(A) Control [^{14}C] aspartic acid
(B) *in situ* pancreatic ribonuclease hydrolysis of [^{14}C] aspartyl-$tRNA^{Asp}$ ($\sim$ 200 pmol)
(C) *in situ* T_1 ribonuclease hydrolysis of [^{14}C] aspartyl-$tRNA^{Asp}$ ($\sim$ 200 pmol)
(D) aminoacylation assay of adenosine in the presence of $tRNA^{Asp}$ NCC
(E) aminoacylation assay of adenosine in the presence of $tRNA^{Asp}$ NC
(F) aminoacylation assay of adenosine in the presence of $tRNA^{Asp}$ N
(G) aminoacylation assay of CCA in the presence of $tRNA^{Asp}$ NCC
(H) aminoacylation assay of CCA in the presence of $tRNA^{Asp}$ NC
(I) aminoacylation assay of CCA in the presence of $tRNA^{Asp}$ N
(J) aminoacylation assay of CCA in the presence of $tRNA^{Asp}$ NCC, followed by pancreatic ribonuclease hydrolysis
(K) aminoacylation assay of CCA in the presence of $tRNA^{Asp}$ NC, followed by pancreatic ribonuclease hydrolysis
(L) aminoacylation assay of CCA in the presence of $tRNA^{Asp}$ N, followed by pancreatic ribonuclease hydrolysis
(★) [^{14}C] aspartic acid; (✵) [^{14}C] aspartyl-adenylate; (◇) [^{14}C] aspartyl-adenosine; (—) [^{14}C] aspartyl-CCA; (○) [^{14}C] aspartic acid contamination.

15 E. Arginine system:

(A) *in situ* pancreatic ribonuclease hydrolysis of [^{14}C] arginyl-$tRNA^{Arg}$ ($\sim$ 200 pmol)
(B) *in situ* T_1 ribonuclease hydrolysis of [^{14}C] arginyl-$tRNA^{Arg}$ ($\sim$ 200 pmol). (Note that yeast $tRNA^{Arg}_3$ exists under two different forms: one with G73, which gives [^{14}C] Arg-CCA upon T_1 hydrolysis, and the other one with C73 which gives [^{14}C]Arg-CCCCA upon T_1 hydrolysis)
(C) control [^{14}C]-arginine
(D) aminoacylation mixture deprived of $tRNA^{Arg}$, adenosine and CCA
(E) aminoacylation assay of adenosine in the presence of $tRNA^{Arg}$ NCC
(F) aminoacylation of CCA in the presence of $tRNA^{Arg}$ NCC
(G) same experiment as (F) followed by pancreatic ribonuclease hydrolysis
(H) aminoacylation assay of CCA in the presence of $tRNA^{Arg}$ N
(I) same experiment as (H) followed by pancreatic ribonuclease hydrolysis. (Note that in tracks (B), (F) and (H) the supplementary [^{14}C] arginyl-adenosine spot results from a partial degradation of [^{14}C] arginyl-CCA
(★) [^{14}C] Arginine; (☆) [^{14}C] arginyl-adenylate; (◇) [^{14}C] arginyl-CCA; (◁) [^{14}C] arginyl-CCCCA; (—) [^{14}C] arginyl-adenosine.

(lanes 2, 3, 6) as shown by the formation of phenylalanyl-adenosine (after pancreatic ribonuclease hydrolysis in the case of CCA). It is interesting to note that the efficiency of aminoacylation of isolated adenosine is much higher in the presence of $tRNA^{Phe}$ lacking the 3′-adenosine than in the presence of a $tRNA^{Phe}$ lacking the last three bases. This could be interpreted to mean that the two C′s of the 3'-terminal sequence play an important role in the catalytic site activation of phenylalanyl-tRNA synthetase. Kinetic measurements allowed us to estimate the affinities for adenosine and CCA-trinucleotide, both in the range of 2 to 4 mM, as well as the maximal velocities for these two substrates, which were respectively found to be 0.1 sec^{-1} and 0.03 sec^{-1} to be compared to the acylation rate for native $tRNA^{Phe}$ (6 sec^{-1}).

In the aspartic acid system, the behaviour of the enzyme, in the presence of $tRNA^{Asp}$ deprived of one, two or three 3′-terminal nucleotides, is identical to that observed in the phenylalanine system. However, CCA appears to be a slightly better substrate than free adenosine (Bacha *et al.*, 1982).

In the valine system, a similar activation of the catalytic centre of valyl-tRNA synthetase was observed, leading to the acylation of isolated CCA in the presence of $tRNA^{Val}$ lacking the CCA-end. The main difference with the two above systems was the lack of activity toward isolated adenosine (Figure 15 B), once more suggesting a particular importance of the C residues. This observation can be correlated to the results obtained in the ultra-violet cross-linking studies (Renaud *et al.*, 1979 a; Renaud *et al.*, 1981 a), which showed that an efficient covalent binding of the RNase T_1 3′-terminal oligonucleotide and of the terminal adenosine took place in the valine system.

The arginine system was chosen because of its peculiarity in forming the aminoacyl-adenylate only in the presence of the cognate tRNA, like the glutamate system (Kern and Lapointe, 1980 a). Any modification of the terminal adenosine of $tRNA^{Arg}$ has been found to hinder the triggering of adenylate formation (Kern and Lapointe, 1980 b). As in the valine system, only the combination of $tRNA^{Arg}$ lacking CCA and free CCA-oligonucleotide was found to trigger the catalytic site activation. A complete correlation was observed in the triggering of the arginyl-adenylate formation in the presence of tRNA^{Arg} lacking all or part of the CCA-end and the complementary fragments (free adenosine, CA or CCA) (Bacha *et al.*, 1982). The above results show that, as in the valine system, the contiguity of the CCA-end is of primordial importance for a correct activition of the catalytic centre. This activation in the arginine system extends to the amino acid adenylation and not only to the tRNA acylation step. The importance of the interaction between arginyl-tRNA synthetase and tRNA^{Arg} in the CCA region is further assessed by the observation that an efficient cross-linking of the CCA-oligonucleotide part of the tRNA^{Arg} to the enzyme is observed upon U.V. irradiation (40%), whereas the terminal adenosine does not seem to be cross-linked ($\approx$ 2%). Another observation also supports the importance of the two C residues in CCA in the arginine system: the Michaelis constant for CCA in arginyl-adenylate formation in the presence of tRNA^{Arg} lacking CCA is 0.5 mM, showing a 4-10 fold higher affinity as compared to that observed in the phenylalanine system.

It was of course very interesting to study the possible occurrence of catalytic centre activation in non-cognate systems. As shown in Table IV, valyl-tRNA synthetase, which efficiently misaminoacylates tRNA^{Phe} (Giegé *et al.*, 1974), is also activated by tRNA^{Phe} lacking the CCA-end. As in the cognate system, only CCA can be aminoacylated. Two other systems, which are efficient in misaminoacylation, phenylalanyl-tRNA synthetase and tRNA^{Arg} or tRNA^{Tyr}, were also found to be efficient in catalytic centre activation, as will be shown in the next section.

In constrast, phenylalanyl-tRNA synthetase, which is unable to aminoacylate tRNA^{Asp} (Giegé *et al.*, 1974) does not show catalytic centre activation upon addition of these tRNAs deprived of their CCA-end (Table IV). Similarly, arginyl-tRNA synthetase, which does not aminoacylate tRNA^{Phe}, is not activated by the latter.

In the phenylalanine system, some information could be gained upon one individual element involved in the mutual adaptation of tRNA and aminoacyl-tRNA synthetase. Indeed wybutine

TABLE IV. *Comparison of aminoacylation and catalytic site activation in cognate and non-cognate systems*

Synthetase	tRNA	Amino-acyla-tion	Catalytic site activation in presence of tRNA lacking all or part of CCA-end	Residue 3′ to anticodon in native tRNA (N)	Modifica-tions
Phe RS	$tRNA^{Val}$	—	—	A	—
Phe RS	$tRNA^{Asp}$	—	—	m^1G	—
Phe RS	$tRNA^{Arg}$	+	+	t^6A	—
Phe RS	$tRNA^{Tyr}$	+	+	i^6A	—
Val RS	$tRNA^{Phe}$	+	+	Y	—
Arg RS	$tRNA^{Phe}$	—	—	Y	—
Asp RS	$tRNA^{Phe}$	—	—	Y	—
Phe RS	$tRNA^{Phe}$ modified	+[1]	—	Y	Y excised by acidic treatment
Phe RS	$tRNA^{Tyr}$ modified	—	—	i^6A	$KMnO_4$ treatment
Phe RS	$tRNA^{Tyr}$ modified	+	+	i^6A	iodination

V_{max} reduced by a factor of 4.

appears to be of particular importance for this process. As shown in Figure 13 and Table III, the excision of wybutine from $tRNA^{Phe}$ hinders the immobilization of formycin in the catalytic centre of phenylalanyl-tRNA synthetase. This is accompanied by a sevenfold decrease of the acylation velocity of $tRNA^{Phe}_F$ (Maelicke *et al.*, 1974) (≈ 4 fold decrease for native $tRNA^{Phe}$). The critical role of wybutine is confirmed (Figure 15 C) by the fact that a $tRNA^{Phe}$ lacking simultaneously all or part of the CCA-end and the wybutine, is no longer able to trigger the catalytic site activation of phenylalanyl-tRNA synthetase. It is interesting to observe that two other non-cognate tRNAs ($tRNA^{Arg}$ and $tRNA^{Tyr}$) which were found to be efficient both in misamino-acylation and catalytic centre activation with phenylalanyl-tRNA synthetase (Table IV), also have, adjacent to the anticodon, an hypermodified bulky residue (t^6 A and i^6 A, respectively) some-what similar to the wybutine residue. In contrast, $tRNA^{Val}$

and $tRNA^{Asp}$, which do not contain an hypermodified base adjacent to the anticodon, cannot be misaminoacylated and do not promote the activation of the catalytic centre of phenylalanyl-tRNA synthetase (Table IV).

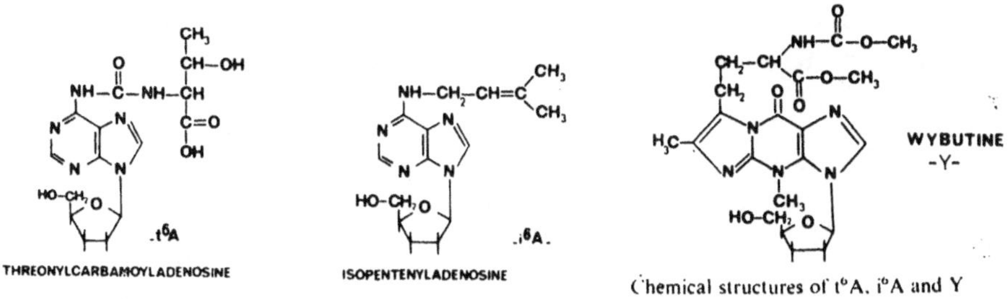

Chemical structures of t^6A, i^6A and Y

Since specific modifications of the i^6A residue in $tRNA^{Tyr}$ have been described (Kline *et al.*, 1969), it was interesting to study the effect of these modifications on the aminoacylation of $tRNA^{Tyr}$ by phenylalanyl-tRNA synthetase as well as on the catalytic centre activation of this enzyme. Table IV shows that permanganate treatment of $tRNA^{Tyr}$, which affords a mixture of dihydroxylated i^6A (70%) and adenine (30%) in place of i^6A, completely suppresses the ability of $tRNA^{Tyr}$ to be aminoacylated by phenylalanyl-tRNA synthetase or to trigger the activation of the catalytic centre of this enzyme. Upon iodine treatment of $tRNA^{Tyr}$ (Kline *et al.*, 1969), which affords a three-ring structure in place of the i^6A residue, no modification of the $tRNA^{Tyr}$ properties is observed (the modified $tRNA^{Tyr}$ can still be misaminoacylated by phenylalanyl-tRNA synthetase and still promotes the activation of the catalytic centre of the latter). In order to understand these observations, it is important to remember that, in order to explain the chemical reactivity of the i^6A residue in $tRNA^{Tyr}$, Hall (1970) was led to assume the structure (A) for the isopentenyl-adenosine. Upon iodine treatment of $tRNA^{Tyr}$, ring closure occurs in i^6A leading to compound (B), almost identical to form (A), which explains why the biological properties of the modified tRNA are identical to those of the native $tRNA^{Tyr}$.

In contrast, permanganate treatment, which results in the disappearance of the allylic double bond and thus destabilizes the three-ring structure, results in the complete disappearance of the misaminoacylation of tRNA^{Tyr} and of the activation of the catalytic centre of phenylalanyl-tRNA synthetase. The last results therefore suggest that the three-ring structure of the wybutine residue is the most important factor in the requirement of this hypermodified base for the triggering of the activation of the catalytic centre of phenylalanyl-tRNA synthetase.

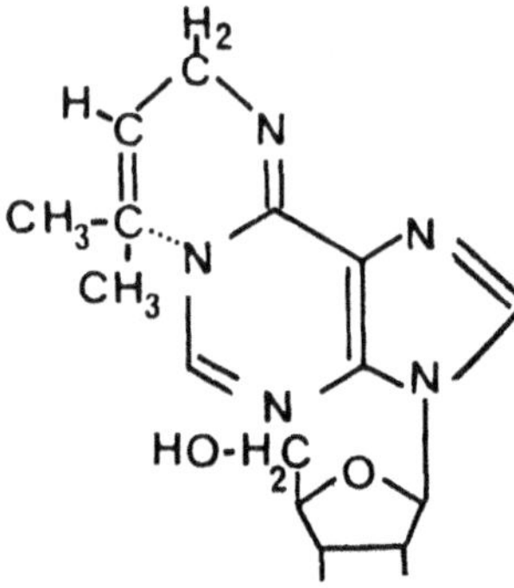

A

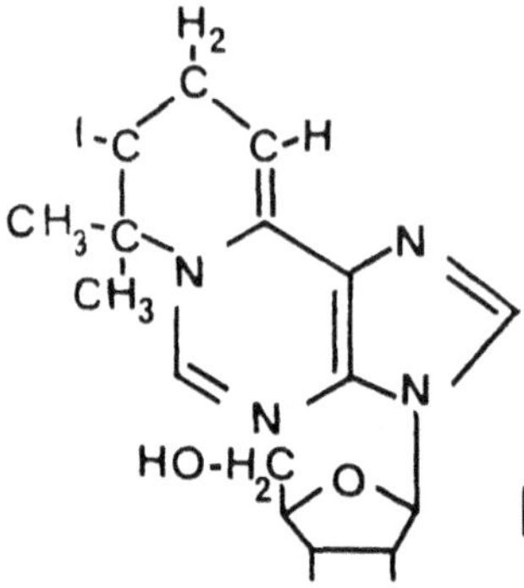

B

Mutual adaptation in the tRNA-aminoacyl-tRNA synthetase systems: a possible key for the aminoacylation specificity: Although no individual element of a tRNA molecule was found strictly indispensable for the interaction with the enzyme (Ebel *et al.*, 1979), most of them appear to be involved in the adaptation process (Krauss *et al.*, 1976; Favre *et al.*, 1979; Ehrlich *et al.*, 1980; Lefevre *et al.*, 1981) which leads to the optimal efficiency of the catalytic mechanism. The induced activation demonstrated above could be responsible for the requirement of the tRNA molecule in the expression of the amino acid activation encountered in the glutamine and arginine systems (Kern and Lapointe, 1980 a, b; Fersht *et al.*, 1978).

Such a multi-step adaptation process of tRNAs and aminoacyl-tRNA synthetases may have been selected so as to allow a fine tuning of the rate of the aminoacylation reaction. Indeed, it has been observed in the phenylalanine system that structural modifications (i.e. excision of the wybutine residue), which prevent the induced fit from occurring, usually result in a substantial lowering of the maximal rate of aminoacylation (Lefevre *et al.*,

1981). Since it is well known that the specificity of the aminoacylation reaction relies more on the maximal velocity of the reaction than on a preferential affinity for the tRNA (Ebel *et al.*, 1973; Bonnet and Ebel, 1975; Renaud *et al.*, 1979 b), this multi-step adaptation process appears to be an important mechanism ensuring the fidelity of the reaction.

IV. Interaction between aminoacyl-tRNA and elongation factor Tu

Elongation factor Tu promotes the A-site specific binding of the aminoacyl-tRNAs (aa-tRNA) to the ribosome via a ternary aa-tRNA.EF-Tu.GTP complex. In the presence of GTP, the protein binds acylated forms of non-initiator tRNAs (Gordon, 1968) for which it has a high affinity, but it discriminates against initiator tRNA (Schulman *et al.*, 1974), non-acylated tRNAs (Gordon, 1967; Schulman *et al.*, 1974; Pingoud *et al.*, 1982) and aminoacyl-tRNAs modified in their 3′-end (Ofengand and Cheng, 1972; Thang *et al.*, 1972; Schulman *et al.*, 1974). Although the aminoacylated 3′-end of tRNA is an important recognition feature, there are several lines of evidence that EF-Tu not only binds the 3′-end of tRNA, but also the amino acid acceptor limb of the L-shaped tRNA structure (Jekowsky *et al.*, 1977).

Here we report (i) experiments obtained with a ribonuclease extracted from the venom of the cobra *Naja oxiana* for partial digestion of the complex (Boutorin *et al.*, 1981). This enzyme has a high specificity for double-stranded regions and makes nicks mainly in the anticodon and acceptor stems of tRNAs (Favorova *et al.*, 1981). The principle of this approach is that regions of tRNA which are cut by the ribonuclease in the naked state become resistant when they are shielded by the protein in the complex; (ii) the effect of EF-Tu on the conformation of aminoacylated $tRNA^{Phe}$ upon ternary complex formation. For this purpose we investigated the chemical stability of tRNA and the accessibility of phosphates towards ENU, in complexed Phe-$tRNA^{Phe}$ and in free $tRNA^{Phe}$ using end-labelling and rapid sequencing gel methodologies (Riehl *et al.*, 1983).

The results with cobra venom nuclease confirm previous conclusions from other groups (Weissbach *et al.*, 1971; Thang *et al.*, 1972; Ringer and Chladek, 1975; Jonak *et al.*, 1980), suggesting an important role of the 3′-terminal region of tRNA molecules in the interaction with EF-Tu: in all the tRNAs we have tested, this region is protected by EF-Tu against digestion.

The region of *E.coli* phenylalanyl-tRNA^{Phe} protected by EF-Tu is shown in Figure 16. These data are in agreement with all

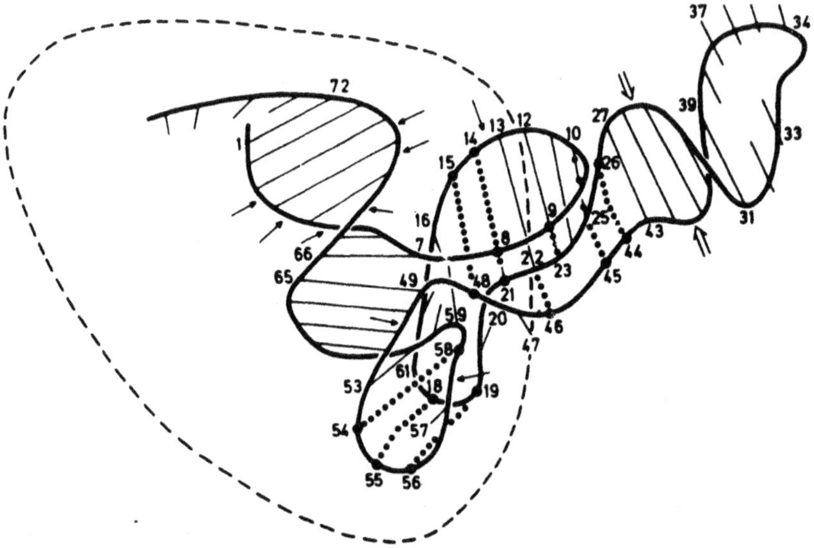

Fig. 16. The structure of yeast tRNA^{Phe} (Robertus *et al.*, 1974). The arrows indicate cleavage positions in the *E.coli* Phe-tRNA^{Phe} molecule by cobra venom ribonuclease. Positions, intensified when tRNA is digested in the complex with EF-Tu. GTP, are indicated by double arrows. The part of the molecule protected by EF-Tu is enclosed by a broken line.

previous results and demonstrate that EF-Tu not only binds to the amino acid acceptor end of the elongator tRNAs but also along the combined amino acid and T-helix. In contrast to the results of Jekowsky *et al.* (1977), who did not observe a protection of the T-loop against T_1 ribonuclease digestion, we also observed a protection of this loop against cobra venom ribonuclease attack. However, this does not necessarily mean that the T-loop is involved in complex formation. Protection in this region could arise from sterical hindrance: cobra venom ribonuclease has a molecular weight of 16,000 whereas that of T_1-ribonuclease is only 11,000.

From these data, it can be concluded that the interaction between aminoacyl-tRNA and EF-Tu mostly consists of an interaction between the protein and that part of the tRNA, including the CCA-end and the amino acid acceptor and T-helices.

The mapping of the complexed tRNA using ENU as a probe unexpectedly did not allow us to define contact points between EF-Tu and the phosphate groups of the aminoacylated tRNA. This result, however, is not contradictory to the nuclease mapping experiments. Because of the small size of ENU, phosphate residues could be readily more accessible to this reagent, even when they are covered by the Tu factor, than to nucleases which are much more bulky. In addition, phosphates at positions 9, 10, 11, 19, 49, and especially 58-60, are already protected by the tertiary structure of the tRNA (Vlassov *et al.*, 1981), and phosphates located near the 3′-and 5′-termini could not be tested because of technical limitations.

This study brings evidence for 6 discrete conformational changes in tRNA upon ternary complex formation. One of them, occurring at position 53, takes place in the T-stem, a region which has been shown to be covered by EF-Tu. This change could be visualized due to the small size of the chemical probe which is not hindered by the factor in reaching this phosphate. The 5 other stuctural alterations occur in the D- and anticodon loops at positions 18, 20, 21 and 34, 36, respectively. In that case, the appearance or disappearance of cuts in complexed tRNA reflects a changed flexibility in its ribose-phosphate backbone. These results clearly indicate that the T-region, the D- and anticodon loops undergo a conformational rearrangement upon ternary complex formation. In the particular case of the D-loop, the occurrence of conformational change would explain the contradictory results obtained in ribonuclease T_1 accessibility studies and in kethoxal modification experiments (Jekowsky *et al.*, 1977; Bertram and Wagner, 1982). Concerning the anticodon loop, these results agree well with spectroscopic (Beres and Lucas-Lenard, 1973; Weygand-Durasevic *et al.*, 1981) and accessibility mapping (Boutorin *et al.*, 1981; Wikman *et al.*, 1982) experiments.

V. Conclusion

Among the various functions of transfer ribonucleic acids in living cells, their best understood role is their participation in ribosome-mediated protein synthesis. These functions lead tRNAs to many interactions with different proteins and nucleic acids. With aminoacyl-tRNA synthetases, enzymes which attach the correct amino acid to the 3′-end of their cognate tRNAs, the molecular recognition must be *highly* specific. The same is true for the decoding of the genetic code at the messenger RNA level. This is not the case with the elongation factor, which carries the aminoacylated tRNAs to the ribosome. One way to solve the problem of adaptability is to vary the three-dimensional structure via conformational changes induced and/or controlled by the reaction. In this review we have presented experimental results on yeast $tRNA^{Asp}$ and $tRNA^{Phe}$ which support this hypothesis. We would like to stress the fact that this information could only be obtained by a strong interpenetration of two approaches: biochemical and crystallographic.

REFERENCES

Arnott, S., Hukins, D.W.L. and Dover, S.D., 1972, Biochem. Biophys. Res. Commun., *48*, 1392.
Artymiuk, P.J., Blake, C.C.F., Grace, D.E.P., Oatley, S.I., Philipps, D.C. and Sternberg, N.J.E., 1979, Nature, *283*, 563.
Bacha, H., Renaud, M., Lefevre, J.F. and Remy, P., 1982, Eur. J. Biochem. *127*, 87.
Baltzinger, M., Fasiolo, F. and Remy, P., 1979, Eur. J. Biochem. *97*, 481.
Barciszewski, H., Romby, P., Ebel. J.P. and Giege, R., 1982, FEBS Lett. *150*, 459.
Beres, L. and Lucas-Lenard, J., 1973, Biochemistry, *12*, 3998.
Bertram, S. and Wagner, R., 1982, Biochem. Internatl. *4*, 117.
Bonnet, J., Ebel, J.P., Dirheimer, G., Shershueva, L.P., Krutilina, A.I., Venkstern, T.V. and Baiev, A.A., 1974, Biochimie (Paris) *56*, 1211.
Bonnet, J. and Ebel, J.P., 1975, Eur. J. Biochem. *58*, 193.
Boutorin, A.S., Clark, B.F.C., Ebel, J.P., Kruse, T.A., Petersen, H.U., Remy, P. and Vassilenko, S., 1981, J. Mol. Biol. *152*, 593.
Boutorin, A.S., Remy, P., Ebel, J.P. and Vassilenko, S., 1982, Eur. J. Biochem., *121*, 87.

Dietrich, A., Giegé, R., Commarmond, M.B., Thierry, J.C. and Moras, D., 1980, J. Mol. Biol., *138*, 129.
Ebel, J.P., Giegé, R., Bonnet, J., Kern, D., Befort, N., Bollack, C., Fasiolo, F., Gangloff, J. and Dirheimer, G., 1973, Biochimie (Paris), *55*, 547.
Ebel, J.P., Renaud, M., Dietrich, A., Fasiolo, F., Keith, G., Favorova, O., Vassilenko, S., Baltzinger, M., Ehrlich, R., Remy, P., Bonnet, J. and Giegé, R., 1979, in Transfer RNA: Structure, Properties and Recognition, eds. Schimmel, P.R., Söll., D. and Abelson, J.N. (Cold Spring Harbor Laboratory, Cold Spring Harbor, NY) p. 325.
Ehrlich, R., Lefevre, J.F. and Remy, P., 1980, Eur. J. Biochem. *103*, 145.
Favre, A., Ballini, J.P. and Holler, E., 1979, Biochemistry *18*, 2887.
Favorova, O.O., Fasiolo, F., Keith, G., Vassilenko, S.K. and Ebel, J.P., 1981, Biochemistry, *20*, 1006.
Fersht, A.R., Gangloff, J. and Dirheimer, G., 1978, Biochemistry *17*, 3740.
Florentz, C., Briand, J.P., Romby, P., Hirth, L., Ebel, J.P. and Giegé, R., 1982, Embo J., *1*, 269.
Frauenfelder, H., Petsko, G.A. and Tsernoglou, D., 1979, Nature, *280*, 558.
Gangloff, J., Keith, G., Ebel, J.P. and Dirheimer, G., 1971, Nature New Biol., *230*, 125.
Giegé, R., Kern, D., Ebel, J.P., Grosjean, H., De Henau, S. and Chantrenne, H., 1974, Eur. J. Biochem. *45*, 351.
Giegé, R., Moras, D. and Thierry, J.C., 1977, J. Mol. Biol., *115*, 91.
Giegé, R., Dietrich, A., Jacrot, B., Zaccai, G., Moras, D., Thierry, J.C., Bacha, H., Remy, P., Renaud, M., Gangloff, J., Kern, D. and Ebel, J.P., 1980, in 7th A. Katzir Katchalsky Conference on Structural Assembly in Biology of Macromolecules (Sussman, Traut and Yonath Eds) p. 667.
Giegé, R., Lorber, B., Ebel, J.P., Thierry, J.C. and Moras, D., 1980, C.R. Séances Acad. Sci. (Paris) Série D, *291*, 393.
Gordon, J., 1967, Proc. Natl. Acad. Sci. USA, *58*, 1574.
Gordon, J., 1968, Proc. Natl. Acad. Sci. USA *59*, 179.
Grosjean, H., De Henau, S. and Crothers, D.M., 1978, Proc. Natl. Acad. Sci. USA, *75*, 610.
Hall, R.H., 1970, Progr. Nucleic. Acid Res. Mol. Biol. *10*, 57.
Huong, P.V., Audry, E., Giegé, R., Moras, D., Thierry, J.C. and Commarmond, M.B., 1984, Biopolymers *23*, 71.
Jack, A., Ladner, J.E. and Klug, A., 1976, J. Mol. Biol. *108*, 619.
Jank, P., Shinda-Okada, N., Nishimura, S. and Gross, H.J., 1977, Nucleic Acids Res. *4*, 1999.
Jekowsky, E., Schimmel, P.R. and Miller, D.L., 1977, J. Mol. Biol., *114*, 451.
Jonak, J., Smrt, J., Holy, A. and Rychlik, I., 1980, Eur. J. Biochem. *105*, 315.
Jones, T.A., 1978, Appl. Cryst., *11*, 268.
Kern, D., Giegé, R. and Ebel, J.P., 1972, Eur. J. Biochem., *31*, 148.
Kern, D. and Lapointe J., 1980 a, Eur. J. Biochem. *106*, 137.
Kern, D. and Lapointe, J., 1980 b, Biochemistry *19*, 3060.
Kim, S.H., Suddath, F.L., Quigley, G.J., Mc Pherson, A., Sussman, J.L., Wang, A.H.L., Seeman, N.C. and Rich, A., 1974, Science *185*, 435.
Kline, L.K., Fittler, F. and Hall, R.H., 1969, Biochemistry *8*, 4361.
Konnert, J.H. and Hendrickson, W.A., 1980, Acta Cryst. *A 36*, 344.
Krauss, G., Riesner, D. and Maas, G., 1976, Eur. J. Biochem., *68*, 81.
Kusmierek, J.T. and Singer, B., 1976, Biochim. Biophys, Acta, *142*, 536.
Ladner, J.E., Jack, A., Robertus, J.D., Brown, R.S., Rhods, D., Clark, B.F.C. and Klug, A., 1975, Proc. Natl. Acad. Sci. USA, *72*, 4414.
La Rossa, R. and Söll. D., 1980, " Transfer RNA " (Altmann, S. ed.) p. 136, Cell Monograph. series 2, The MIT Press., Cambridge and London.
Lavery, R., Pullman, A. and Pullman, P., 1980 a, Theoret, Chim. Acta (Berl.) *57*, 233.
Lavery, R., Pullman. A., Pullman, B. and de Oliveira, M., 1980 b, Nucleic Acids Res., *8*, 5095.
Lefevre, J.F., Ehrlich, R., Kilhoffer, M.C. and Remy, P., 1980, FEBS Lett. *114*, 219.

Lefevre, J.F., Bacha, H., Renaud, M., Ehrlich, R., Gangloff, J., Von der Haar, F. and Remy, P., 1981, Eur. J. Biochem. *117*, 439.

Leonard, N.J., Mc Donald, J.J., Henderson, R.E.L. and Reichman, M.E., 1971, Biochemistry, *10*, 3335.

Lorber, B., Giegé, R., Ebel, J.P., Berthet, C., Thierry, J.C. and Moras, D., 1983, J. Biol. Chem., *258*, 8429.

Maelicke, A., Sprinzl, M., Von der Haar, F., Khwajh, T.A. and Cramer, F., 1974, Eur. J. Biochem. *43*, 617.

Maxam, A.M. and Gilbert, W., 1980, Methods in Enzymology, *65*, 499.

Moras, D., Commarmond, M.B., Fischer, J., Weiss, R., Thierry, J.C., Ebel, J.P. and Giegé, R., 1980, Nature, *286*, 669.

Moras, D., Lorber, B., Romby, P., Ebel, J.P., Giegé, R., Lewitt-Bentley, A. and Roth, M., 1983, J. Biomole. Str. Dyns. 1, 209.

Ofengand, J. and Cheng, C.M., 1972, J. Biol. Chem. *247*, 2049.

Ofengand, J., 1977, in Molecular Mechanisms of Protein Biosynthesis, eds. Weissbach, H. and Petska, S. (Academic Press, New York) p. 1.

Peattie, D.A., 1979, Proc. Natl. Acad. Sci. USA, *76*, 1760.

Peattie, D.A. and Gilbert, W., 1980, Proc. Natl. Acad. Sci. USA, *77*, 4679.

Pingoud, A., Block, W., Wittinhofer, A., Wolf, H. and Fischer, E., 1982, J. Biol. Chem., *257*, 11261.

Quigley, G.J., Wang, A., Seeman, N.C., Suddath, F.L., Rich, A., Sussman, J.L. and Kim, S.H., 1975 a, Proc. Natl. Acad. Sci. USA, *72*, 4866.

Quigley, G.J., Seeman, N.C., Wang, A.H.T., Suddath, F.L. and Rich, A., 1975 b, Nucl. Acids. Res. *2*, 2329.

RajBhandary, U.L. and Chang, S.H., 1968, J. Biol. Chem., *243*, 598.

Ravel, J.M., Shorey, R.L. and Shive, W., 1977, Biochem. 1979 b, Biophys, Res. Commun. *29*, 68.

Renaud, M., Dietrich, A., Giegé, R., Remy, P. and Ebel, J.P., 1979 a, Eur. J. Biochem. *101*, 475.

Renaud, M., Ehrlich, R., Bonnet, J. and Remy, P., 1979, Eur. J. Biochem. *100*, 157.

Renaud, M., Bacha, H., Dietrich, A., Remy, P. and Ebel, J.P., 1981 a, Biochim. Biophys. Acta, *653*, 145.

Renaud, M., Bacha, H., Remy, P. and Ebel, J.P., 1981 b, Proc. Nat. Acad. Sci. USA, *78* 1606.

Rich, A. and Schimmel, P.R., 1977, Nucleic. Acids Res. *4*, 1649.

Riehl, N., Giegé, R., Ebel, J.P. and Ehresmann, B., 1983, FEBS Lett. *154*, 42.

Rigler, R., Pachmann, U., Hirsch, R. and Zachau, H.G., 1976, Eur. J. Biochem. *65*, 307.

Rigler, R. and Wintermeyer, W., 1983, Ann. Rev. Biophys. Bioeng., *12*, 475.

Ringer, D. and Chladek, S., 1975, Proc. Nat. Acad. Sci. USA, *72*, 2950.

Robertus, J.D., Ladner, J.E., Finch, J.T., Rhodes, D., Brown, R.S., Clark, B.F.C. and Klug, A., 1974, Nature (London) *250*, 546.

Roe, B., Sirover, M. and Dudock, B., 1973, Biochemistry *12*, 4146.

Rosa, J.J., Rosa, M.O. and Sigler, P.B., 1979, Biochemistry, *18*, 637.

Schimmel, P.R., Söll, D. and Abelson, J.N., 1979, Editors of " Transfer RNA Structure, Properties and Recognition ", Cold Spring Harbor Lab. Ed.

Schimmel, P.R., and Redfield, A.G., 1980, Ann. Rev. Biophys, Bioeng., *9*, 181.

Schulman, L.H., Pelka, H. and Sundari, R.M., 1974, J. Biol. Chem. *249*, 7102.

Söll, D. and Schimmel, P.R., 1974, in The Enzymes, ed. Boyer P.O. (Academic Press New York), Vol. 10, 3rd Ed., p. 494.

Starzyk, R., Koontz, S. and Schimmel, P., 1982 Nature, *298*, 136.

Stout, C.D., Mizuno, H., Rao, S.T., Swaminathan, P., Rubin, J., Brennan, T. and Sundaralingam, M., 1978, Acta Crys. *B 54*, 1529.

Sussman, J.L., Holbrook, S.R., Warrant, R.W., Church, G.M. and Kim, S.H., 1978, J. Mol. Biol. *123*, 607.

Thang, H.N., Dondon, L., Thang, D.C. and Rether, B., 1972, FEBS Lett., *26*, 145.
Thiyagarayan, P. and Ponnuswamy, P.K., 1979, Biopolymers, *18*, 2233.
Vlassov, V.V., Giegé, R. and Ebel, J.P., 1980, FEBS Lett., *120*, 12.
Vlassov, V.V., Giegé, R. and Ebel, J.P., 1981, Eur. J. Biochem. *119*, 51.
Vlassov, V.V., Kern, D., Romby, P., Giegé, R. and Ebel, J.P., 1983, Eur. J. Biochem. *132*, 537.
Von der Haar, F. and Cramer, F., 1978, Biochemistry *17*, 4509.
Weissbach, H., Redfield, B. and Brot. N., 1971, Arch. Biochem. Biophys. *145*, 676-684.
Westhof, E., Dumas, P. and Moras, D., 1983, J. Biomole. Str. Dyns. *1*, 337.
Weygand-Durasevic, I., Kruse, T.A. and Clark, B.F.C., 1981, Eur. J. Biochem. *116*, 59.
Wikman, F.P., Siboska, G.E., Petersen, H.U. and Clark, B.F.C., 1982, Embo J., *1*, 1095.
Willick, G.E. and Kay, C.M., 1976, Biochemistry *15*, 4347.
Wrede, P., Woo, N.H. and Rich, A., 1979, Proc. Natl. Acad. Sci. USA, *76*, 3289.
Wübbeler, W., Lossow, C., Fittler, F. and Zachau, H.G., 1975, Eur. J. Biochem. *59*, 405.
Zaccai, G., Morin, P., Jacrot, B., Moras, D., Thierry, J.C. and Giegé R., 1979, J. Mol. Biol. *129*, 483.

ANTIBODY STRUCTURE AND SPECIFICITY

DAVID R. DAVIES

Laboratory of Molecular Biology
National Institutes of Health
NIADDK
Bethesda MD 20205 USA

ABSTRACT

The three-dimensional structure of the antibody molecule is described with particular reference to the refined structures of two mouse antibody Fabs, McPC603, and J539. The molecular basis of the binding of phosphorylcholine to M603 and of galactan to J539 can be understood in terms of the amino acid side chains with which they come in contact. Sequence analysis of other antibodies specific for phosphorylcholine and for galactan reveal that for each hapten the antibodies form a relatively restricted class with what seem to be closely related binding sites.

INTRODUCTION

Specificity in biological systems is nowhere more exquisitely displayed than in the immune system, where it is coupled with diversity in such a way that as many as 10^7 different antibody specificities can be generated by a single individual. As a result of the investigations of the last two decades we can understand a little about the structural basis of antibody specificity and a lot at the level of the gene about the generation of antibody diversity. In this paper I shall summarize some of the results from three-dimensional structure investigations of antibody molecules by X-ray diffraction, coupled with amino acid sequence determination and I shall try to relate these to some of the questions from other fields with particular emphasis on specificity.

This area has recently been reviewed by Davies and Metzger (1983).

The basic three-dimensional structure

Antibody function is mediated by a molecule whose structure consists of two distinct regions—one that carries a recognition site for antigenic determinants, and a second through which the antibody reacts with receptors of a variety of effector systems. Detailed structural analyses have provided an explicit molecular basis for antibody specificity while the analysis of gene structure has yielded explanations for the generation of antibody diversity.

The three-dimensional structure of the antibody molecule can be summarized by Fig. 1, showing the structure of the human IgG cryoglobulin Dob (Silverton *et al.*, 1977), in which the salient features of the molecule can be clearly seen. The molecule consists of four polypeptide chains, two heavy (H) and two light (L) in identical pairs. It is a glycoprotein with, in the case of IgG, a branched chain of complex carbohydrate attached to Asn 297 of each heavy polypeptide chain.

Mild proteolysis by papain will split the molecule into three fragments, two Fabs that bind antigen, and one Fc that is used in the generation of effector functions (Porter, 1959). Most of the X-ray analysis to date has been performed on Fab or Fc fragments or on L chain or VL dimers. Although crystals of intact molecules have been obtained there have been problems associated with the crystals: in Kol (Marquart *et* al., 1980) and Zie (Ely *et al.*, 1978) the Fc part of the molecule is disordered and does not contribute significantly to the observed scattering. In both Dob and Mcg (Steiner & Lopes, 1979; Fett *et al.*, 1973) the molecule has a 15 residue deletion in the hinge, presumably reducing the flexibility of the molecule and enabling the Fc to be fixed in the structure. Thus our knowledge of the three-dimensional structure of the antibody molecule has to be a sum of its parts. However, the checks that can be made on this structure are reassuring and indicate that the composite picture is basically correct.

Fig. 1 clearly illustrates the domain structure of the molecule (Edelman *et al.*, 1969) each chain being divided into discrete globular domains of approximately 110 amino acid residues. While the carboxy terminal domains are constant in antibodies

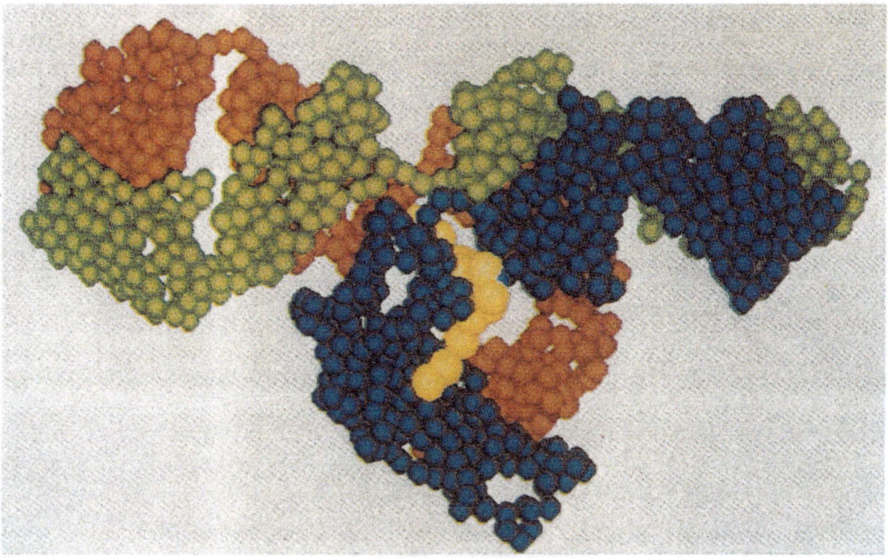

FIG. 1. A space-filling Cα model of the Dob cryoglobulin. The heavy chains are blue and red, the light chains are green and the carbohydrate is yellow.

of a given class or subgroup, the first domains at the amino termini vary considerably from one antibody to another. An analysis of this variability showed that in these variable domains there are several hot spots of variability termed hypervariable regions on both VH and VL (Kabat & Wu, 1970).

Fig. 2 shows the overall structure of the Fab of McPC 603, a mouse myeloma protein with specific binding to phosphorylcholine, and illustrates the clustering of six of these seven hypervariable regions at the tip of the Fab to form the complementarity determining surface. A characteristic feature of the structure is the strong lateral interaction between pairs of domains on opposite chains with relatively weak interactions between domains on a single chain, suggesting that the functional units consist of pairs of domains, i.e. VH : VL and CH1 : CL.

The Fc part of the molecule is shown in Fig. 3 (taken from Deisenhofer, 1981). The two CH3 domains have structures closely related to the CH1 and CL domains of the Fab, and have a similar mode of interaction However, there is no protein-protein contact between the two CH2 domains; instead the complex car-

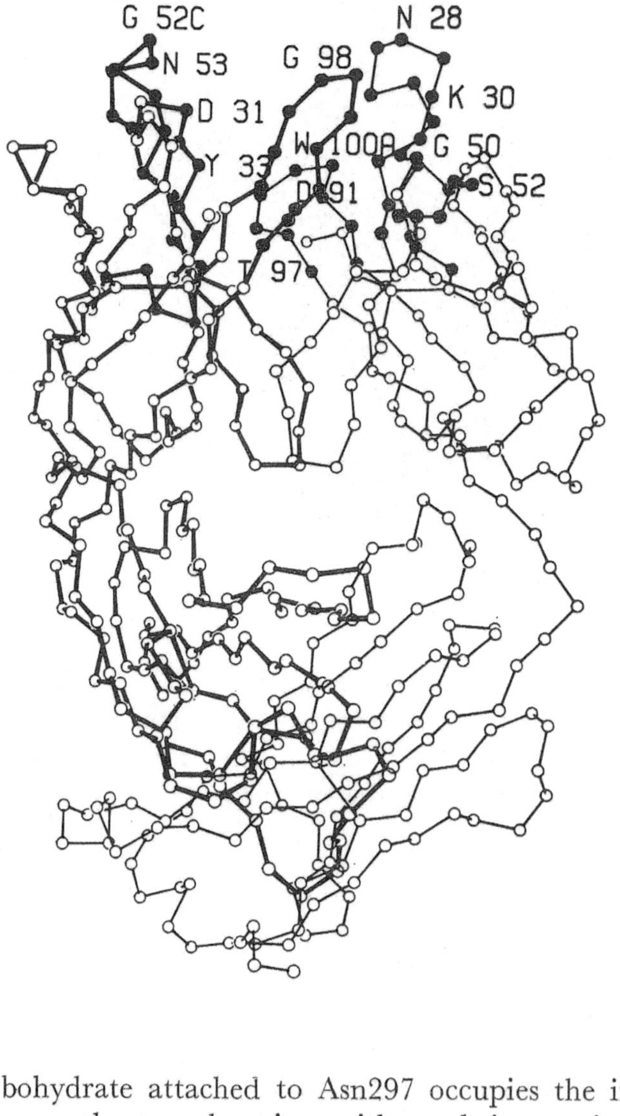

Fig. 2. A Cα drawing of the Fab of McPC603. The two variable domains are above and the two constant domains below. Thick lines represent the heavy chain and thin lines the light. The six hypervariable regions are represented by solid circles.

bohydrate attached to Asn297 occupies the interface region between the two domains, with weak interactions between the two carbohydrate groups (Deisenhofer, 1981).

The Dob immunoglobulin illustrated in Fig. 1 is abnormal in that it has a 15 residue deletion of the hinge region of the heavy chain that links the Fab to the Fc. Most antibodies have a hinge consisting of from about a dozen residues in IgGl to over 60 in IgG3. The hinge is rich in proline and in cysteine, and Marquart

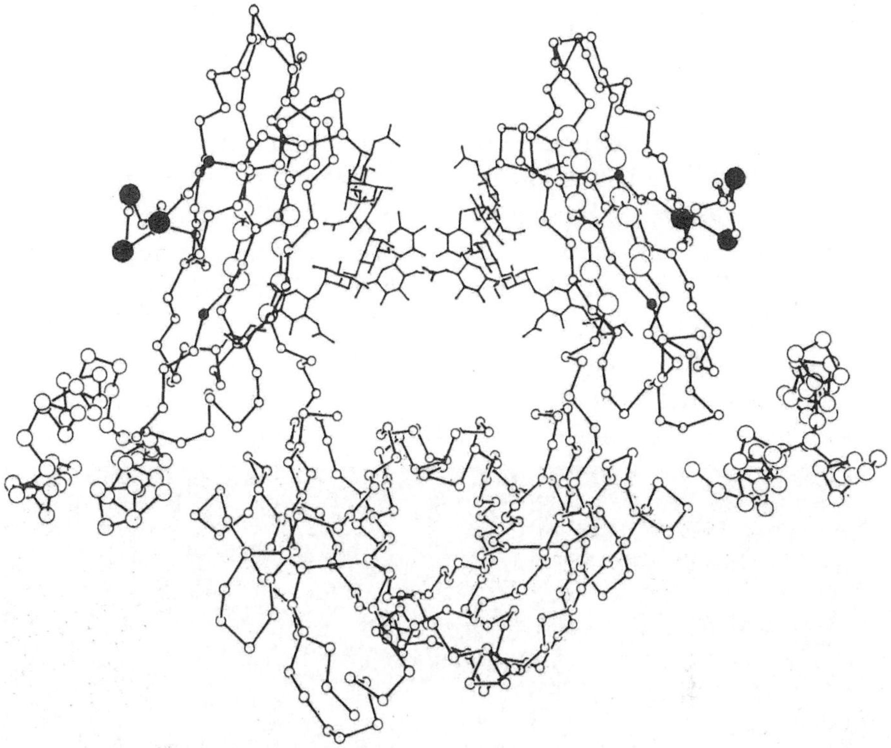

FIG. 3. The Fc of human IgG bound to fragment B of protein A from *staphylococcus aureus*. Based on coordinates of Deisenhofer (1981). The complex carbohydrate occupies the space between the two upper (CH2) domains. The large and solid circles are the α-carbons of those residues that have been proposed to bind to Clq (Boragado *et al.*, 1982; Burton *et al.*, 1980; Prystowsky *et al.*, 1981).

et al., (1980) in an X-ray study of a human myeloma protein Kol observed that the two strands each have a polyproline-like structure joined by disulfide bridges.

Klein *et al.* (1980) have demonstrated that the proteins Dob and Lec, both of which have a 15 residue deletion of the hinge, could not bind C1q nor activate complement. The role played by the hinge is not entirely clear, but it most probably acts as a spacer in separating the part of the Fc that binds the C1q from the Fab. In molecules lacking the hinge access to this region would be restricted.

ANTIBODY COMBINING SITES AND THE MOLECULAR BASIS OF SPECIFICITY

McPC603

Crystals of proteins usually contain approximately 50% solvent of crystallization which occupies the spaces between neighboring protein molecules. Haptens can be diffused through the solvent channels to the protein surface, where they may attach to appropriate binding sites. Difference electron density maps can then be used to determine the location and disposition of the hapten on the protein surface.

Phosphorylcholine has been diffused into crystals of McPC603, and Fig. 4 shows the location of the binding site determined in

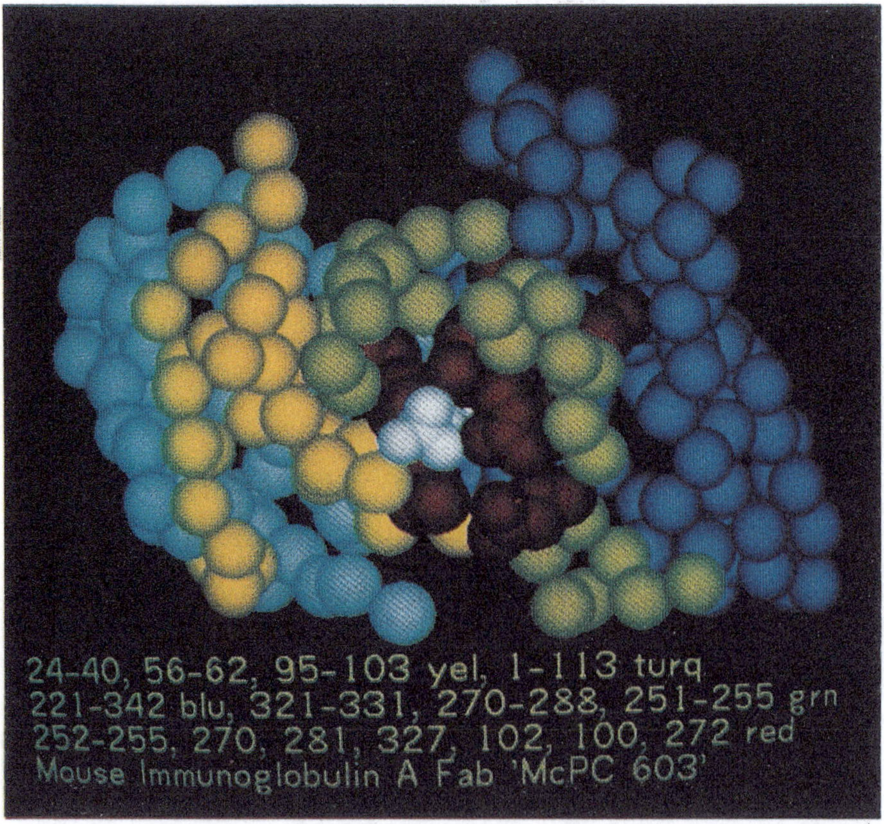

FIG. 4. The combining site of McPC603 with phosphorylcholine bound (white). The yellow and green are the α-carbons of the hypervariable residues. The red are the side chains of those residues that interact with phosphorylcholine.

this way with the bound hapten. The binding site consists of a pocket in a cavity located in the middle of the hypervariable surface. The phosphate group is on the surface, exposed to solvent while the choline moiety is buried in the pocket. Only a small part of the total available hypervariable surface is involved in the binding of this hapten, the remainder being presumably ca-

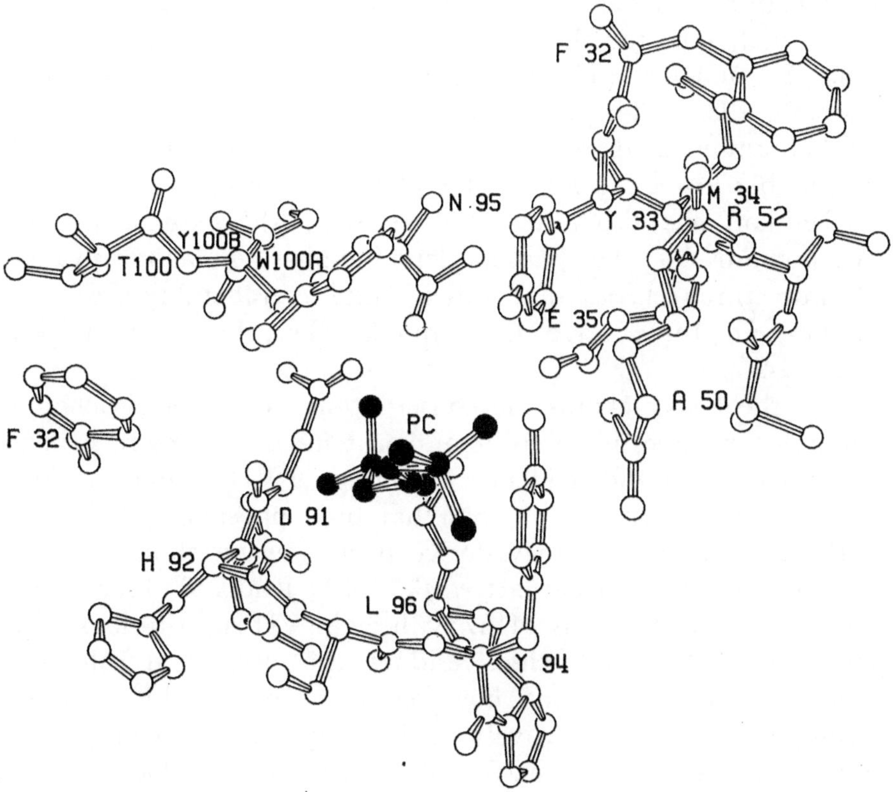

FIG. 5. The combining site of McPC603 with phosphorylcholine bound (black), showing the locations of the contacting residues.

pable of associating with the rest of the antigen. In Fig. 5 the residues involved directly in hapten binding are shown. They include the third hypervariable region of the light chain (L3), particularly the side chains of Tyr94, Pro95 and Leu96; residues

Tyr33 from H1, and Arg52 from H2, which form hydrogen bonds to the oxygens of the phosphate; and finally, Trp100a which forms a lid to the pocket. At the back of the pocket the positive charge on the choline is partly neutralized by the side chain of Asp97 of the light chain.

Leu96 plays a key role in the pocket and it is interesting to note that the codon for this residue occurs at the first position of the J region minigene, a position that is highly variable (Kabat *et al.*, 1982). In the three regions examined, it appears that the same JL minigene (J5) is used on each occasion (reviewed by Rudikoff, 1983). The use of any other JL region would substitute another amino acid for the Leucine in this position and would presumably lower the affinity for phosphorylcholine.

Rudikoff has also noted that although the sequences of the VL domains of the phosporylcholine-binding mouse proteins fall into three classes that differ quite considerably from one another, all three have the sequence Tyr, Pro, Leu in positions 94-96.

For the heavy chains the conservation of phosphorylcholine-contacting residues in the different proteins is even more striking. A comparison of the sequence encoding 19 VH domains of myeloma and hybridoma antibodies that bind phosphorylcholine led to the conclusion that they derive from one germline T15 gene (Crews *et al.*, 1981; Gearhart *et al.*, 1981; Rudikoff, 1983). The most divergent protein is M167 with eight VH substitutions. The phosphorylcholine-contacting residues are present in all these sequences. The D regions, however, show considerable variation, accompanied by only relatively small changes in phosphorylcholine affinity, consistent with the fact that, with the possible exception of Trp100b, the CDR3 of the heavy chain does not play an important role in defining the binding pocket.

These data, and in particular the conservation in both the light and the heavy chains of the phosphorylcholine-contacting residues, also suggest that the mode of phosphorylcholine binding in these other antibodies closely resembles that observed by X-ray diffraction analysis in McPC603.

Mutations Affecting Binding

Although the residues that are in direct contact with hapten and form the immediate combining site are obviously very important for antibody specificity, it is clear that other residues can also play a role in defining this specificity. For example, Rudikoff *et al.* (1982) have examined mutants of S107, a mouse myeloma protein with phosphorylcholine-binding capability that is closely related to McPC603. They find that phosphocholine-binding capability is lost when Asp35 in the heavy chain is changed to Ala. Although not in direct contact with the hapten, this residue plays a double role in making a hydrogen bond to the hydroxyl group of Tyr 94 L, thus presumably anchoring this side chain at the side of the pocket, as well as helping to bind this part of VH to VL.

Conformational Change

The question of how antibodies signal to effector systems their attachment to antigen is one that has led to a number of hypotheses. One of the mechanisms proposed for effector function activation involves an allosteric change upon antigen binding. Such changes could be detected through changes in the crystal structure, such as those observed when deoxyhemoglobin is oxygenated which results in a shattering of the crystals. When crystals of McPC603 are soaked in solutions containing phosphorylcholine, no significant conformational change occurs in the protein. There are small adjustments of the amino acid side chains in contact with the hapten, but no other change of any significance. A similar absence of conformational change is observed when FAB NEW binds to the neutral vitamin K_1OH hapten (Amzel *et al.*, 1974.) Thus, although such a change cannot be rigorously excluded, X-ray diffraction does not provide support for a conformational change upon hapten binding.

J539 FAB

The crystal structure of J539, an antigalactan Fab from Balb/c mice, has been determined and refined at 2.7A resolution. (Suh, Bhat, Cohen, Davies, in preparation). The VH : VL of J539 is very similar to the corresponding part of McPC603, the principal differences occurring in those hypervariable loops that contain insertions and deletions relative to each other. The effect on the overall appearance of the combining site is, however, quite striking. Instead of a large concavity centered by a pocket, J539 has a central cavity from which leads a cleft. It would appear that the galactan must bind in this cleft, but we have not yet attempted to construct a detailed model.

The Elbow Bend and Flexibility

Although the hinge has long been recognized as a source of segmental flexibility between the FAB and the FC (Yguerabide *et al.*, 1970), a comparison of the structures of the four Fabs and two intact molecules has suggested another region of flexibility in the " switch " part of the molecule connecting the VH : VL region to the CH : CL. Flexibility in this region can be measured by the " elbow bend " or the angle between the two local symmetry axes relating VH to VL and CH1 to CL (Marquart *et al.*, 1980), which varies from approximately 135° in McPC603 and in New (Amzel & Poljak, 1979), through 145° for J539 and Dob (Silverton *et al.*, 1977) to 170° for Kol (Marquart *et al.*, 1980). Although this kind of flexibility cannot be directly deduced from a single crystal structure determination, the observed spread of bend values from six separate structures strongly suggests that the switch is a very flexible region of the molecule.

DISCUSSION

Although the immune system can produce 10^7 or more different combining specificities, there are at present only 4 high resolution structures of antibody combining sites with three of these having known specificity. They are McPC603 and J539, two mouse IgA,k antibodies specific for phosphorylcholine and galactan, respectively; Fab New, human IgG,k specific for a hydroxy derivative of Vitamin K (Amzel *et al.*, 1974); and Kol, a human IgG, that has no known specificity (Marquart *et al.*, 1980).

What generalizations can be made from these few structures about the nature of antibody combining sites? In McPC603 the hapten binds in a pocket. In New the menadione ring of VitKOH binds in a narrow groove with the phytyl chain on the surface. In the galactan binding J539 there is a broad groove that presumably represents the site where the hapten binds. In Kol the groove of New is filled by several aromatic side chains and this site interacts strongly in the crystal with the C-terminal part of another Fab in what could be a prototype antibody/protein complex.

For small antigenic determinants one might expect by analogy with enzymes a pocket or groove to provide the specificity. The sizes of combining sites are comparable to those of some enzymes (Givol, 1979). Lysozyme for example has a groove that accommodates a hexasaccharide, and this is believed to represent the upper limit in size for antibodies to this kind of antigen (Kabat, 1982). Another example would be the aspartyl proteases where an extended groove can be used to bind hexapeptide inhibitors with considerable avidity (Bott *et al.*, 1982). Based on the X-ray analyses it would appear that for antibodies the location of this groove would be between the third hypervariable loops of the light and heavy chains. However, it is not very likely that every antibody would have this kind of feature since complementarity of two surfaces can be achieved without necessarily invoking grooves or pockets and significant contributions to the free energy of interaction can come from the exclusion of water from these surfaces. The Kol protein provides an example of this kind of interaction.

These generalizations about antibody structure may be important in the modeling of combining sites. The first attempt to construct a specific model for a combining site was for MOPC315, a dinitrophenol-binding myeloma protein (Padlan *et al.*, 1976). The principles used in constructing the model were to accept the framework part of the structure from known structures (in this case McPC603) and to attach to this the loops containing the hypervariable residues. To a first approximation the shape of these loops was taken from corresponding loops of the same size that had been determined by X-ray diffraction. Adjustments were then made in the structure to remove poor nonbonded contacts. This model building has continued with other proteins (reviewed by Davies & Metzger, 1983). However, caution should be used in dealing with the results since even for this restricted problem in protein folding there exist many solutions. A preliminary comparison of predicted models of J539 with the observed X-ray structure indicates that large errors of as much as 5 or 6 A in the Cα positions can readily occur, presumably because it is hard to analyze the effects of simultaneously altering amino acid residues on the six hypervariable loops.

In order to improve the understanding of combining site structure and of the interactions with different classes of antigen it is clear that new structure investigations will be needed, particularly of antibody-antigen complexes. For example, there are no structures so far of protein-antibody complexes, and there are a number of questions concerning this interaction that remain unanswered. The degree of flexibility in both the antigen and in the antibody combining site and the way in which this plays a role in forming the complex is unknown. Some rather unexpected results have come from the recent use of small peptides as synthetic antigens where a considerable fraction of the antibodies produced can also bind to the protein containing that peptide sequence (Niman *et al.*, 1983). The explanation for this phenomenon is not clear, and for an examination of the questions arising from these results further X-ray diffraction investigations will be needed of the different complexes with peptide and with protein. The basic question of conformational change in the antibody as a result of antigen binding also can only be answered by a direct study of the structures of complexes as well as of the isolated antibodies. Conformational changes, if they occur, might be induced by the larger energy of interaction with larger antigens while failing to be triggered by small haptens.

REFERENCES

Amzel, L.M. and Poljak, R.J., 1979, Ann. Rev. Biochem. *48*; 961.
Amzel, L.M., Poljak, R.J., Saul, F., Varga, J.M. and Richards, F.F., 1974, Proc. Natl. Acad. Sci. USA *71*; 1427.
Boragado, R., Lopez de Castro, J.A., Juare, C., Albar, J.P., Garcia Pards, A., Ortiz, F. and Vivanco-Martinex, F., 1982, J. Immunol. *19*; 579.
Bott, R.R., Subramian, E. and Davies, D.R., 1982, Biochemistry *21*; 956.
Burton, D.R., Boyd, J., Braumpton, A.D., Easterbrook-Smith, S.W., Emanuel, E.J., Novotny, J., Rademacher, T.W., Van Schravendijk, M.R., Sternberg, M.J.E. and Dwek, R.A., 1980, Nature *288*; 388.
Crews, S., Griffin, J., Huang, H., Calame, K. and Hood, L., 1981, Cell *25*; 59.
Davies, D.R. and Metzger, H., 1983, Ann. Rev. Immunol. *1*; 87.
Deisenhofer, J., 1981, Biochemistry *20*; 2361.
Edelman, G.M., Cunningham, B.A., Gall, W.E., Gottlieb, P.D., Rutishauser, U. and Waxdal, M.J., 1969, Proc. Natl. Acad. Sci. USA *63*; 78.
Ely, K.R., Colman, P.M., Abola, E.E., Hess, A.C., Peabody, D.S., et al., 1978, Biochemistry *17*; 820.
Fett, J.W., Deutsch, H.F., Smithies, O., Colman, P. M. and Lattman, EE., 1973, Immunochemistry *10*; 115.
Gearhart, P.J., Johnson, N.D., Douglas, R. and Hood, L., 1981, Nature *291*; 29.
Givol, D., 1979, Int. Rev. Biochem. *287*; 301.
Kabat, E.A., 1982, Pharm. Rev. *34*; 23.
Kabat, E.A. and Wu, T.T., 1970, J. Exp. Med. *132*; 211.
Kabat, E.A., 1982, Pharmacological Reviews *34*; 23.
Klein, M., Haeffner-Cavaillon, N., Isenman, D.E., Rivat, C., Navia, M.A. and Davies, D.R., 1981, Proc. Natl. Acad. Sci. USA *78*; 524.
Marquart, M., Deisenhofer, J., Huber, R. and Palm, W., 1980, J. Mol. Biol. *141*; 369.
Niman, H.L., Houghten, R.A., Walker, L.E., Reisfeld, R.A., Wilson, I.A., Hogle, J.M. and Lerner, R.A., 1983, Proc. Natl. Acad. Sci. USA *80*; 4949.
Padlan, E.A., Davies, D.R., Pecht, I., Givol, D. and Wright, C., 1976, Cold Spring Harbor Symp. Quant. Biol. *XL*; 627.
Porter, R.R., 1959, Biochem. J. *73*; 119.
Prystowsky, M.B., Kehoe, J.M. and Erickson, B., 1981, Biochemistry *20*; 6349.
Rudikoff, S., 1983, Contemp. Top. Mol. Immunol. *9*; 169.
Silverton, E.W., Navia, M.A. and Davies, D.R., 1977, Proc. Natl. Acad. Sci. USA *74*; 5140.
Steiner, L.A. and Lopes, A.D., 1979, Biochemistry *18*; 4045.
Yguerabide, J., Epstein, H.F. and Stryer, L., 1970, J. Mol. Biol. *51*; 573.

SPECIFICITY OF IONOPHORE-CATION INTERACTION

YU. A. OVCHINNIKOV AND V. T. IVANOV

Shemyakin Institute of Bioorganic Chemistry USSR Academy of Sciences
Moscow, USSR

ABSTRACT

Structure, ion binding properties and mechanisms of induced ion transport are considered for various groups of ion carriers. Contribution of geometrical and electronic factors to the experimentally observed ionic selectivity is discussed for selected ionophores and in general terms. Molecular mechanisms underlying the channel mediated ion transport in amphotericin B, gramicidin A and alamethicin I families are presented and analyzed in view of their ionic selectivity.

Molecular origins of ionic specificity in the living nature have been attracting considerable interest for long. In the mid-60's a novel phenomenon was discovered of selective transmembrane ion transport induced by a number of low molecular natural products, mostly antibiotics, called ionophores. The discovery prompted further search for novel members of that class, both natural and artificial, as well as detailed study of their ion binding structure and molecular mechanism of action. The crucial role of conformation in selecting the particular ion has been demonstrated in addition to the obvious importance of the nature of liganding functional groups. Environment and the membrane system also had a notable impact on ionic selectivity. Metal binding properties of various ionophores will be briefly considered in this paper as well as major factors affecting their selectivity and transport properties.

1. Transmembrane ion carriers

a) *Valinomycin and its analogs*

Historically, valinomycin was the first ionophore, on the example of which the induced membrane transport has been observed (Moore and Pressman, 1964). This macrocyclic depsipeptide (Fig. 1a) exerts its action in a variety of artificial

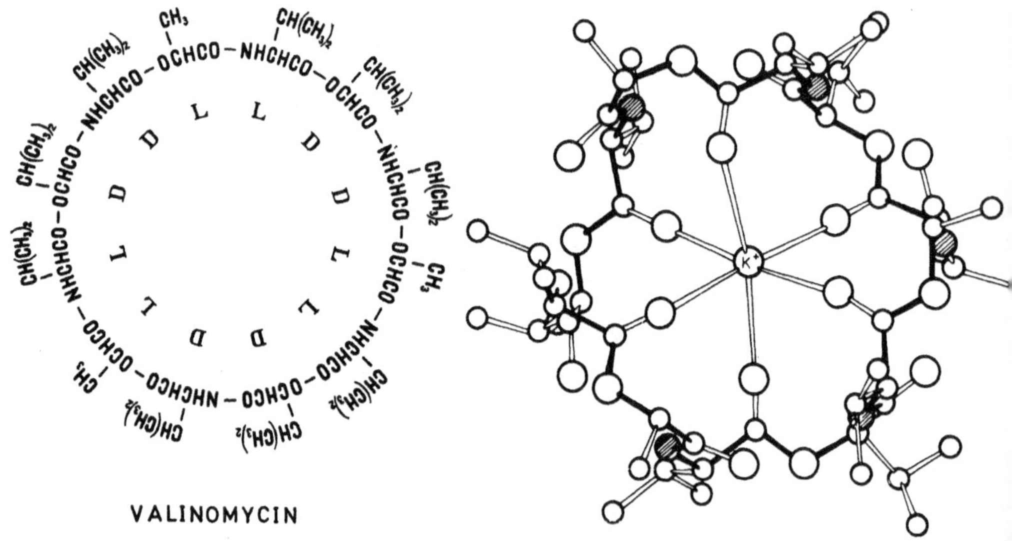

Fig. 1. Formula of valinomycin (a) and three-dimensional structure of its K-complex (view along the symmetry axis) (b).

and biological membranes (see Ovchinnikov *et al.*, 1974, and references therein; Pressman, 1976). A unique feature of the induced ionic current is its extreme K/Na selectivity, reaching in some cases 10^4. In 1967 Shemyakin *et al.* discovered the ion binding capacity of the antibiotic which was as selective as the induced transport and which therefore lies at the basis of the transport phenomenon. In addition to K and Na, valinomycin also binds other alkali, alkaline earth, a few transition metal ions and ammonium. As seen from Table 1 the maximal stability was found with K and Rb. Cs-Complex is slightly less stable and other cations show much weaker binding. Binding constants fall sharp-

TABLE 1. *Stability Constants* (M^{-1}) *of Valinomycin Complexes at 25° C**

Solvent	Li^+	Na^+	K^+	Rb^+	Cs^+	NH_4^+	Ag^+	Tl^+	Mg^{2+}	Ca^{2+}	Sr^{2+}	Ba^{2+}
Hexane (Rose and Henkens, 1974)	—	$> 10^6$	$> 10^6$	—	—	—	—	—	—	—	—	—
Dioxane (Rose and Henkens, 1974)	—	$\sim 10^3$	$> 5\times 10^4$	—	—	—	—	—	—	—	—	—
Acetonitrile (Rose and Henkens, 1974)	—	$< 10^3$	3.2×10^5	—	—	—	—	—	—	—	—	—
(Brezina *et al.*, 1974)	—	350	5.6×10^6	7.9×10^6	1.0×10^6	—	—	7.9×10^5	—	—	—	—
Isopropanol (Rose and Henkens, 1974)	—	180	2.5×10^6	—	—	—	—	—	—	—	—	—
Ethanol (Andreev *et al.*, 1971; Shemyakin *et al.*, 1969)	—	< 50	2.6×10^6	2.9×10^6	65×10^6	—	—	—	—	—	—	—
		25 (Rose and Henkens, 1974)	3.8×10^5 (Rose and Henkens, 1974)			4.2×10^4		(Cornelius *et al.*, 1974)				
99.5% aqueous ethanol (Moschler *et al.*, 1971)	—	—	1.2×10^6	—	—	—	—	—	—	—	—	—
Methanol (Grell *et al.*, 1972)	5	4.7	0.8×10^5	1.8×10^5	2.6×10^4	47	8000	5400	5	500	170	2200
(Cornelius *et al.*, 1974)	—	2300	6.2×10^4	1.31×10^5	2.1×10^4	—	—	2.3×10^4	—	—	—	—
(Rose and Henkens, 1974)	—	5.9	2.95×10^4	—	—	—		—	—	—	—	—
(Wipf *et al.*, 1968; Haynes *et al.*, 1971)	—	12	0.8×10^4	—	—	—	—	—	—	—	—	—
Andreev *et al.*, 1971)	—	—	2.7×10^4	—	—	—	—	—	—	—	—	—
Methanol-water, 9:1 (Grell *et al.*, 1975)	—	—	4700	—	—	—	—	—	—	—	—	—
Methanol-water, 7:3 (Grell *et al.*, 1975)	—	—	93	—	—	—	—	—	—	—	—	—
Methanol-water, 1:1 (Grell *et al.*, 1975)	—	—	10.5	—	—	—	—	—	—	—	—	—
Water (Feinstein and Felsenfeld, 1971)	—	0.1	2.3	5.9	0.12	—	—	—	—	—	—	—

* The values for the bivalent cation complexes are low, owing to neglect of (anion·metal)$^+$ pair formation.

ly on increasing solvent polarity. For instance, with K^+ the binding constants in water and ethanol differ by six orders of magnitude (see below, Table 2).

The ligand : cation ratio in valinomycin complex is usually 1:1. However, with Ca^{2+} and Ba^{2+} 1:2 and/or 2:1 stoichiometries were found (Devarajan and Easwaran, 1981; Devarajan *et al.*, 1980; Ivanov, 1975; Vishwanath and Easwaran, 1982). Formation of the 2:1 sandwich complex with potassium preliminarily communicated by Ivanov (1975) so far did not find further confirmation.

Formation of (valinomycin · cation)$^+$ · anion$^-$ ion pairs has not been studied yet in a systematic way. Such complexes are amply documented for some Na (Golikov and Vichutinsky, 1981; Davis and Tosteson, 1975; Eastman, 1974; Steinrauf *et al.*, 1982) and Ba (Devarajan and Easwaran, 1981) salts. Golikov and Vichutinsky (1981) also found ion pairs with the K^+-complex which proved much less stable.

Of fundamental importance for understanding the ion binding and membrane properties of valinomycin was conformational analysis of the free and ion-bound antibiotic (see pp. 563-573 of the review by Ovchinnikov and Ivanov, 1982; Hamilton *et al.*, 1981). With K^+, Rb^+, Cs^+ and Tl^+, both in solution and in the crystal, a C_3-symmetric structure is formed, resembling a bracelet with hydrophobic periphery and a molecular cavity lined with six ester carbonyl oxygens. Each of these atoms interacts with the unsolvated cation placed at the center of the cavity (Fig. 1b). The size of the cavity fits best the size of potassium and rubidium. The amide NH and CO groupings participate in a perfect system of intramolecular hydrogen bonds fixing six β-turns which in turn form a cylindric structure of six condensed 10-membered rings protecting the internal coordination sphere from the environment. Spectacular progress of electronic microscopy allowed direct observation of that structure on electron micrographs (Ottensmeyer *et al.*, 1978, 1978-79).

Being quite rigid, the bracelet structure cannot shrink so as to adapt the size of the inner cavity to the smaller than potassium size of sodium (and of course lithium). Therefore valinomycin

TABLE 2. *Stability Constants of Equimolar Potassium Complexes of Valinomycin and its Analogs* (M^{-1})

Nº	Compound	CH_3CN	EtOH	96% EtOH	80% EtOH	MeOH	64% MeOH	H_2O
1.	cyclo [-(D-Val-Lac-Val-D-Hyi)$_3$-] Valinomycin *	320000 5600000	630000 2600000	130000 (Andreev *et al.*, 1971)	8000 (Andreev *et al.*, 1971)	8000 27000 29500 62000 80000	160	2,3
2.	cyclo [-(D-Val-Lac-Val-D-Hyi)$_2$-] Octa-valinomycin (Fonina *et al.*, 1980)	—	—	< 10	—	—	—	—
3.	cyclo [-(D-Val-Lac-Val-D-Hyi)$_4$-] Hexadeca-valinomycin (Shemyakin *et al.*, 1969)	—	—	∾50	—	—	—	—
4.	cyclo [-(D-Ile-Lac-Ile-D-Hyi)$_3$-] Isoleucinomycin (Fonina *et al.*, 1980)	—	—	180000	—	—	—	—
5.	cyclo [-(D-Ile-Lac-Ile-D-Hyi)$_2$-] Octa-isoleucinomycin (Fonina *et al.*, 1980)	—	—	< 10	—	—	—	—
6.	cyclo [-(D-Ile-Lac-Ile-D-Hyi)$_4$-] Hexadeca-isoleucinomycin (Fonina *et al.*, 1980)	—	—	< 100	—	—	—	—
7.	cyclo [-(D-Val-Hyi-Val-D-Hyi)$_3$-] Meso-Hyi-valinomycin (Ivanov *et al.*, 1974)	—	> 10	—	—	880000	—	—
8.	cyclo [-(D-Val-MeAla-Val-D-Hyi)$_3$-] (Vinogradova *et al.*, 1974; Ivanov *et al.*, 1980)	—	$> 10^1$	—	40000	—	—	—
9.	cyclo [-(D-Val-Lac-Val-D-Pro)$_3$-] (Avotin *et al.*, 1978; Balashova *et al.*, 1983)	—	—	35000	—	—	—	—
10.	cyclo [-(D-Val-Pro-Val-D-Pro)$_3$-] Prolinomycin (Gizin *et al.*, 1978)	—	—	—	—	—	100000	630
11.	cyclo [-(Val-Gly-Gly-Pro)$_3$-] (Baron *et al.*, 1977)	170000	—	—	—	—	—	—

* For references see Table 1.

is unable to compensate for the higher desolvation energy of sodium with equivalent enhancement of ligand-metal interactions. Such is the source of the 7-8 kcal/mol free energy difference in Na and K binding which in turn gives rise to 10^4 times difference in binding and transport efficiency.

Unlike most antibiotics, valinomycin is a sole naturally occurring representative of this group of depsipeptides. On the other hand, over a hundred synthetic valinomycin analogs have been described differing from the parent substance in the ring size, the nature or configuration of amino/hydroxy acid residues, in CONH → COO, CONH → CONR replacements or in combination of the above modifications (pp. 399-408 in Ovchinnikov and Ivanov, 1982b). Comparative studies allowed to trace regularities connecting the chemical formula, three-dimensional structure, stability of the complexes and ion transporting capacity in this extensive series. Tables 2 and 3 include some analogs

TABLE 3. *Stability Constants (K, M^{-1}) of Complexes of Valinomycin and its Analogs in 96% Aqueous Ethanol (Fonina et al., 1980)*

Nº	Compound	Na^+	K^+	Rb^+	Cs^+
1.	Valinomycin	∾10	130000	240000	63000
2.	Octa-valinomycin	∾10	< 10	< 10	< 10
3.	Hexadeca-valinomycin	< 50	50–100	∾100	500
4.	Isoleucinomycin	< 50	180000	80000	42000
5.	Octa-isoleucinomycin	∾10	< 10	< 10	< 10
6.	Hexadeca-isoleucinomycin	< 50	50–100	∾100	100–1000

whose ion binding properties require a special comment. Analogs (2, 3, 5 and 6), being generally quite poor ion binders, nevertheless instead of K prefer Na (analogs 2 and 5) or Cs (analogs 3 and 6), in accord with the smaller or larger, than in valinomycin, ring size. Moreover, hexadeca-isoleucinomycin (6) selectively transfers voluminous organic cations into the organic phase (Table 4) while hexadeca-valinomycin selectively carries these cations across lipid bilayers (Eisenman and Krasne, 1975). Potassium 1:1 complexes of analogs (7,8 and 10) closely resemble

TABLE 4. *Constants of Bulk Extraction ($k_{w/o}$) from Water into CH_2CL_2 of Picrates of Univalent Ions by Valinomycin and its Analogs (Fonina et al., 1980)*

Nº	Compound	K^+	Rb^+	Cs^+	NH_4^+	Me_4N^+	Acetyl-choline
1.	Valinomycin	430 *	530 *	150 *	670	—	57
3.	Hexadeca-valinomycin	< 10 *	< 10 *	< 10 *	—	—	—
4.	Isoleucinomycin	1700 *	2300 *	510 *	1400	—	62
6.	Hexadeca-isoleucinomycin	40	120	270	10	550	1450

* Dinitrophenolate

valinomycin complexes in their structures (Hamilton *et al.*, 1980; Ivanov *et al.*, 1974, 1980), these analogs being at the same time considerably more powerful ion binders (Table 2). The peptide analog (11) is endowed with very high stability of 1:1 complexes with Mg^{2+}, Ca^{2+}, Ba^{2+} and Li^+. It also shows a pronounced trend to form " sandwich " complexes with 2:1 ligand:ion ratio as well as " reverse sandwiches " with two metal ions bound to one peptide molecule (Li^+, Mg^{2+}, Ca^{2+}, Ba^{2+}, all data for acetonitrile, Baron *et al.*, 1977; Easwaran *et al.*, 1979). Analog (9) also forms a $(9)_2 \cdot K^+$ " sandwich " in methanol while in chloroform the same analog forms a usual $(9) \cdot K^+$ complex. In other words, not only binding constants but even stoichiometry of ion binding proved solvent dependent in this case (Balashova *et al.*, 1983).

Free prolinomycin (10) is too hydrophylic to enter the lipid matrix which is not the case for valinomycin and the K^+-complexes of both depsipeptides. This circumstance gives rise to a dramatic difference in the mechanisms of ion transfer. A key step in the valinomycin mediated transfer is heterogeneous complexation of valinomycin in the interface and the ion in the aqueous phase (Fig. 2a, b). Only a much less efficient pathway (c) is allowed for prolinomycin which includes ion binding in water and passage by the charged complex of both interfaces (Benz *et al.*, 1976).

In summary, ion selectivity in the valinomycin series is clearly dependent upon geometrical fit of the cavity and the ion size

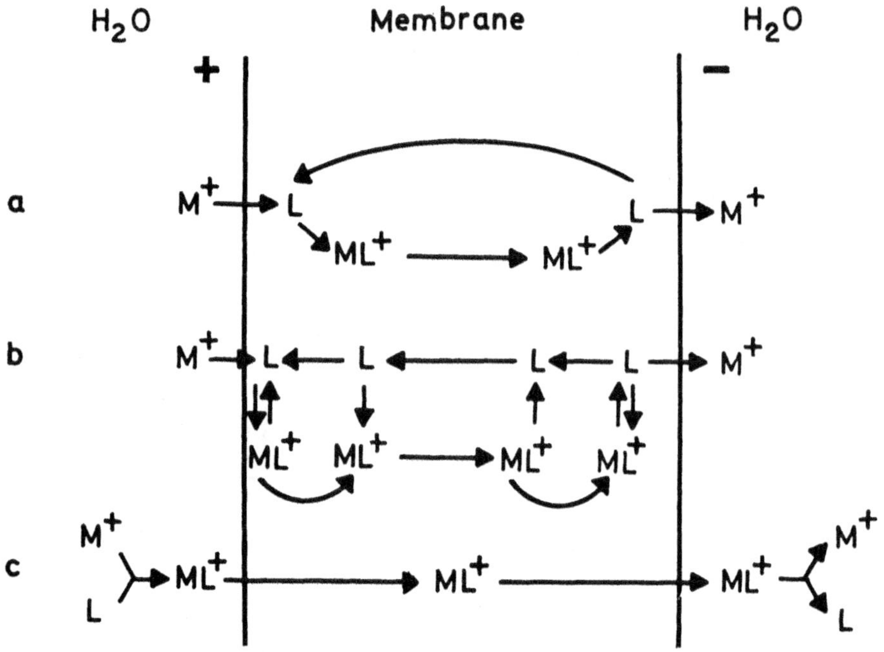

FIG. 2. Scheme of alkali metal ion transport induced by valinomycin (a, b) and prolinomycin (c). In most cases valinomycin induced transport is satisfactorily described by scheme (a) (Läuger, 1980), although with bilayers made of the total lipid of brain white matter a more complicated scheme (b) had to be accepted (p. 290 in Ovchinnikov *et al.*, 1974).

(analogs 2-6). Replacement of the ester moieties by amides enhances the affinity to alkaline earth cations and the trend to form complexes with other than 1:1 stoichiometries (compounds 9-11). A more detailed analysis would be premature at this stage due to complexity of the system.

b) *Enniatins*

Similarly to valinomycin, enniatins are also built of alternating amino and hydroxy acid residues (Fig. 3). Their dominating property as ion binding species is a broad spectrum of action and low ionic selectivity. In solution enniatins form complexes with alkali, alkaline earth metals, ammonium and some transition metal ions - Ag^+, Tl^+, Mn^{2+}, Zn^{2+}, Cd^{2+} (see Ovchinnikov *et al.*, 1974, pp. 12-16 and references therein). Stability constants of these complexes are relatively low. For example, they vary in the $10\text{-}10^3$ range for enniatin B in methanol, in 96% ethanol they may be 2-3 times higher. In addition to equimolar complexes, enniatins B, C and beauvericin form complexes in ratio macro-

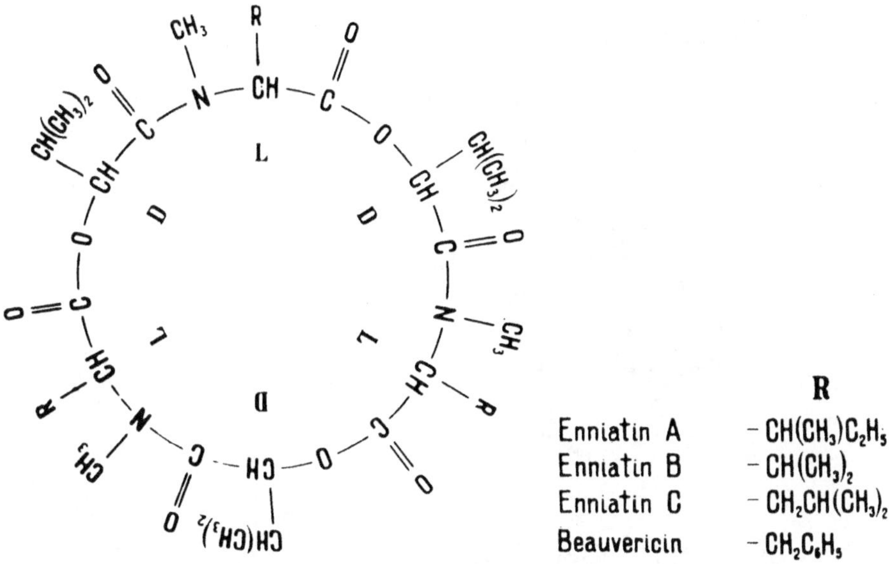

FIG. 3. Enniatin antibiotics.

cycle: cation = 2:1 (and possibly also 3:2), the second binding constant being 1-2 orders of magnitude lower than the first. Stability of the " sandwich " complexes falls in the order

$$K > Cs > Na > Li$$

Even more " exotic " stoichiometry was found in the crystalline complex of beauvericin with barium picrate ($BaPic_2$), viz. $[(\text{beauvericin})_2\ Ba_2 \cdot Pic_3]^+$ (Braden *et al.*, 1980).

Figs 4-6 demonstrate the principal structures proposed for complexes of various stoichiometries, each of which can be considered as a potential ion carrying structure. Conformation of the (enniatin B · K)$^+$ complex (Fig. 4) was derived from the low resolution X-ray analysis (Dobler *et al.*, 1969). The characteristic C_3-symmetric disc-like shape of that structure, the alternating up-down orientation of N-methyl amide and ester carbonyls as well as pseudo equatorial orientation of C^{α} substituents has been further substantiated for enniatin complexes in solution, the free enniatins in polar solvents (Ovchinnikov *et al.*,

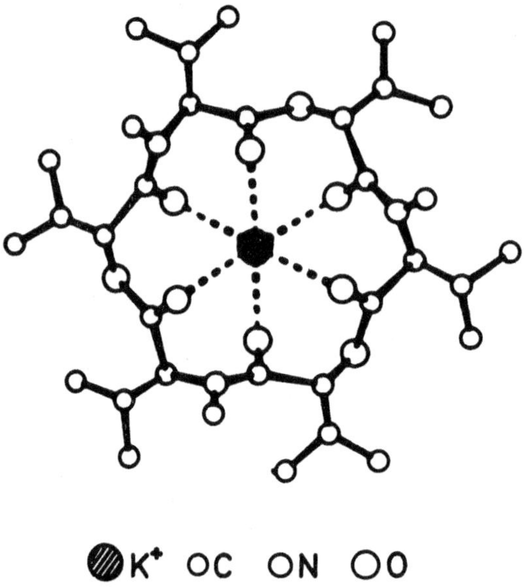

FIG. 4. Conformation proposed for the crystalline complex (enniatin B) · KNCS.

1974) and in the crystal (Tishchenko *et al.*, 1976), and for beauvericin and the above mentioned barium complex in the crystalline state (Braden *et al.*, 1980; Geddes and Akrigg, 1976). At the same time, as pointed out by Hamilton *et al.* (1975), location of the bound cation in the center of the molecular cavity requires further proof. It well might be that even in the equimolar complex the ion is placed at the disc periphery, as in Fig. 5a, interacting with only amide or only ester carbonyls and probably retaining part of the solvent sheath. In the " sandwich " (Fig. 5b) and in the " stack " (Fig. 5c) the cation is more efficiently screened from the solvent and the counterion. Studies in solution do not allow to choose between structures in Figs 4 and 5a. In all cases when X-ray structures of metal ion complexes with cyclic peptides related to the enniatin family had been established, the peripheral location of the bound ion was found, namely:

the rubidium complex of the LDLLDL enniatin B diastereomer (Tishchenko *et al.*, 1977), (cyclo[-(Pro-Gly)$_3$-])$_2 \cdot Mg^{2+}$, (cyclo [-(Pro-Gly)$_3$-])$_2 \cdot Ca^{2+}$ (Kartha *et al.*, 1982), (cyclo[-(Gly-Pro--Pro)$_2$-]) $\cdot Mg^{2+}$ (Karle and Karle, 1981), (cyclo [-(Pro-Gly)$_4$-])$_2 \cdot Rb_2^+$ (Chin *et al.*, 1977), the beauvericin complex shown in Fig. 6. This argument indirectly speaks in favour of the structural type shown in Fig. 5a.

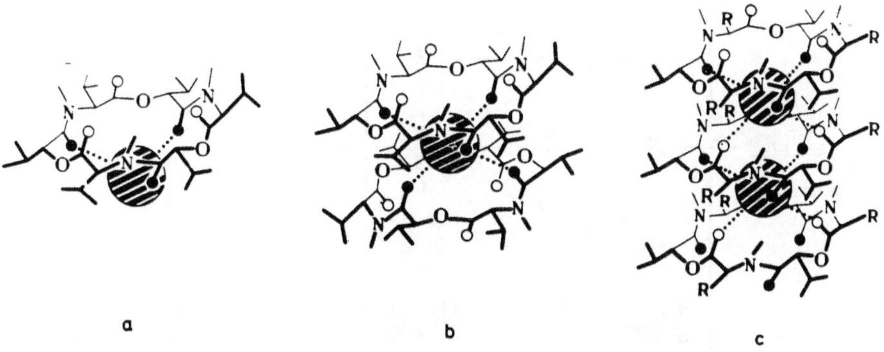

FIG. 5. Conformations proposed for enniatin B complexes of various stoichiometry.

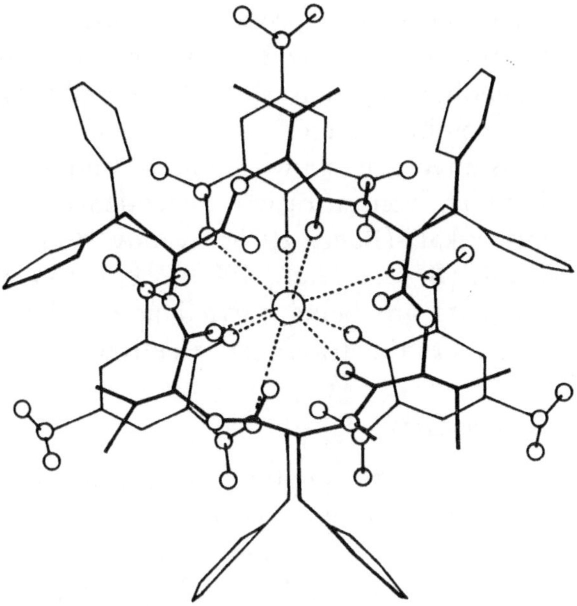

FIG. 6. Conformation of the crystalline beauvericin$_2 \cdot$ (BaPic$_2$)$_2$ complex.

In dioleylphosphatidylcholine bilayers enniatins A and B induce Li, Na, K, Rb and Cs conductivities proportionally to the ionophore concentration (Benz, 1977). However, with another lipid (total from the brain white matter) a cubic dependence was obtained with Cs, a square dependence — with K and only with Na conductivity was (as usually) proportional to ionophore concentration (Ivanov *et al.*, 1973). These data point to a unique ion-dependent mechanism of ionophore action, shown in Fig. 7.

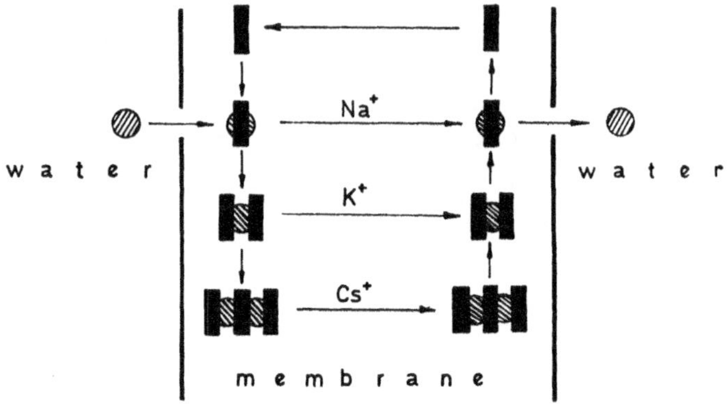

FIG. 7. Scheme of enniatin B induced alkali metal ion transport.

According to this proposal the major contribution to the overall ion transfer is provided by complexes of variable stoichiometry, depending on their stability, lipophylic properties, mobilities in the membrane, etc.

Beauvericin shows high, second or third, power of conductivity *vs.* concentration dependence with all lipids tested both with alkali (Benz, 1978; Ivanov *et al.*, 1973) and alkaline earth metals (Braden *et al.*, 1980). It is thought that in the latter case ion pairs are being carried across the membrane (Prince *et al.*, 1974), i.*e.* the transport selectivity becomes anion-dependent.

c) *Antamanide*

This cyclic peptide (Fig. 8a) isolated from the poisonous mushroom *Amanita phalloides* causes preventive action against the toxins of the same mushroom, the phalloidins. Being a weak ionophore, antamanide still presents an interesting example of

Na, Ca/K binding selectivity (see Ovchinnikov and Ivanov, 1982, and references therein). The ion in antamanide complexes is conventionally placed into the molecular cavity where it is held with four metal-carbonyl bonds. In addition, one solvent molecule (acetonitrile in Fig. 8b) remains tightly bound to the complexed ion. A hypothesis is put forward (p. 561 in Ovchinnikov and Ivanov, 1982b) according to which antamanide binds to the lipid or the protein components of the biomembrane *via*

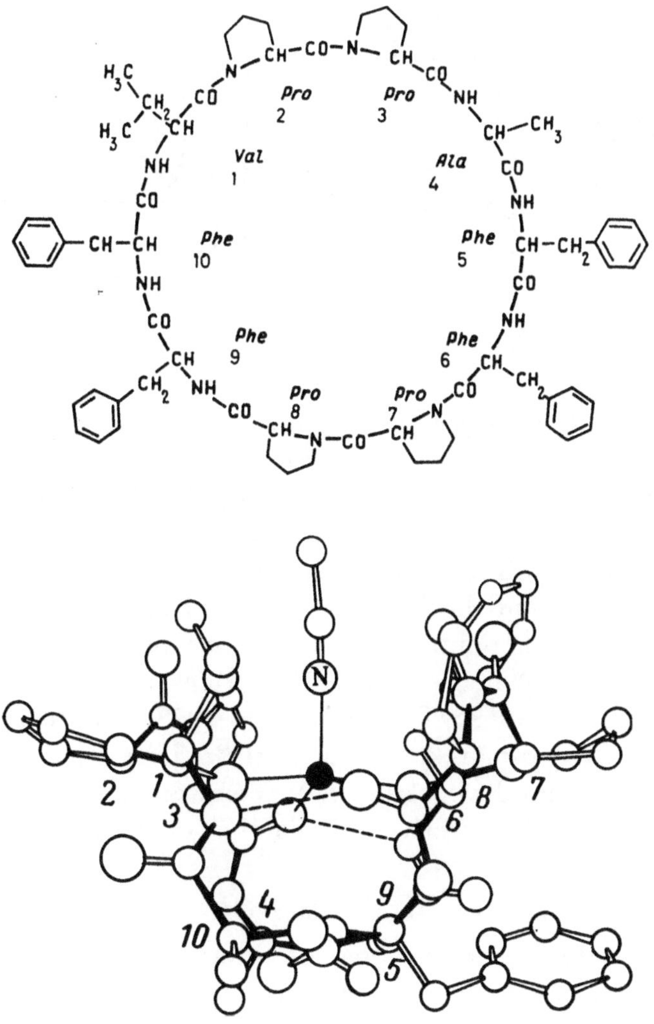

FIG. 8. Formula of antamanide (a) and three-dimensional structure of its crystalline Li⁺-complex (Karle *et al.*, 1973) (b).

the metal ion which serves in this case as a link between the interacting species. Such modification of the membrane surface makes it less permeable to phalloidins which explains the antitoxic effect (Fig. 9).

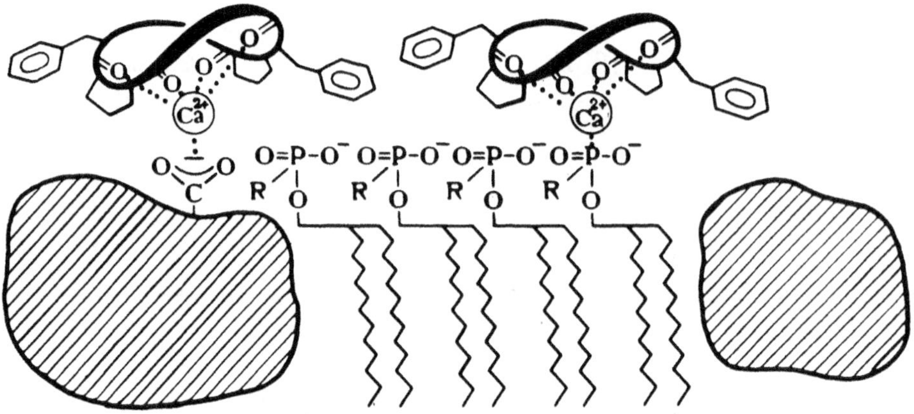

FIG. 9. Hypothetic principle of antamanide interaction with a biomembrane (hatched areas refer to protein globule).

d) *Nactins*

Nonactin is a parent member of a family of closely related ionophores called macrotetrolides (i.e. macrocyclic lactones) or nactins (Keller-Schierlein and Gerlach, 1968; Ovchinnikov *et al.*, 1974, p. 47, and references therein). Chemical formulae of better characterized nactins are shown in Fig. 10.

Nactins bind in solution the ions of alkali and alkaline earth metal ions, thallium, ammonium and a number of substituted ammonium derivatives (Asher *et al.*, 1977; Briggs and Hinton, 1979; Eisenman *et al.*, 1969; Phillies *et al.*, 1975; Prestegard and Chan, 1969, 1970; Simon and Morf, 1973; Ueno and Kishimoto, 1979; Ueno and Kyogoku, 1979; Vishwanath and Easwaran, 1981). The ionic selectivity of nactins is characterized by the sequence

$$NH_4 > Tl > K > Rb > Na > Cs > Ca > Ba > Li$$

In K/Na selectivity nactins occupy an intermediate position between valinomycin and enniatins. A striking property of nactins as ionophores is their exceptionally high affinity to ammonium ions (Eisenman *et al.*, 1969; Scheler and Simon, 1970; Szabo *et al.*, 1969). In no case does the stoichiometry of binding in nactin series deviate from the classical 1:1 ratio.

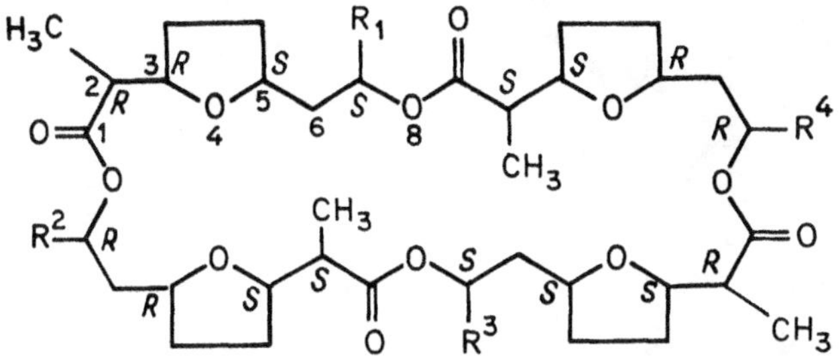

	R^1	R^2	R^3	R^4
Nonactin	CH_3	CH_3	CH_3	CH_3
Monactin	CH_3	CH_3	CH_3	C_2H_5
Dinactin	CH_3	C_2H_5	CH_3	C_2H_5
Trinactin	CH_3	C_2H_5	C_2H_5	C_2H_5
Tetranactin	C_2H_5	C_2H_5	C_2H_5	C_2H_5

FIG. 10. Macrotetrolide antibiotics.

Nonactin was the first antibiotic ionophore where a number of fundamental features of ionophore complexation have been elucidated in the course of the X-ray analysis of the crystalline nonactin · KNCS complex (Fig. 11, Dobler *et al.*, 1969; Kilbourn *et al.*, 1967). To such features belong of course sequestering of the bare ion in the middle of the molecular cavity, participation of oxygen atoms in ion binding and hydrophobic nature of the molecular periphery. Quite similar structures have all the complexes of nonactin with K, Na (Dobler and Phizakerley, 1974) and ammonium (Neupert-Laves and Dobler, 1976) and of tetranactin with Na, K, Rb, Cs (Sakamaki *et al.*, 1976, 1977) and ammonium (Nawata *et al.*, 1975, 1977). Four ester carbonyls and four ether oxygens arrange themselves around the bound ion approximately at the cube apexes. The size of the cavity varies in accord with the size of the bound ion at the expense of slight alterations of rotational angles. Notable distortion of the cubic symmetry was found in Na^+-complexes which explains their relatively low stability. Ammonium complexes are stabilized

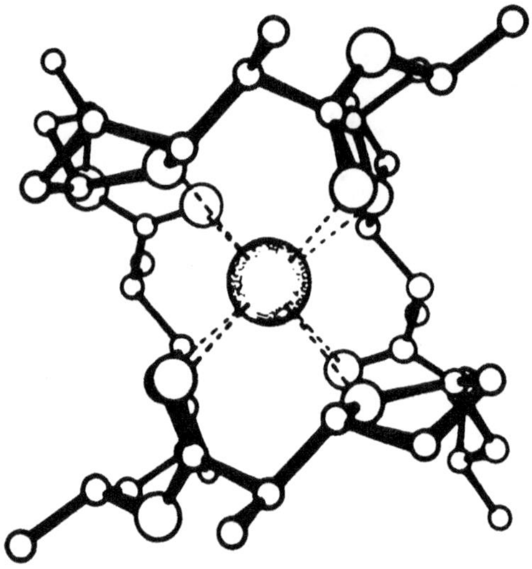

FIG. 11. Conformation of the crystalline complex nonactin · K^+

by hydrogen bonds with ether oxygens (Fig. 12) apparently responsible for the extreme ammonium selectivity (Umeyama *et al.*, 1980). Spectral studies of nactin complexes with alkali metal ions (Asher *et al.*, 1977; Phillies *et al.*, 1975; Prestegard and Chan, 1969, 1970; Ueno and Kyogoku, 1979), Tl^+ (Asher *et al.*, 1977; Briggs and Hinton, 1979; Phillies *et al.*, 1975), Ca^{2+} (Vishwanath and Easwaran, 1981), Ba^{2+}, NH_4^+, NH_3OH^+ and $C(NH_2)_3^+$ (Asher *et al.*, 1977; Phillies *et al.*, 1975) found no significant differences from the structures presented in Figs 11 and 12. The characteristic "tennis ball seam" conformation of the naction complexes is replaced by flatter structures on ion removal (Anteunis and Bruyn, 1977; Dobler, 1972; Kyogoku *et al.*, 1975; Nawata *et al.*, 1974, 1980; Prestegard and Chan, 1969, 1970).

Similarly to valinomycin, nactins serve as popular tools for studying transport mechanisms across bilayers (Ciani *et al.*, 1973; Eisenman and Krasne, 1975; Hladky *et al.*, 1974; Szabo *et al.*, 1973), thick layers (Lev *et al.*, 1973; Wipf *et al.*, 1970) and in two

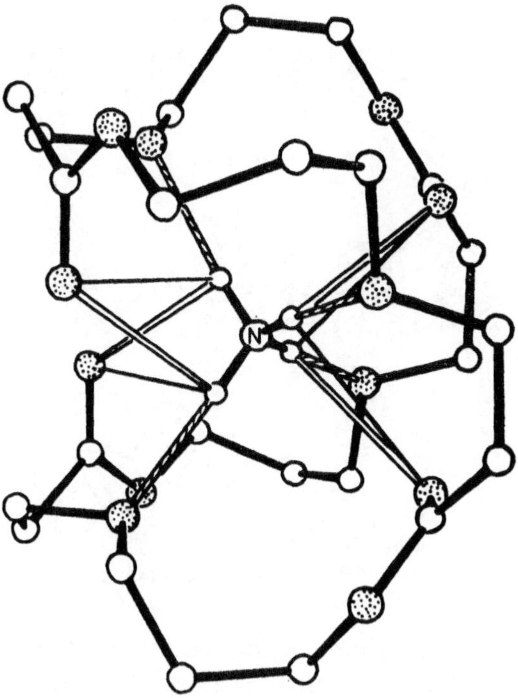

FIG. 12. Hydrogen (hatched) and ion-dipole bonds in the crystalline tetranactin · NH_4^+ complex (only the macrocyclic skeleton is shown for clarity).

phase systems (Eisenman *et al.*, 1969; Haynes and Pressman, 1974; Szabo *et al.*, 1969). In all cases nactins behave as typical cage carriers with the above shown selectivity order.

e) *Carboxylic ionophores*

Polyether antibiotics comprise a seemingly heterogeneous, already large and still rapidly growing family of the so-called carboxylic ion carriers. Among the popular members of that group are the Na selective monensin, the K selective nigericin and the Ca selective calcimycin (known also as antibiotic A 23187) (Fig. 13). Chemistry and biology of carboxylic ionophores are covered in the recent monograph (Westley, 1982-83).

In spite of the wealth of accumulated data general structure-functional regularities are not easily discernible in this area. One of the few suggestions of that sort, advanced by Duax *et al.* (1983), provides a clue to the dramatic difference in monensin

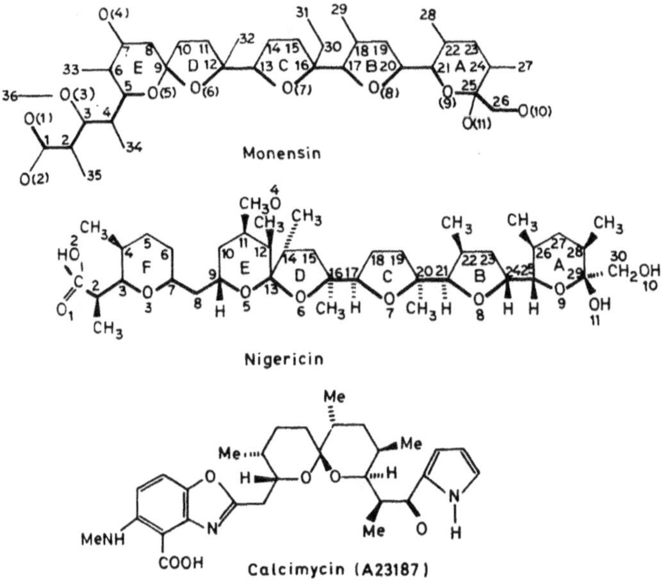

FIG. 13. Chemical formula of selected carboxylic ionophores.

and nigericin selectivities: Na/K for the former and K/Na for the latter, in both cases ca. 10-fold (Lutz *et al.*, 1970). Comparing the crystalline structures of Na complexes (Fig. 14) the authors noted that the overall number of liganding oxygens is 6 in monensin and 5 in nigericin. At the same time judging from the

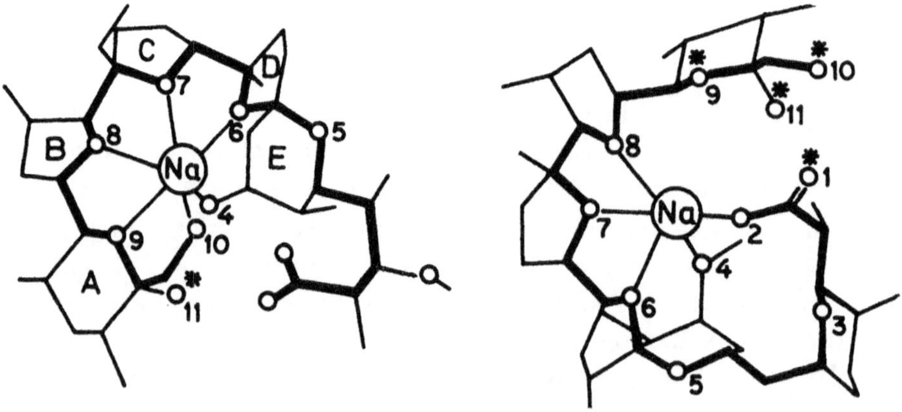

FIG. 14. Three-dimensional structure of the sodium salts of monensin (a) and nigericin (b). Potential (after minor conformational rearrangement) metal binding 0-atoms are marked with the asterisk.

potential total number of oxygens in the coordination sphere, the number of actually binding oxygens with K^+ might reach 7 with monensin and 8 with nigericin. Such a change in the balance of ion binding oxygens explains the shift in selectivity.

Calcimycin contains a unique heterocyclic moiety which together with the carboxylated anion directly interacts with the bound calcium in the crystalline "sandwich" complex (Fig. 15). Theoretical treatment of the origins of ionic selectivity in this case still remains to be carried out (Chaney *et al.*, 1976; Smith and Duax, 1976).

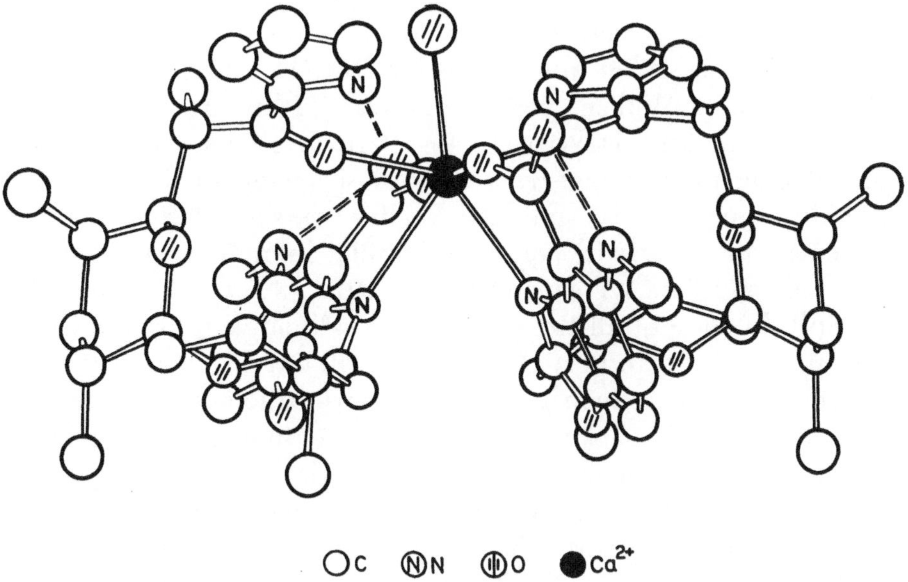

FIG. 15. Conformation of the crystalline complex of calcimycin with Ca^{2+}.

f) *Crown ethers*

The vast class of synthetic macrocyclic polyethers provides a variety of approaches to selective ion binding and transport. Many hundreds and possibly thousands of such chemicals have already been prepared, and their number continues to grow with a high rate (Bradshaw and Scott, 1980; Izatt and Christensen, 1979). Already in 1967 Pedersen in his pioneering work has shown that many crown ethers form stable complexes with a broad circle of cations. The first generation of crown ethers (monocyclic, containing exclusively oxygens as metal binding atoms) was followed by nitrogen and sulfur containing ligands and by bi- and tricyclic cryptands of Lehn *et al.* (1973, 1978, 1979, 1983). Introduction of additional binding moieties and steric barriers at strategic sites has led to topo-selective ligands imitating enzymes in some cases (Chao and Cram, 1976; Cram and Cram, 1978; Lehn *et al.*, 1973, 1978, 1979, 1983).

Crown ethers have found application in enantiomeric separation of chiral cations (Cram and Trueblood, 1981; Tolley *et al.*, 1982), transmembrane ion transport (McLanghlin *et al.*, 1972), building of ion selective electrodes (Kimura *et al.*, 1979; Ryba and Petranek, 1973) and even for isotope enrichment of Na (Knöchel and Wilken, 1981) and Ca (Heumann and Schefer, 1980; Jepson and DeWitt, 1979). Crown ethers are also indispensable agents in phase-transfer catalysis owing to their chemical stability, availability and the ability to solubilize in organic solvents the otherwise quite insoluble inorganic salts (Starks and Liotta, 1978; Weber and Gokel, 1977).

Still, in spite of all the progress the valinomycin-like selectivity, not speaking of transport efficiency, is still out of reach in the field of crown ethers. Such features of crown ethers as relative flexibility of the cyclic backbone and incomplete screening of the bound cation from the solvent and the counterion may be responsible for that fact. DeJong and Reinhoudt (1981) demonstrate that as with other ionophores the maximal binding capacity requires perfect fit of the size of the bound ion to the cavity size (Fig. 16). Nonetheless such an approach when applied in a straightforward way often fails to produce correct predictions. Neglect of solvent interactions is the main cause of such difficulties (see e.g. Michaux and Reisse, 1982).

g) *Factors contributing to ionic selectivity*

Let us consider in more detail various factors contributing to the overall free energy of complexation. It is well known that desolvation energy $\Delta F_{desolv.}$ of ion M_i monotonously decreases with the increase in the ion size. When a macrocycle interacts with the ion, flatter dependence is expected for small ions (steric hindrance does not allow perfect fit of the cavity size) and steeper dependence — for large ions (too large to " squeeze in "). As a

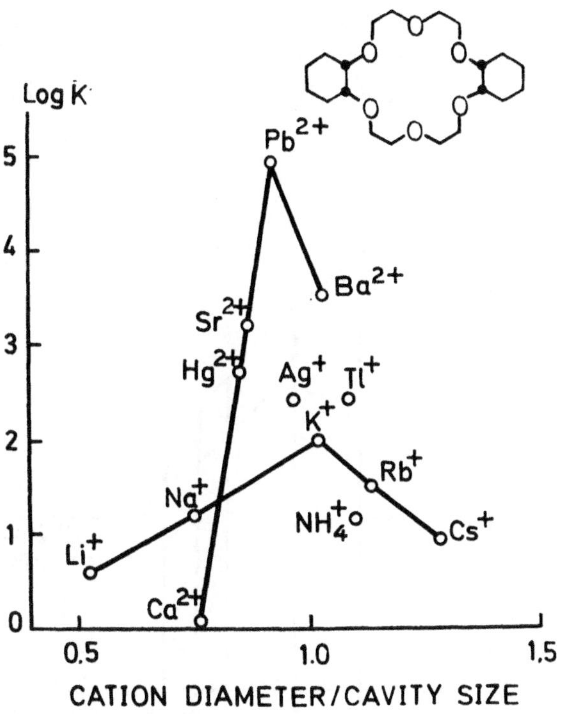

Fig. 16. Relation between logK and the ratio of cation diameter to cavity size for (cis-syn-cis)-dicyclohexyl-18-crown-6 in water at 25° C.

result, the free energy difference between these two processes which provide by far the largest contribution to the overall binding energy (i.e. to the complex stability) is expected to reach the maximum in the area of geometrical fit (Fig. 17). This, we believe, is the origin of the empirical observation discussed above on the example of valinomycin depsipeptides (Tables 2 and 3) and crown ethers (Fig. 16). Other factors should be considered when more detailed analysis is attempted, in particular, the energy of conformational transitions of the ligand which inevitably accompany ion binding ($\Delta F_{conf.}$) and the term which reflects solvation differences of the free ligand and of the complex ($\Delta F_{solv.}$) (p. 229 in Ovchinnikov *et al.*, 1974):

$$\Delta F_{compl.} = \Delta F_{ion\text{-}dipole} + \Delta F_{desolv.} + \Delta F_{conf.} + \Delta F_{solv.}$$

In other words, ionic selectivity $\Delta(\Delta F_{compl.})^{M_1}_{M_2}$ is dependent upon the differences in conformational energies of the two states, $\Delta(\Delta F_{conf.})^{M_1}_{M_2}$, and upon the differences in solvent interactions, $\Delta(\Delta F_{solv.})^{M_1}_{M_2}$. From the data presented above it also follows

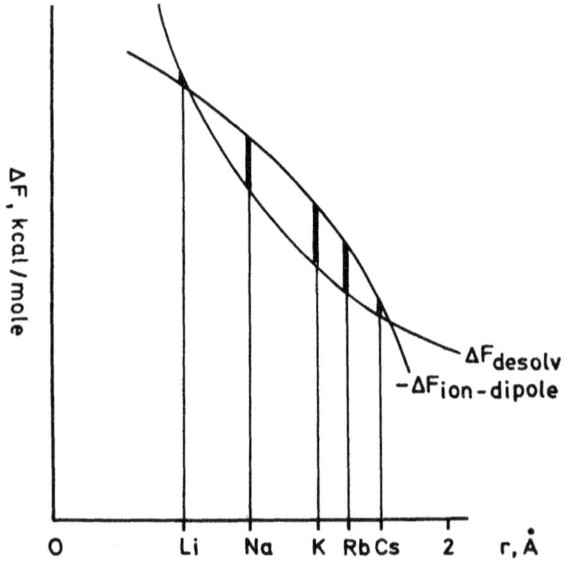

FIG. 17. Dependence of the free energy of desolvation ($-\Delta F_{desolv.}$) and the free energy of ion-dipole interaction of the desolvated cation with the macrocyclic ligand ($\Delta F_{ion\text{-}dipole}$) from the size of the cation.

that the real situation is often further complicated by variable stoichiometry of complexation.

It should be noted that electronic properties of the ion binding moieties of the ligand are not easily interpretable at present state of theoretical treatment. At the same time in a number of cases this factor plays a crucial role. Proton carriers often used as uncouplers of oxidative phosphorylation in mitochondria can serve as an obvious example. Usually these are weak organic acids which penetrate the membrane and reversibly dissociate into the lipophylic anion and the proton. The widely used proton carriers are e.g. dinitrophenol and carbonyl-cyanide-p-trifluoromethoxy-phenylhydrazone (so-called FCCP):

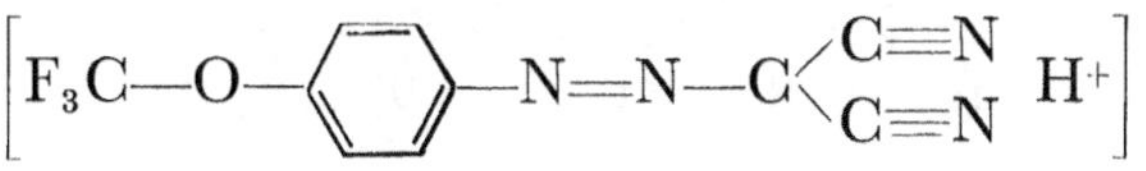

Replacements 0 → S in crown ethers bring about stronger binding of transition metal ions, such as Ag^{+} and Pb^{2+}, due to increase in covalent interactions, complexation with alkali metal ions being rather insensitive to such replacements (p. 21 in DeJong and Reinhoudt, 1981).

It is also tempting to ascribe the Ca selectivity of calcimycin to participation of the unusual binding atom, the heterocyclic nitrogen in the coordination sphere (see Fig. 15).

2. Channel formers

Although natural ionophores continue to stimulate a great deal of biological research it is becoming increasingly clear that the normal, physiological ion transport across biological membranes is performed by channel forming proteins. To such proteins belongs the light dependent proton pump, bacteriorhodopsin, the acetylcholine receptor of postsynaptic nerve membranes, ion-conducting units of excitable membranes, transport ATP-ases performing active transport of protons, the Na, K or Mg, Ca antiport at the expense of ATP hydrolysis, etc. (see Ovchinnikov, 1981 and references cited therein). All these proteins form supramolecular aggregates spanning the membrane matrix and containing a molecular pore capable of accommodating the moving ion. Molecular structure of these aggregates is one of the key problems of modern membranology. In view of extreme complexity of the problem considerable effort is spent on studies of model channel formers which successfully imitate particular functional aspects of their protein prototypes. Being at the same time of much lower molecular weight they proved more amenable to detailed physicochemical study. Typical representatives of the three best studied families of model channel formers are shown below.

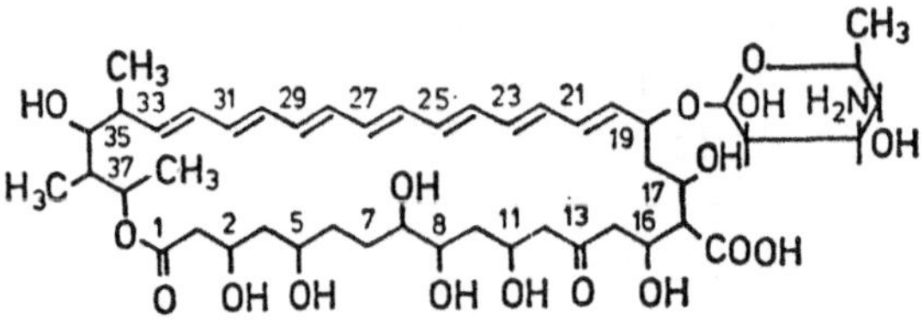

AMPHOTERICIN B

L L D L D L D L D L D L D L

OHC-VAL-GLY-ALA-LEU-ALA-VAL-VAL-VAL-TRY-LEU-TRY-LEU-TRY-LEU-TRY-NHCH$_2$CH$_2$OH

GRAMICIDIN A

AC-AIB-PRO-AIB-ALA-AIB-ALA-GLN-AIB-VAL-AIB-GLY-LEU-AIB-PRO-VAL-AIB-AIB-GLU-GLN-PHOL

ALAMETHICIN I

a) *Macrolide antibiotics*

Macrolide (or polyene) antibiotics form transmembrane pores filled with water and permeable both, to anions and — less readily — to cations. Nonelectrolytes also pass through the pore if their effective radius is less than 4 Å (for instance, glucose is too large to penetrate the pore) (Finkelstein and Holz, 1973). Cholesterol must be present in the membrane in order to form the pore. Moreover, the macrolide, e.g. amphotericin B, has to be added from both sides of the membrane if lipid bilayer is used as the test system (Ermishkin *et al.*, 1977). The number of pores is highly concentration dependent (exponential dependence of the 4-12 order).

In accordance with the above mentioned anionic selectivity, ion currents through the channel formed by amphotericin B are anion dependent and cation independent (Fig. 18) (Ermishkin *et al.*, 1977).

A plausible model has been advanced which accords with the available experimental evidence (Finkelsein *et al.*, 1973; DeKruyff and Demel, 1974). According to that model (Fig. 19) the channel (pore) consists of two end-to-end joined cylinders

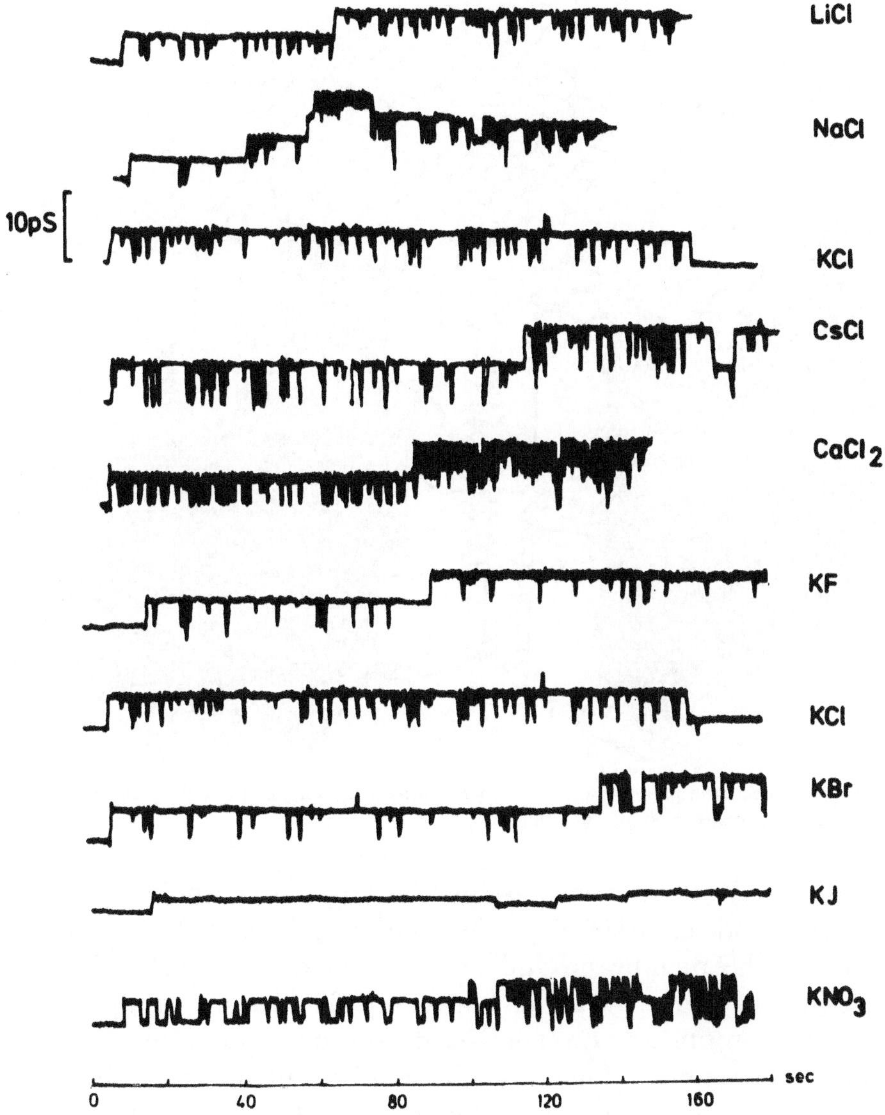

FIG. 18. Time dependence of amphotericin B induced membrane conductivity at 200 mV potential in 2M salt solutions. Concentration of amphotericin B $(0.5\text{-}5.0)\cdot 10^{-7}$ M, total brain phospholipids.

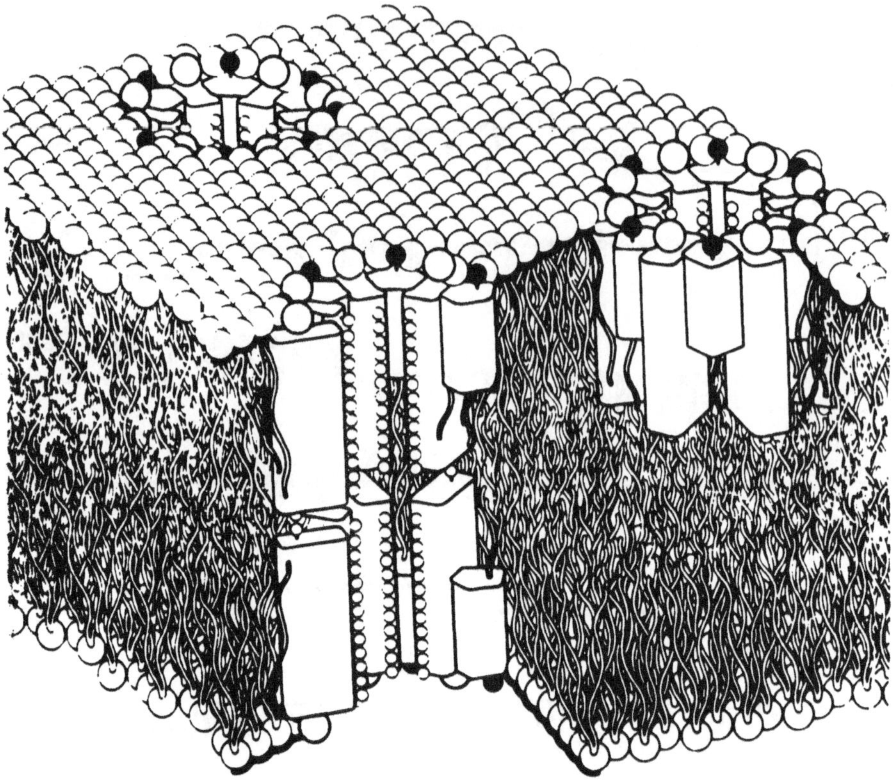

FIG. 19. Schematic presentation of the pores formed by amphotericin B in a lipid (the half-pore and a cross-section of the full pore).

(half-pores), formed in turn by the alternating molecules of antibiotic and cholesterol. The half-pores assemble at opposite sides of the membrane and are connected with each other with the help of hydrogen bonds between C-35 hydroxy groups. Other hydroxy groups are directed inside the pore providing its hydrophilic character. Hopefully, this model will be of use in further unravelling of the structure-functional relationship of macrolide antibiotics.

b) *Gramicidin A*

This linear peptide antibiotic is, beyond doubt, the best so far characterized model channel former. In a great variety of biological membranes and lipid bilayers it forms narrow pores permeable to alkali metal ions, protons, ammonium and impermeable bivalent metal ions and to anions (see p. 310 in Ovchinnikov *et al.*, 1974, and references therein). Conductivity of gramicidin A channels in various lipid bilayers follows the order

$$H > K, Rb, Cs, NH_4 > Na > Li$$

Membrane thickness only slightly affects the single channel conductivity, but the channel mean life-time sharply falls in thicker bilayers (Table 5).

TABLE 5. *Conductance (at 100 mV and 23° C) and Mean Duration of the Single Channel of Gramicidin A as a Function of Membrane Thickness (Hladky and Haydon, 1972)*

Membrane-Forming Lipid Solution	Hydrocarbon Thickness (Å)	Single-Channel Conductance in M/2 NaCl ($\Omega^{-1} \times 10^{11}$)	Mean Duration of Single Channel (sec)
gmpo[1] + *n*-hexadecane	26	1.7	∾60
gmo[2] + *n*-hexadecane	31	1.7	2.2
gmo + *n*-tetradecane	40	1.7	1.3
gmo + *n*-decane	47	1.7	0.4
phsa[3] + gmo + *n*-decane	∾64	1.7 0.8	∾0.03

[1] glyceryl monopalmitoleate
[2] glyceryl monooleate
[3] polyhydroxystearic acid

Veatch *et al.* (1975) and Veatch and Stryer (1977) have shown that the gramicidin channel is built of two antibiotic molecules. Two principal structures were proposed for this dimer. According to Ramachandran and Chandrasekharan (1972a, b), Urry (1971) and Urry *et al.* (1971) the channel is a single stranded, so-called π_{LD}-helix wherein the monomers are connected in a head-to-head manner with 4-10 intermolecular hydrogen bonds. In the alternative model of Veatch *et al.* (1974) the channel is a double helix called $\pi\pi_{LD}$-helix, in which all facing each other amino acid residues of the two monomers are connected with intermolecular hydrogen bonds. Some of these structures, which seem to accord with the known properties of the channel, are depicted in Fig. 20.

Both models possess similar geometrical parameters, both have hydrophobic outer surface and the axial cavity lined with

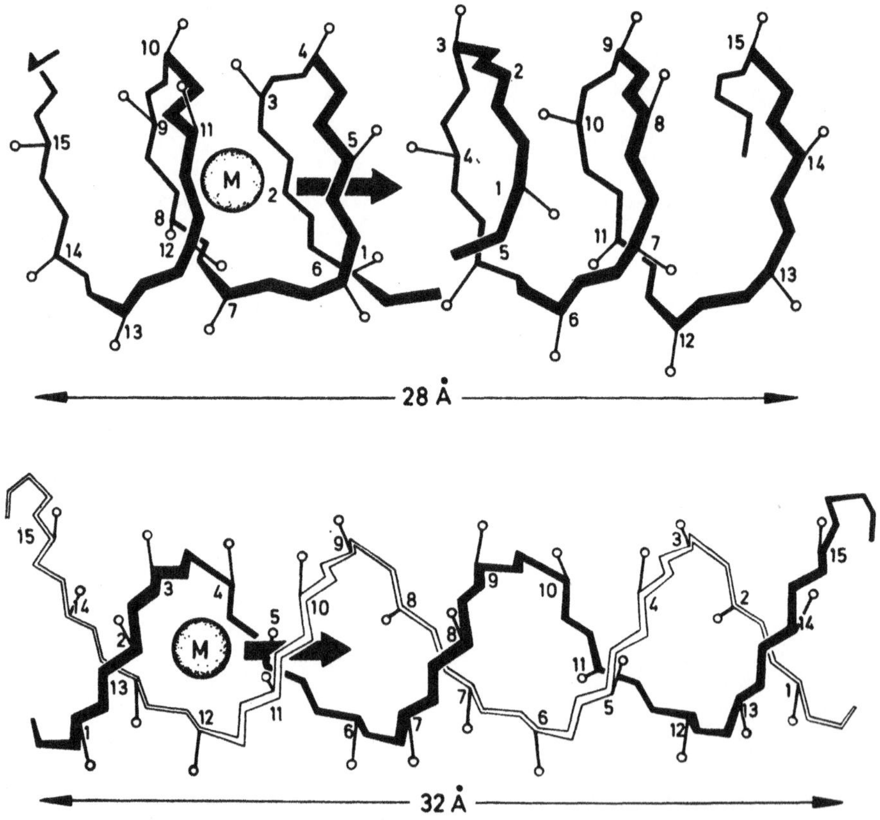

FIG. 20. Left-handed $\overrightarrow{\pi_{LD}^{6.3}}\,\overleftarrow{\pi_{LD}^{6.3}}$ (a) and $\downarrow\uparrow\pi\pi_{LD}^{5.6}$ (b) helical dimers of gramicidin A (the hydrogen bonds and the amino acid side chains are omitted for clarity).

polar amide moieties and suitable for accommodating the partially solvated metal ion or hydoxonium, H_3O^+.

With the aid of spectral studies we were able to show that in solution and in lipid membranes both helical types are present and that the double helices dominate the equilibrium (Ivanov, 1983; Ivanov and Sychev, 1982; Ovchinnikov and Ivanov, 1982;

Sychev and Ivanov, 1982)[1]. A "zipper" or "sliding" mechanism was hypothetically suggested (Urry *et al.*, 1975) and experimentally supported (Ivanov and Sychev, 1982; Ovchinnikov and Ivanov, 1982a) for interconversion of single and double stranded helices (Fig. 21). A model was developed of gramicidin A function (Fig. 22) according to which both helical types contribute to the overall

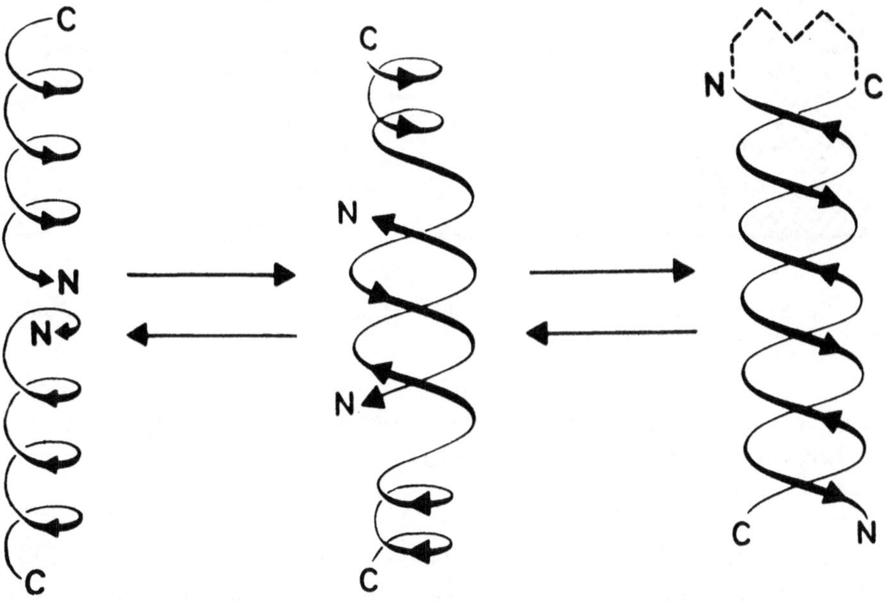

FIG. 21. "Zipper" mechanism of antiparallel double helix formation by gramicidin A.

ionic current proportionally to their molar fraction in the given membrane system (Ivanov and Sychev, 1983). Single stranded helices can dissociate into the monomers that results in switching off of the channel (b → a). The double stranded dimer (c) can fully or partially "unscrew" giving rise to the conducting dimer (b) or, if "narrow" helices are formed at the ends, to non-con-

[1] In a recent paper Urry *et al.* (1983) doubted that conclusion. However, the experimental data cited in that paper were obtained with the high gramicidin: lipid ratio, i.e. at conditions where high molecular aggregates of the antibiotic are formed. In these aggregates gramicidin A has quite different conformation than in the dimer. Double helical conformation of gramicidin A was unequivocally proved for a dioxane solution by two-dimensional NMR spectroscopy (Arseniev *et al.*, 1983).

ducting dimer (d). Both processes, b → a and c → d, are coupled with fluctuation of membrane thickness which therefore is a driving force of switching the channel. The model is in accord with diminishing channel life-times in thicker membranes (Table 5). Further support came from studies of N-succinyl (n = 2) and

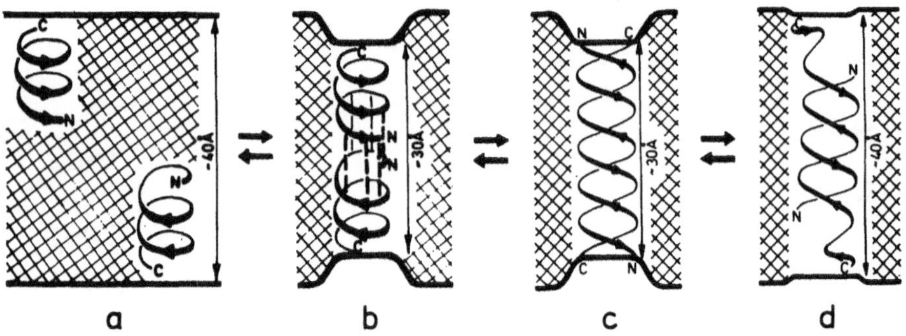

FIG. 22. Switching on and off of the single stranded (a, b) and double stranded (c, d) helical dimers of gramicidin A.

N-suberyl (n = 6) gramicidin A derivatives having additional carboxylic function at N-terminus:

$$HOOC(CH_2)_nCO\text{-Val-}\ldots\text{-Trp-}NHCH_2CH_2OH$$

It was found (Ivanov, 1983; Ivanov and Sychev, 1982, 1983; Ovchinnikov and Ivanov, 1982a) that in lipid membranes (liposomes) these derivatives have higher than in the parent peptide content of single stranded helices, apparently due to additional carboxyl-to-carboxyl hydrogen bonds (Fig. 23). In good agreement with that fact, kinetic measurements of Apell *et al.* (1979) revealed increased contribution of bimolecular reactions to the switching off process (transition b → a).

Interesting results, also consistent with the above concept, were obtained with bis-gramicidin A analogs, wherein the dimeric structure is fixed by covalent bridge in a N-terminus to N-terminus (head-to-head), N-terminus to C-terminus (head-to-tail) or C-terminus to C-terminus (tail-to-tail) manner.

$$HOCH_2CH_2NH\text{-Trp-}\ldots\text{-Val-}CO(CH_2)_nCO\text{-Val}\ldots$$
$$\ldots\text{-Trp-}NHCH_2CH_2OH$$

Head-to-head, n = 2, 3, 4, 6 or 7

OHC-Val-. . .-Trp-$NHCH_2CH_2$-$OCO(CH_2)_nCO$-Val- . . .
. . .Trp-$NHCH_2CH_2OH$

Head-to-tail, n = 2, 3, 4, 6 or 7

OHC-Val-. . .-Trp-$NHCH_2CH_2$-$OCO(CH_2)_nCOO$-CH_2CH_2NH-
Trp-. . .-Val-CHO

Tail-to-tail, n = 2, 3 or 7

All three types of bis-derivatives proved efficient channel formers, both in single channel and noise analysis experiments (Fonina *et al.*, 1981; Irkhin, 1982; Ivanov and Sychev, 1982;

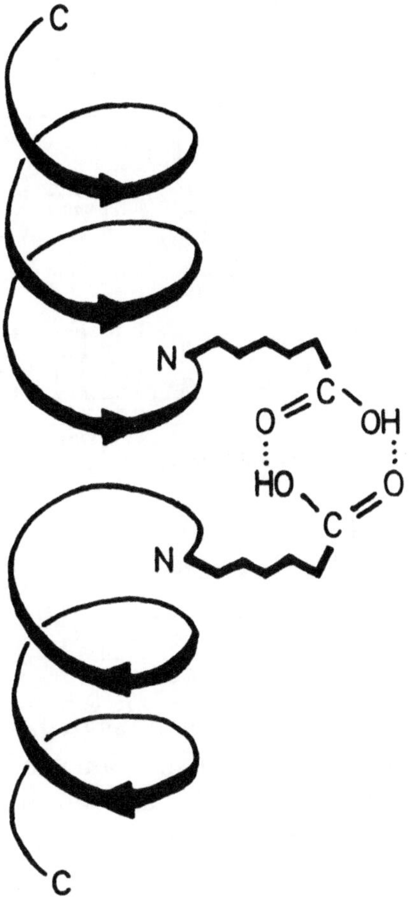

FIG. 23. Stabilization of single stranded helices in carboxyl containing gramicidin A analogs.

Ovchinnikov and Ivanov, 1982a). Selected results are shown in Fig. 24 and Table 6. Spectral studies have shown that both in solution and in the membrane all analogs preferably form single stranded helical structures if intermolecular association is avoided (e.g. at single-channel membrane measurements). At higher

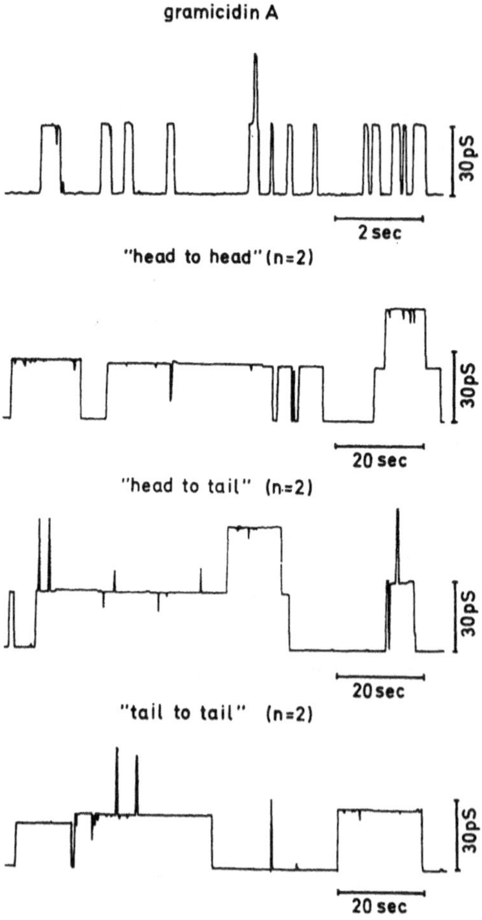

FIG. 24. Single-channel conductance recordings of gramicidin A and its bis-derivatives. Glycerolmonooleate/cholesterol (2/1 molar ratio) membranes were used with an area of $6 \cdot 10^{-4}$ cm^2 in 0.5 M KCl at 21° C. The applied voltage was 100 mV, the feed-back resistor of the amplifier was $10^9\ \Omega$.

concentrations (e.g. at noise analysis measurements) aggregates are formed which are predominantly double-helical. Since neither of the b → a and c → d transitions (Fig. 7) are possible for monomeric bis-derivatives, life-times of the channels formed by these dimers are much longer than for gramicidin A at the same

TABLE 6. *Channel Parameters of Lipid Bilayers Modified by Gramicidin A and its Bis-derivatives*

Compound	Single channel measurements [1]			Noise analysis [2]	
	Minimal active concentration,	Single channel conductance,	Mean channel life-time,	Single channel conductance,	Mean channel life-time,
	M	pS	Sec	pS	Sec
Gramicidin A	10^{-13}	30.2	0.2	18	0.45
Head-to-head, n = 2	10^{-15}	24.2	12.6	—	—
Head-to-head, n = 4	10^{-16}	24.6	10.1	—	—
Head-to-head, n = 7	10^{-11}	24	∾9	—	—
Head-to-tail, n = 2	10^{-13}	24.3	9.7	—	—
		30.3	∾0.2	12	0.55
Tail-to-tail, n = 2	10^{-13}	24.8	7.5	—	—
		29.4	∾0.2	12	1.42

[1] Conditions see in caption to Fig. 24.

[2] Membranes having an area 6×10^{-4} cm^2 were formed from dioleoyl-phosphatidylcholine/cholesterol 1.8/0.2 (percentage) solution in n-decane. The KCl concentration was 1 M, T = 21°C. Mean macroscopic conductance was up to $10^{-2}\,\Omega^{-1}\,cm^{-2}$.

condition ($\sim$10 sec instead of 0.2 sec). With aggregates of bis-derivatives the same regularities are observed as with gramicidin A itself, hence the similarity of respective life-times observed in noise-analysis experiments (Table 6). It should be also noted that selectivity sequences found for all bis-derivatives closely reproduce the above shown sequence of gramicidin A (Fonina *et al.*, 1981; Irkhin, 1982). This is considered as a consequence as well as a proof of close similarity of the pore parameters for all types of junction of polypeptide chains.

c) *Alamethicin I*

Alamethicin I represents a large family of membrane active antibiotics (alamethicin, suzukacillins, antiamoebins, emerimicins, trichotoxin, hypelcins, etc, see e.g. the review by Nagaraj and Balaram, 1981). As seen from the formula, the N-terminus of alamethicin is acetylated and the C-terminus is occupied with phenylalaninol (Phol); a number of unusaul amino acid residues (Aib, α-amino isobutyric acid or α-methyl alanine) is also present. A fundamental property of alamethicin induced conductivity in lipid bilayers is a strong potential dependence. In the absence of potential the conductivity is zero, but starting from certain level each 4-8 mV increment increases *e*-fold the number of channels. In this respect alamethicin is a valuable model of the ionic channels of excitable membranes.

Single-channel recordings (Fig. 25) reveal another unique feature of the alamethicin channel, viz., existence of a number of interconverting conducting states, differing from each other in their conductivity (see the review by Ehrestein and Lecar, 1977, and references therein). The lower state resembles the gramicidin channel in a number of parameters, such as conductivity (19 pS for K^+, cf. Table 6) and impermeability to bivalent ions and anions. Conductivity of other states varies from 280 to 6200 pS (for K^+), and they all conduct alkali, alkaline earth cations as well as various anions (Hanke and Boheim, 1980); the pore diameter reaches 15 Å in the highest conducting state (Boheim, 1974).

There is no doubt that the multistate nature of the alamethicin channel is due to formation of aggregates of increasing molecular weight, each of which contributes to the overall conductivity and has its own set of properties. A number of proposals have been suggested in an attempt to relate the structure of ala-

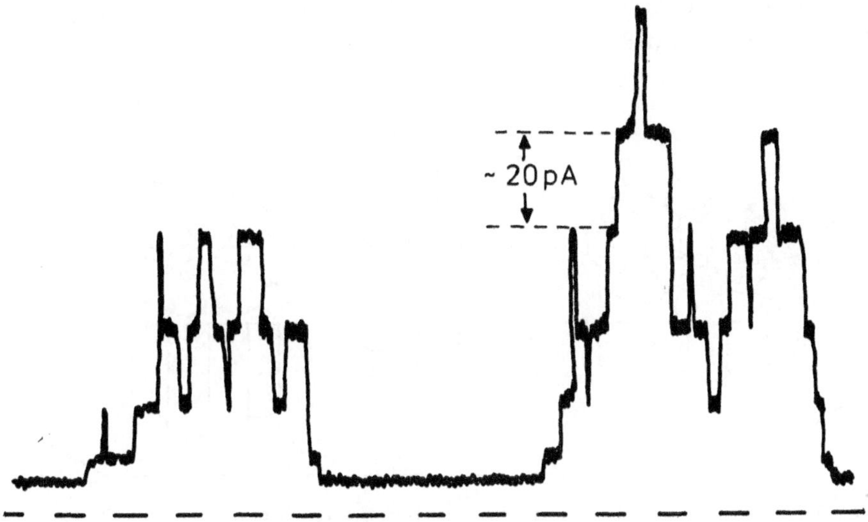

FIG. 25. Typical multiple step fluctuations of the alamethicin channel in a lipid bilayer. Recording time is 0.1-10 sec, depending on the lipid.

methicin with its channel forming properties. Two of them will be mentioned below. Boheim *et al.* (1983) consider that in the absence of applied potential alamethicin, which is predominantly α-helical, forms aggregates where monomers are arranged in pairs of alternating (up-down) orientation of the α-helical dipole (which is quite large). This state is non-conducting. When voltage is applied a uniform orientation of α-helices in the membrane is forced (flip-flop hypothesis), giving rise to the observed set of conducting aggregates.

X-ray studies of Fox and Richards (1982) have shown that at least in the crystalline state alamethicin I is indeed a α-helix, except for the Pro-containing segment 11-14, where a bend is observed. The authors further suggested that alamethicin I has a similar conformation in the aggregates which can vary in size. A decamer is shown in Fig 26; a driving force of aggregation is formation of hydrogen bonded network between the Gln carboxamido moieties. In the absence of potential only the 1-11 segment is merged into the membrane while the bend and the C-terminal helix are placed at the interface. When potential is applied the structure is straightened up (the middle part of Fig. 26) and shifted along the electrical field to the opposite side

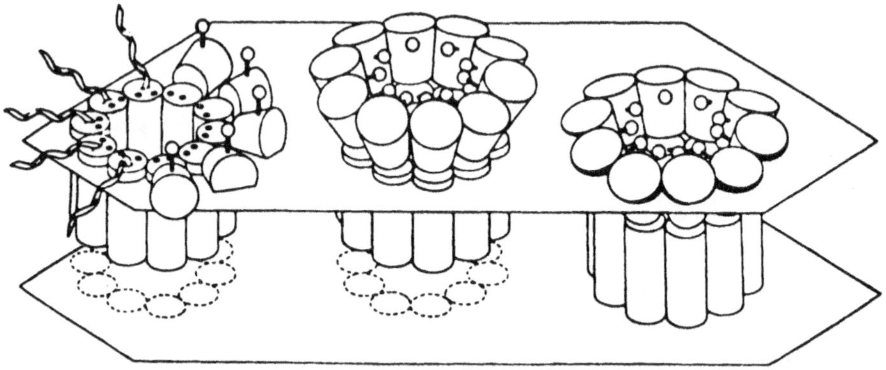

FIG. 26. Voltage-gating channel model. The cork-shaped elements represent the C-terminal region of alamethicin, bent away from the channel axis. The spheres represent the carboxyl functions of Gln residues.

of the membrane. The channel is switched on. The same motive is utilized in this model as with gramicidin A: in one way or another the peptide must reach both surfaces of the bilayer in order to open the channel. It well might be that this principle is operative with other model channels.

In conclusion it should be noted that even the simplest, model channel formers pose formidable problems in the course of their structure-functional studies. Nonetheless the present day thinking of selectivity and transport in biological systems is to a large extent influenced by the results of such studies. There are all grounds to expect further development of this exciting field in the near future.

REFERENCES

Andreev, I.M., Malenkov, G.G., Shkrob, A.M. and Shemyakin, M.M., 1971, Mol. Biol. (Moscow), *5*, 614.

Anteunis, M.J.O. and De Bryun, 1977, Bull. Soc. Chim. Belg., *86*, 445.

Apell, H.J., Bamberg, E. and Alpes, H., 1979, J. Membr. Biol., *50*, 271.

Arseniev, A.S., Bystrov, V.F., Ivanov, V.T. and Ovchinnikov, Yu. A., 1984, FEBS Lett., *1*, 165, 51.

Asher, I.M., Phillies, G.D.J., Kim, B.J. and Stanley, H.E., 1977, Biopolymers, *16*, 157.

Avotin, G.Ya., Fonina, L.A., Ivanov, V.T. and Ovchinnikov, Yu.A., 1978, Bioorg. Khim., *4*, 581.

Balashova, T.A., Fonina, L.A., Gurevich, A.Z., Starovoitova, N.V., Senyavina, L.B., Avotin G.Ya., Ivanov, V.T. and Ovchinnikov, Yu.A., 1983, Bioorg. Khim. *9*, 883.

Balashova, T.A. et al., 1983, Bioorg. Khim. (in press).

Baron, D., Pease, L.G. and Blant, E.R., 1977, J. Amer. Chem. Soc., *99*, 8299.

Benz, R., 1978, J. Membr., *43*, 367.

Benz, R., Gisin, B.F., Ting-Beall, H.P., Tosteson, D.C. and Läuger, D., 1976, Biochim. Biophys. Acta, *445*, 665.

Boheim, G., 1974, J. Membr. Biol., *19*, 277.

Boheim, G., Hanke, W. and Jung, G., 1983, Biochim. Biophys. Acta, (in press).

Braden, B., Hamilton, J.A., Sabesan, M.N. and Steinrauf, L.K., 1980, J. Amer. Chem. Soc., *102*, 2704.

Bradshaw, J.S. and Scott, P.E., 1980, Tetrahedron, *26*, 461.

Brenzina, M., Hofmanova-Matejkova, A. and Koryta, J., 1974, Biophys. Chem., *2*, 264.

Briggs, R.W. and Hinton, J.F., 1979, J. Magn. Res., *33*, 363.

Cornelius, G. Grätner, W. and Haynes, D.H., 1974, Biochemistry, 13, 3052.

Chaney, M.O., Jenes, N.D. and Debono, M., 1976, J. Antib., *29*, 424.

Chao, Y. and Cram, D.J., 1976, J. Amer. Chem. Soc., *98*, 1015.

Chin, Y.H., Brown, L.D. and Lipskomb, W.N., 1977, J. Amer. Chem. Soc., *99*, 4799.

Ciani, S.M., Eisenman, G., Laprade, R. and Szabo, G., 1973, in: Membranes — A Series Advances. V. 2. Eisenman, G. Ed., N.Y.: Marcel Dekker, p. 61.

Cram, D.J. and Cram, J.M., 1978, Acc. Chem. Res., *11*, 8.

Cram, D.J. and Trueblood, K.N., 1981, Topics in Current Chemistry, *98*, 43.

Davis, D.G. and Tosteson, D.C., 1975, Biochemistry, *14*, 3962.

DeJong, F. and Reinhoudt, 1981, Advances in Physical Organic Chemistry. V. 17. London-N.Y.: Acad. Press.

DeKruyff, B. and Demel, R.A., 1974, Biochim. Biophys. Acta, *339*, 57.

Devarajan, S. and Easwaran, K.P.K., 1981, Biopolymers, *20*, 891.

Devarajan, S., Nair, C.M.K., Easwaran, K.P.K. and Vijayan, M., 1980, Nature, *286*, 640.

Dobler, M., 1972, Helv. Chim. Acta, *55*, 1371.

Dobler, M., Dunitz, J.D. and Kilbourn, B.T., 1969, Helv. Chim. Acta, *52*, 2573.

Dobler, M., Dunitz, J.D. and Krajewski, J. ,1969, J. Mol. Biol., *42*, 603.

Dobler, M. and Phizackerley, R.D., 1974, Helv. Chim. Acta, *57*, 664.

Duax, W.L., Smith, G.D. and Van Roey, P., 1983, in: Physical Chemistry of Trans-membrane Ion Motions. Spach, G. Ed., Amsterdam: Elsevier, p. 217.

Eastman, M.P., 1974, Chem. Commun., p. 789.

Easwaran, K.P.K., Pease, L.G. and Blant E.R., 1979, Biochemistry, *18*, 61.

Ehrenstein, G. and Lecar, H., 1977, Ann. Rev. Biophys., *10*, 1.

Eisenman, G., Ciani, S. and Szabo, G., 1969, J. Membr. Biol., *1*, 295.

Eisenman, G. and Krasne, S.J., 1975, in Fox, MTP Int. review of science, Biochemistry series one. V. 2. London: Butterworth, p. 27.

Ermishkin, L.N., Kasumov, K.M. and Potseluev, V.M., 1977, Biochim. Biophys. Acta, *470*, 357.

Feinstein, M.B. and Felsenfeld, H., 1971, Proc. Natl. Acad. Sci. USA, *68*, 2037.

Finkelstein, A. and Holz, R., 1973, in: Membranes — A Series Advances. V. 2. Eisenman, G. Ed., N.Y.: Marcel Dekker, p. 378.

Fonina, L.A., Avotin, G.Ya., Balashova, T.A., Starovoitova, N.V., Senyavina, L.B., Savelov, I.S., Bystrov, V.F., Ivanov, V.T. and Ovchinnikov, Yu.A., 1980, Bioorg. Khim., *6*, 1285.

Fonina, L.A., Demina, A.M., Hlavacek, Ya., Sychev, S.V., Irkhin, A.I., Melnik, E.I. and Ivanov, V.T., 1981, in: Peptides 1980, Brunfeldt, K. Ed., Copenhagen: Scriptor, p. 469.

Fox, R.O. and Richards, F.M., 1982, Nature *300*, 325.

Geddes, A.J. and Akrigg, D., 1976, Acta Crystallogr., Sect. B., *B32*, 3164.

Gizin, B.F., Ting-Beall, H.P., Davis, D.G., Grell, E. and Tosteson, D.C., 1978, Biochim. Biophys. Acta *509*, 201.

Golikov, A.G. and Vichutinski, A.A., 1981, DAN SSSR, *260*, 219.

Greel, E. Funk, T. and Eggers, F., 1972, in: Molecular Mechanisms of Antibiotic Action on Protein Biosynthesis and Membranes. Munoz, E., Garcia-Ferrandiz, F. and Vazquez, D. Eds., Amsterdam: Elsevier, p. 646.

Grell, E., Funk, T. and Eggers, F., 1975, in Membranes. V. 3. Eisenman, G. Ed., N.Y.: Marcel Dekker, p. 2.

Hamilton, J.A., Sabesan, M.N. and Steinrauf, L.K., 1980, Acta Crystallogr., Sect. B., *B36*, 1052.

Hamilton, J.A., Sabesan, M.N. and Steinrauf, L.K., 1981, J. Amer. Chem. Soc., *103*, 5880.

Hamilton, J.A., Steinrauf, L.K. and Braden, B., 1975, Biochem. Biophys. Res. Commu., *64*, 151.

Hanke, W. and Boheim, G., 1980, Biochim. Biophys. Acta, 596, 456.

Haynes, D.H. and Pressman, B.C., 1974, J. Membr. Biol., *18*, 1.

Haynes, D.H., Pressman, B.C. and Kowalsky, A., 1971, Biochemistry, *10*, 852.

Hladky, S.B., Gordon, L.G.M. and Haydon, D.A., 1974, Ann. Rev. Phys, Chem., *25*, 11.

Hladky, S.B. and Haydon, D.A., 1972, Biochim. Biophys. Acta, *274*, 294.

Heumann, K.G. and Schiefer, H.P., 1980, Angew. Chem. *92*, 406.

Irkhin, A.I., 1982, Dissertation, Moscow.

Ivanov, V.T., 1975, Ann. N.Y. Acad. Sci., *264*, 221.

Ivanov, V.T., 1983, in: Peptides 1982. Blaho, K. and Malon, P. Eds, Berlin-N.Y.: De Gruyter, p. 73.

Ivanov, V.T., Evstratov, A.V., Sumskaya, L.V., Melnik, E.I., Chumburidze, T.S., Portnova, S.L., Balashova, T.A. and Ovchinnikov, Yu.A., 1973, FEBS Lett., *36*, 65.

Ivanov, V.T., Fonina, L.A., Senyavina, L.B., Ovchinnikov, Yu.A., Chervin, I.I. and Yakovlev, G.I., 1980, Bioorg. Khim., *6*, 1008.

Ivanov, V.T., Sanafaryan, A.A., Chervin, I.I., Yakovlev, G.I., Fonina, L.A., Senyavina, L.B., Sychev, S.V., Vinogradova, E.I. and Ovchinnikov. Yu.A., 1974, Izv. AN SSSR (ser. khim.) (Bull. Acad. Sci. USSR, ser. chim.), p. 2310.

Ivanov, V.T. and Sychev, S.V., 1982, in: Biopolymer Complexes. Snatzke, G. and Bartmann W. Eds, N.Y.: Wiley, p. 107.

Ivanov, V.T. and Sychev, S.V., 1983, Proc. VIII Amer. Pept. Symp., Pierce Chem. Comp., Rockford, Illinois, (in press).

Izatt, R.M. and Christensen, J.J., 1979, Progress in Macrocyclic Chemistry. N.Y.: Wiley.

Jepson, B.E. and DeWitt, R., 1976, J. Inorg. Nucl. Chem., *38*, 1175.

Karle, I.L. and Karle, J., 1981, Proc. Natl. Acad. Sci. USA. *78*, 681.

Karle, I.L., Karle, J., Wieland, T., Burgermeister, W., Faulstich, H. and Witkop, B., 1973, Proc. Natl. Acad. Sci. USA, *70*, 1836.

Kartha, G., Varughese, K.I. and Aimoto, S., 1982, Proc. Natl. Acad. Sci. USA, *79*, 4519.

Keller-Schierlein, W. and Gerlach, H., 1968, in:Forschritte der Chemie organischer Naturstoffe. Bd. 25. Zechmeister, L. Ed., p. 161.

Kilbourn, B.T., Dunitz, J.D., Pioda, L.A.R. and Simon, W., 1967, J. Mol. Biol., *30*, 559.

Kimura, K., Maeda, T., Tamura, H. and Shamo, T., 1979, J. Electroanal. Chem., *95*, 91.

Knöchel, A. and Wilken, R.D., 1981, J. Amer. Chem. Soc., *103*, 5707.
Kyogoku, Y., Ueno, M., Akutsu, H. and Nawata, Y., 1975, Biopolymers, *14*, 1049.
Läuger, D., 1980, J. Membr. Biol., *57*, 163.
Lehn, J.-M., 1973, Structure and Bonding, *16*, 1.
Lehn, J.-M., 1978, Pure Appl. Chem., *50*, 871.
Lehn, J.-M., 1979, Pure Appl. Chem., *51*, 979.
Lehn., J.-M., 1983. in: Physical Chemistry of Transmembrane Ion Motions, Spach, G. Ed., Amsterdam: Elsevier, p. 181.
Lev, A., Malev, V. and Osipov, V., 1973, in: Membranes — A Series of Advances. V.2. Eisenman, G. Ed., N.Y.: Marcel Dekker, p. 480.
Lutz, W.K., Wipf, H.K. and Simon, W., 1970, Helv. Chim. Acta, *53*, 1741.
McLaughlin, S.G.A., Szabo, G., Ciani, S. and Eisenman, G., 1972, J. Membr. Biol., *9*, 3.
Michaux, G. and Reisse, J., 1982, J. Amer. Chem. Soc., *104*, 6895.
Moore, C. and Pressman, B.C., 1964, Biochem. Biophys. Res. Commun., *15*, 562.
Möschler, H.J., Weder, H.-C. and Schwyzer, R., 1971, Helv. Chim. Acta, *54*, 1437.
Nagaraj, R. and Balaram, P., 1981, Acc. Chem. Res., *14*, 356.
Nawata, Y., Hayashi, T. and Iitaka, Y., 1980, Chemistry Lett., 315.
Nawata, Y., Sakamaki, T. and Iitaka, Y., 1974, Acta Crystallogr., Sect. B., *B30*, 1047.
Nawata, Y., Sakamaki, T. and Iitaka, Y., 1975, Chemistry Lett., 151.
Nawata, Y., Sakamaki, T. and Iitaka, Y., 1977, Acta Crystallogr., Sect. B., *B33*, 1201.
Neupert-Laves, K. and Dobler, M., 1976, Helv. Chim. Acta, *59*, 614.
Ottensmeyer, F.D., Bazett-Jones, D.P., Henkelman, R.M., Korn, A.P. and Whiting, R.F., 1978-79, Chemica Scripta, *14*, 257.
Ottensmeyer, F.D., Bazett-Jones, D.P., Hewitt, J. and Price, G.B., 1978, Ultramicroscopy, *3* 303.
Ovchinnikov, Yu.A., 1981, Biochem. Soc. Symp., *46*, 103.
Ovchinnikov, Yu.A. and Ivanov, V.T., 1982a, in: Conformation in Biology. Srinivasan, R. and Sarma, R.H. Eds, Guilderland, N.Y.: Acad. Press. p. 155.
Ovchinnikov, Yu.A. and Ivanov, V.T., 1982b, in: The Proteins. V. 5. Neurath, H. and Hill, R.L. Eds, N.Y.: Academic Press, pp. 307-642.
Ovchinnikov, Yu.A., Ivanov, V.T. and Shkrob, A.M., 1974, Membrane Active Complexones. BBA Library, v. 12, Amsterdam: Elsevier.
Ovchinnikov, Yu.A., Ivanov, V.T., Evstratov, A.V., Mikhaleva, I.I., Bystrov, V.F., Portnova, S.L., Balashova, T.A., Tulchinski, V.M. and Meshcheryakova, E.A., 1974, Int. J. Pept Prot. Res., *6*, 465.
Pedersen, C.J., 1967, J. Amer. Chem. Soc., *89*, 2495; 7017.
Phillies, G.D.J., Asher, I.M. and Stanley, H.E., 1975, Science, *188*, 1027.
Pressman, B.C., 1976, Ann. Rev. Biochem., *45*, 501.
Prestegard, J.H. and Chan, S.I., 1969, Biochemistry, *8*, 3921.
Prestegard, J.H. and Chan, S.I., 1970, J. Amer. Chem. Soc., *92*, 4440.
Prince, R.C., Crofts, A.R. and Steinrauf, L.K., 1974, Biochem. Biophys. Res. Commun., *59*, 697.
Ramachandran, G.N. and Chandrasekaran, R., 1972a, Ind. J. Biochem. Biophys., *9*, 1.
Ramachandran, G.N. and Chandrasekaran, R., 1972b, in: Progress in Peptide Research. Lande S. Ed., N.Y.: Cordon & Breach, p. 195.
Rose, M.C. and Henkens, R.W., 1974, Biochim. Biophys. Acta, *372*, 426.
Ryba, O. and Petranek, J., 1973, Electroanal. Chem. *44*, 423.
Sakamaki, T., Iitaka, Y. and Nawata, Y., 1976, Acta Crystallogr., Sect. B, *B32*, 768.
Sakamaki, T., Iitaka, Y. and Nawata, Y., 1977, Acta Crystallogr., Sect. B, *B33*, 52.
Scheler, R.P. and Simon, W., 1970, Chimia, *24*, 372.
Shemyakin, M.M., Ivanov, V.T., Antonov, V.K., Vinogradova, E.I., Shkrob, A.M., Malenkov, G.G., Evstratov, A.V., Ryabova, I.D., Laine, I.A. and Melnik, E.I., 1969, J. Membr. Biol., *1*, 402.

Shemyakin, M.M., Ovchinnikov, Yu. A., Ivanov, V.T., Antonov, V.K., Shkrob, A.M., Mikhaleva, I.I., Evstratov, A.V. and Malenkov, G.G., 1967, Biochem. Biophys. Res. Commun., *29*, 834.

Simon, W. and Morf, W.E., 1973, in: Membranes — A series of Advances. V. 2, Eisenman, G. Ed., N.Y.: Marcel Dekker, p. 329.

Smith, G.D. and Duax, W.L., 1976, J. Amer. Chem. Soc., *98*, 1578.

Starke, C.M. and Liotta, C., 1978, Phase Transfer Catalysis. N.Y.: Acad. Press.

Steinrauf, L.K., Hamilton, J.A. and Sabesan, M.N., 1982, J. Amer. Chem. Soc., *104*, 4085·

Sychev, S.V. and Ivanov, V.T., 1982, in: Membranes and Transport. V.2. Martonosi, A.N. Ed., N.Y.: Plenum Corp., p. 301.

Szabo, G., Eisenman, G. and Ciani, S.M., 1969, J. Membr. Biol., *1*, 346.

Szabo, G., Eisenman, G., Laprade, R., Ciani, S.M. and Krasne, S., 1973, in: Membranes — A Series of Advances. V. 2. Eisenman, G. Ed., N.Y.: Marcel Dekker, p. 179.

Tishchenko, G.N., Karimov, Z., Andrianov, V.I., Vainstein, B.K., Evstratov, A.V., Ivanov, V.T. and Ovchinnikov, Yu.A., 1977, Bioorg, Khim., *3*, 467.

Tishchenko, G.N., Karimov, Z., Vainstein, B.K., Evstratov, A.V., Ivanov, V.T. and Ovchinnikov, Yu.A., 1976, FEBS Lett., *65*, 315.

Tolley, S.T., Bradshaw, J. and Izatt, R.M., 1982, J. Heterocyclic Chem., *19*, 3.

Ueno, M. and Kishimoto, H., 1979, Chemistry Lett., 1487.

Ueno, M. and Kyogoku, Y., 1979, Biopolymers, *18*, 2645.

Umeyama, H., Nakagawa, S., Nomoto, T. and Moriguchi, I., 1980, Chem. Pharm. Bull., *28*, 745.

Urry, D.W., 1971, Proc. Natl. Acad. Sci. USA, *68*, 672.

Urry, D.W., Goodall, M.C., Glickson, J.D. and Mayers, D.F., 1971, Proc. Natl. Acad. Sci, USA, *68*, 1907.

Urry, D.W., Long, M.M., Jacobs, V. and Harries, R.D., 1975, Ann. N.Y. Acad. Sci., *264*, 203.

Urry, D.W., Shaw, R.G., Trapane, T.L. and Prasad, K.U., 1983, Biochem. Biophys. Res. Commun., *114*, 373.

Veatch, W.R., Fossel, E.T. and Blout, E.R., 1974, Biochemistry, *13*, 5249.

Veatch, W.R., Matheis, R., Eisenberg, M. and Stryer, L., 1975, J. Mol. Biol., *99*, 75.

Veatch, W.R. and Stryer, L., 1977, J. Mol. Biol., *113*, 89.

Vinogradova, E.I., Fonina, L.A., Ryabova, I.D. and Ivanov, V.T., 1974, Khim. Prir. Soedin., 278.

Vishwanath, C.K. and Easwaran K.P.K., 1981, Biochemistry, *20*, 2018.

Vishwanath, C.K. and Easwaran, K.P.K., 1982, Biochemistry, *21*, 2612.

Weber, W.P. and Gokel, G.W., 1977, Phase Transfer Catalysis in Organic Synthesis. Berlin: Springer Verlag.

Westley, J.W., 1982-83, Polyether Antibiotics. V. 1, 2. N.Y.: Marcel Dekker.

Wipf, H.-K., Olivier, A. and Simon, W., 1970, Helv. Chim. Acta, *53*, 1605.

Wipf, H.-K., Pioda, L.A.R., Stefanac, Z. and Simon, W., 1968, Helv. Chim. Acta, *51*, 377.

THEORETICAL ANALYSIS OF FACTORS RESPONSIBLE FOR SPECIFICITY IN IONOPHORE-CATION INTERACTIONS

ALBERTE PULLMAN

Institut de Biologie Physico-Chimique
Fondation Edmond de Rothschild
13, rue Pierre et Marie Curie - 75005 Paris
France

ABSTRACT

Techniques developed in our laboratory for calculating intermolecular interactions, based on ab initio computations, have been used for the calculation of the energy of association of metal cations with ionophores. The systems studied involve the interaction of alkali cations with valinomycin, Na^+, K^+, NH_4^+ with nonactin and Mg^{2+}, Ca^{2+} with ionophore A 23187. The delimitation in each case of the different components of the interaction energy and the inclusion of the solvation-desolvation aspects of the interaction enable to account for the selectivity and specificity observed in each of these associations.

1. Introduction

Although the qualitative phenomenological analysis of the factors involved in the selectivity of bioorganic ion carriers has been available for some time (Eisenmann and Krasne 1975; Eisenmann *et al.*, 1975; Morf and Simon 1971 a, b, Morf *et al.*, 1972), the quantitative evaluation of the components of the energy balance has been limited to very simplified models and treatment. It is my aim in this paper to show that, owing to techniques developed recently in our laboratory for calculating intermolecular interactions based on the results of *ab initio* computations (Gresh *et al.*, 1979, A. Pullman *et al.*, 1979), we can not only compute the overall energy of association of a cation with an ionophore but also delimit the relative weights of the different components of this energy, whose interplay, combined with the corresponding desolvation energies of the ions, determines the selectivity or specificity of the different associations. We shall illustrate this situation in the examples of the carriers valinomycine, nonactin and ionophore A 23187 (calcimycine).

2. Methods

The intermolecular energies are computed by means of an additive procedure, elaborated earlier (Gresh *et al.*, 1979) and applied besides ionophores to diverse problems involving binding specificities (Gresh et B. Pullman 1981 a, b, 1983 a, b).

The interaction energy ΔE is evaluated as a sum of four components:

$$\Delta E = E_{MTP} + E_{pol} + E_{rep} + E_{dl}$$

where E_{MTP} is the electrostatic interaction energy computed between the overlap multipole expansions of the charge distributions of the entities in interaction, E_{pol} is the polarization energy, E_{rep} is the sum of bond-bond repulsions, and E_{dl} is a dispersion-like term calibrated in Gresh et al. 1979. To compute E_{MTP} we use a procedure (A. Pullman *et al.*, 1979) established for the calculation of the molecular electrostatic potential of large macromolecules, in which the macromolecule is built from subunits chosen in such a way as to minimize the electrostatic perturbations caused by the subdivision. The multicenter multipolar expansions (up to quadrupoles) of the charge distributions of the constituent fragments, required to compute the electrostatic and the polarization contributions to the binding energy, are derived from *ab initio* SCF computations; the minimal orbital basis set utilized is described in Gresh and Pullman, 1978.

Details on this methodology may be found in the original papers (see also, A. Pullman 1983 a).

In spite of important recent developments of the theoretical calculations on the hydration enthalpies of cations (see e.g. A.

Pullman, 1983 a), we have generally utilized, because of their availability, the experimental values of these enthalpies, for evaluating the corresponding desolvation energies.

3. Results and discussion

A) *Selectivity of valinomycin towards alkali cations* (Gresh *et al.*, 1981; Etchebest *et al.*, 1982; A. Pullman 1983 a, b). The chemical formula of valinomycin is recalled in figure 1. The binding

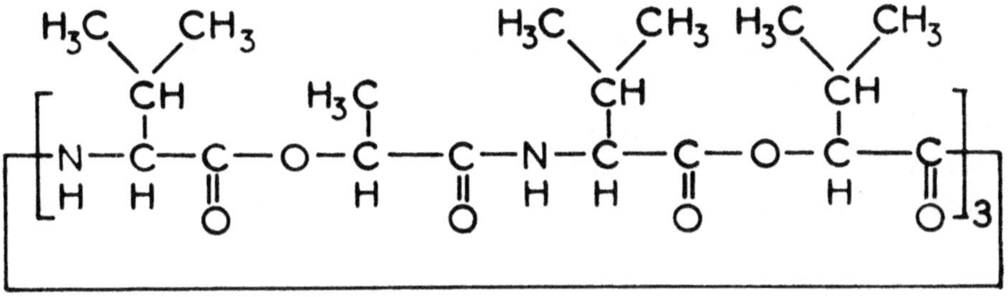

FIG. 1. Molecular structure of valinomycine.

constants measured in ethanol (Shemyakin *et al.*, 1969) and methanol (Grell *et al.*, 1972) for its complexes with the alkali ions follow the order

$$Rb^+ \approx K^+ > Cs^+ \gg Na^+$$

Until very recently, only one of these complexes, that of K^+, had been crystallized and analyzed through X-ray crystallography (Neupert-Laves and Dobler 1975): in the complex the molecule takes up a very characteristic " bracelet-like " structure in which all the amide NH and CO groups form a tight array of successive hydrogen bonds (fig. 2), leaving the six ester carbonyls free to " encage " the ion, three above and three below (fig. 3). Taking into consideration the rigidity conferred on the molecule by the sequence of hydrogen bonds and experimental indications on the permanence of a bracelet-like structure in the other alkali complexes, we have attempted (Gresh *et al.*, 1981) to account for the observed specificity order by considering two components of the energy balance involved in complex formation, namely the inter-

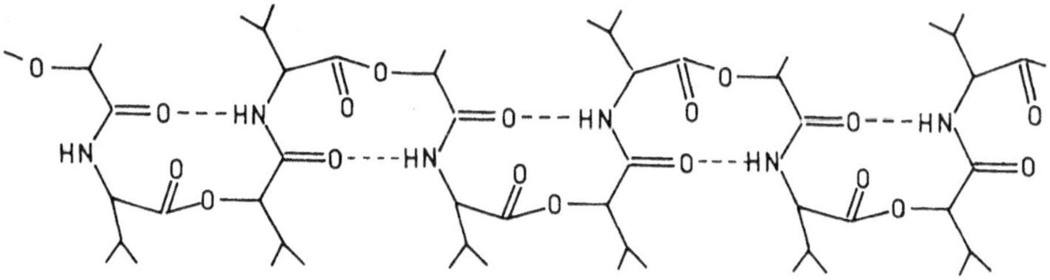

FIG. 2. The array of hydrogen-bonds in the K^+ complex of valinomycine.

action energy between the ions and valinomycin, assumed to keep the geometry of the K^+ complex, and the desolvation energy of the ions. The first term was obtained by the technique indicated in Methods, optimizing the position of each cation inside the cavity by an energy-minimizing procedure. For the energy of desolvation, the negatives of the available experimental values

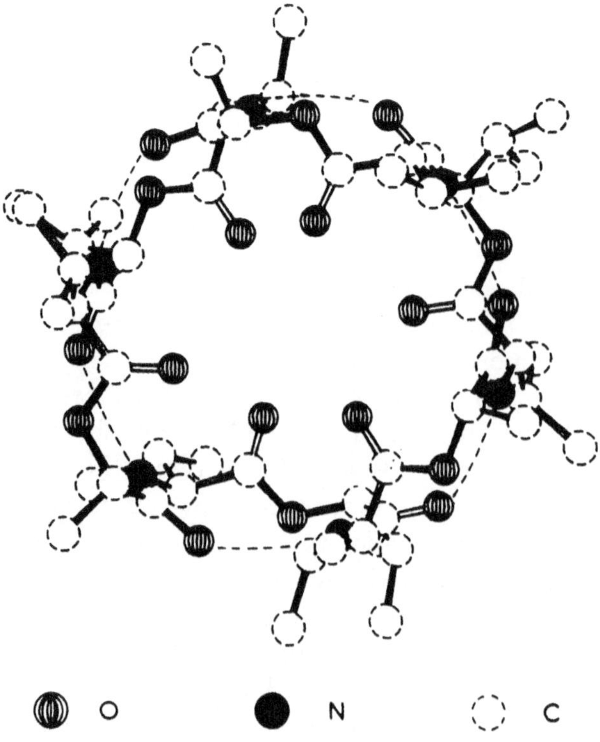

FIG. 3. The "bracelet" structure of the valinomycine-K^+ complex.

of the enthalpies of hydration (Kaufmann *et al.*, 1976) have been used (see Gresh *et al.*, 1981 for a discussion).

The results are indicated in table I, from which it is seen that:

a) In the K^+ complex, the computed optimal distances are in very reasonable agreement with the range of distance (2.69-2.83 Å) observed in the crystal structure, a feature supporting the credibility of the computations.

b) K^+, Rb^+ and Cs^+ prefer an essentially central position in the cavity. On the contrary, Na^+ prefers a position shifted from the center towards the vicinity of two carbonyl oxygens. It is interesting that such a situation was suggested on the basis of infrared measurements of the ester carbonyl frequencies of the Na^+-complex (Ovchinnikov *et al.*, 1972).

c) The optimal values of the *energies of interaction* are in the order $Na^+ > K^+ > Rb^+ > Cs^+$. (Note the favorable binding of Na^+ in spite of its peculiar position in a cavity too large for it). They differ thus greatly from the experimentally observed order.

d) The order of the experimental *desolvation energies* is the same as the order of the interaction energies and thus does not account either, by itself, for the experimental order of preferential binding.

e) *When the two preceding components are taken into account simultaneously, the computed order of the binding energies becomes exactly identical to the selectivity order*, with K^+ very close to Rb^+, Cs^+ somewhat beyond, and Na^+ very much beyond.

Thus, if one assumes that the complexes all adopt a structure similar to that of the K^+ complex, it is possible to account for the specificity of association by taking into consideration solely the complexation and the desolvation energies, the two terms operating in opposite directions, but with different differential effects.

Very recently (Steinrauf *et al.*, 1982) the crystal structure of a valinomycin-sodium picrate complex was established. The configuration of the crystal is very particular: it contains a water molecule inside the cavity of a bracelet-like valinomycin, the Na^+ and picrate ions being situated outside the frame of the antibiotic. The situation is thus distinctly different from that of the similar K^+ complex (Hamilton *et al.*, 1981). We have already indicated in one of our previous publications (Gresh *et al.*, 1981 and references therein) that in the results reported in table I, the energy balance for Na^+ is so unfavourable as compared to the other cations, that the very occurrence of the assumed conformation upon Na^+

TABLE I. *Valinomycin - cation complexes. Equilibrium distances d (Å) to the ester oxygens, energies of interaction ΔE of the cations inside the K^+ cavity, experimental dehydration enthalpies ΔE_E, energy balance δE and difference δ with respect to the best balance. Energies in kcal/mole*

	Na^+	K^+	Rb^+	Cs^+
d_1	3.05	2.08	2.74	2.70
d_2	2.95	2.68	2.73	2.72
d_3	2.81	2.73	2.77	2.75
d_4	2.42	2.62	2.69	2.72
d_5	2.47	2.72	2.68	2.69
d_6	2.67	2.77	2.75	2.78
ΔE	—108.5	—106.4	—102.7	—88.8
ΔE_H [1]	106.0	85.8	79.8	72.0
δE [2]	— 2.5	— 20.6	— 22.9	— 16.8
δ	20.4	2.3	0	6.1

[1] (Kaufmann *et al.*, 1976).
[2] $\delta E = \Delta E + \Delta E_H$.

binding may be questioned. Indeed, the circular dichroism spectrum of valinomycin upon Na^+ binding in methanol is characterized by a negative Cotton effect, whereas the complexes with K^+, Rb^+ and Cs^+ are characterized by a positive effect, and this was taken as evidence for the involvement of a different Na^+-binding conformation upon complexing in methanol. A similar conclusion was arrived at on the basis of ^{13}C and infrared measurements. This situation is markedly different from the one occurring in a non-polar solvent, in which the closed conformation of valinomycin seems involved in the binding of all four alkali cations.

The reasons for the peculiar structure adopted by the Na^+ picrate complex necessitate obviously a special investigation, which is being planned in our laboratory.

B) *The K^+/Na^+ selectivity of nonactin* (Gresh and A. Pullman, 1982). Nonactin is another cyclic carrier with the formula of figure 4.

Like valinomycin, it prefers K^+ to Na^+. In this case the crystal structures of both complexes are known. They both show the molecule wrapped around the ion, using its four carbonyl and its four tetrahydrofuran oxygens to encage the ions in two somewhat different cavities. This knowledge has made possible the study

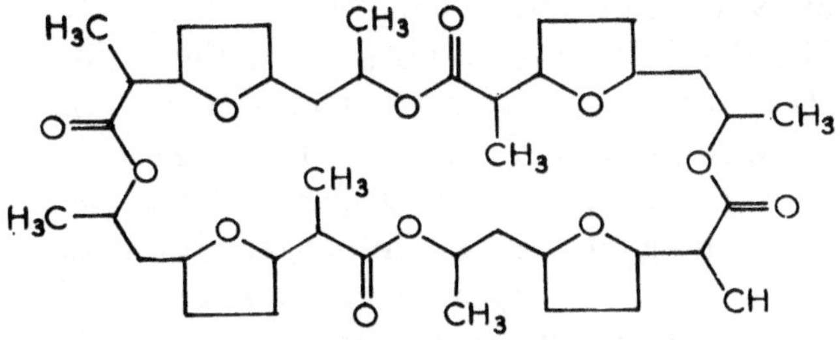

FIG. 4. The chemical formule of nonactin.

in this case (Gresh and Pullman 1982) of the mutual interplay of the *three* most conspicuous components of the energy balance, namely, the binding energy of the ion to its specific cavity, the desolvation energy of the cation and the change in intramolecular energy of the ionophore in going from the geometry corresponding to the K^+-cavity to that of the Na^+-cavity.

The data are summarized in table II. ΔE is computed, as before, by optimization of the position of each ion in its cavity. The desolvation energies are the negatives of the experimental values of the enthalpies of solvation in methanol and the variation

TABLE II. *Contributions to the energy balance in the complexation of K^+ and Na^+ to nonactin*

	Na^+	K^+
ΔE	-151.2	-117.5
ΔE_H [1]	$+104.9$	$+84.2$
$\delta E = \Delta E - \Delta E_H$	-46.3	-33.3
ΔR [2]	$+26.1$	$+9.3$
$\delta' E = \delta E - \Delta R$	-20.2	-24.0

[1] Enthalpies in methanol (Hedwig and Parker 1974).
[2] Repulsion.

in the intramolecular energy of the ionophore is computed as due to the different repulsions between its nucleophilic units in the two geometries corresponding to the two different cavities, this variation being estimated to represent the essential term of the overall difference (for details see Gresh and A. Pullman 1982).

It is seen that the binding energy of Na^+ to nonactin in its specific cavity is appreciably *larger* than that of K^+ in its own cavity. The desolvation energies play in the right direction for *reducing* this difference between the two ions, leaving them, however, in the wrong order. Finally, *it is the ligand-ligand repulsions, appreciably stronger in the* Na^+*-cavity than in the* K^+*-one*, (essentially due to a shortening of the oxygen-oxygen distances), *which bring the overall balance in favour of* K^+ *by a few kcal/mole* and which thus ensure the agreement with experiment.

C) *The* NH_4^+/K^+ *selectivity of nonactin* (Gresh and Pullman, 1982).

Nonactin prefers NH_4^+ to K^+ for complex formation. The crystal structure of the ammonium complex being unknown, the K^+-cavity was used for optimizing the position of the ammonium ion in it. The results are given in table III. A first observation is that, in its optimal position, NH_4^+ is directly hydrogen-bonded to the

TABLE III. *Nonactin complexes. Equilibrium distances d (Å) to the eight oxygens* ([1]) *of* K^+ *and of the hydrogens of* NH_4^+*, optimized binding energies* ΔE*, desolvation energies* ΔE_H *and balance* δE *(kcal/mole)*

	K_T	NH_4^+
d_1	2.75	2.63, 2.59
d_2	2.80	2.58, 2.83
d_3	2.74	2.48, 2.76
d_4	2.80	2.72, 2.60
d_5	2.81	1.90
d_6	2.90	1.99
d_7	2.80	1.94
d_8	2.90	2.00
ΔE	— 117.5	— 131.8
ΔE_H [2]	79.8 / 85.8	89.0
δE	— 38.7 / — 31.7	— 52.8

[1] d_1 to d_4 refer to the ester carbonyls, d_5 to d_8 to the tetrahydrofuran oxygens; for the carbonyl oxygens the two HO distances refer to the two closest hydrogens.
[2] Dehydration energies.

four tetrahydrofuran oxygens, the carbonyl oxygens being situated roughly between two hydrogens at appreciably larger distances, indicating no direct H-bonding in that case. This is in perfect agreement with conclusions drawn on the basis of infrared spectral measurements of the C = O stretching frequencies of this complex in solution (Züst *et al.*, 1973). Interestingly, the strongest binding found in the case of ammonium is reversed with respect to the tendency observed in the K^+ complex, where the shortest distances to oxygens occur with the carbonyl ones.

As to the specificity NH_4^+ versus K^+, it is seen that the binding energies in the cavity already strongly favor ammonium. When the balance is made with the desolvation enthalpies, the overall preference, although decreased, remains largely in favour of ammonium. *This is thus an example where the ordering set by the cation-cavity interactions is so much in favour of the preferred ion that it is not modified by the desolvation energies.*

D) *Selectivity of ionophore A* 23187 *towards* Ca^{2+} *and* Mg^{2+} (Gresh and A. Pullman 1983).

The open-chain carboxylic ionophore A 23187 (fig. 5) transports divalent ions accross membranes (Reed and Lardy 1972; Pfeiffer *et al.*, 1974) with a marked selectivity over monovalent ions.

X-ray crystal data are available for the ionophore (Chaney *et al.*, 1974), as well as for its 2: 1 Ca^{2+} complex (Smith and Daux 1976), and since very recently, for its 2: 1 Mg^{2+} complex (Al-

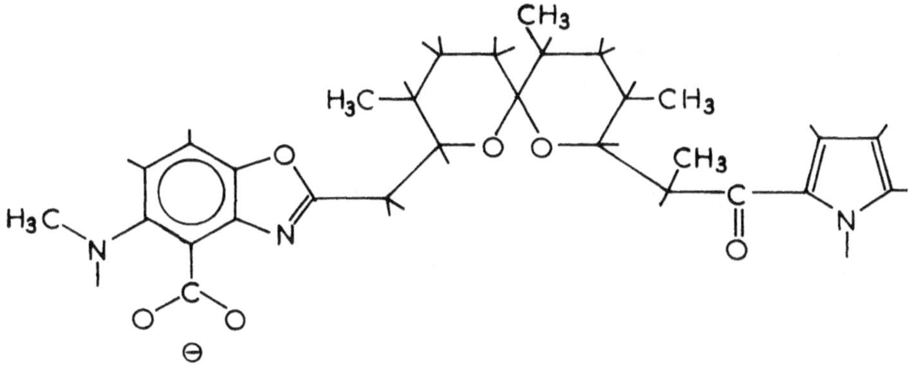

FIG. 5. The molecular structure of A 23187.

leaume and Barrans, 1983). The solution structure of these two complexes was also investigated by NMR spectroscopy (Antenis 1977; Diber and Pfeiffer 1976).

In both 2: 1 crystal complexes, each deprotonated molecule of A 23187 (symbol A^-) binds the cation by one of its carboxylate oxygens, by the aromatic nitrogen of its benzoxazole group and by the carbonyl oxygen of its ketopyrrole moiety. In both complexes also, an additional stabilization occurs through the formation of two hydrogen bonds, involving the NH group of the ketopyrrole ring of one ionophore and the metal-bound anionic oxygen of the benzoxazole ring of the other ionophore. The Ca^{2+} complex differs from the Mg^{2+} complex by the inclusion of one water molecule in the coordination polyhedron of the cation, which is thus seven-fold coordinated.

It should be pointed out that in contrast to the strong divalent/monovalent selectivity of A 23187, its discrimination in favour of Ca^{2+} *versus* Mg^{2+} is small. (C. Krause *et al.*, 1983; Puskin and Gunter 1975; Pfeiffer and Lardy 1976; Pohl *et al.*, 1980; Debons *et al.*, 1981; Nakashina *et al.*, 1982).

The availability of the detailed X-ray coordinates of both complexes has stimulated a search for an extensive enthalpy balance incorporating the binding energy of the cation, the mutual ionophore-ionophore interactions, the variations of intramolecular energies of the individual ionophores, and the dehydration enthalpy of the cation. Particular attention was devoted to the significance of the presence of the water molecule in the Ca^{2+} complex.

The essential results of the computations are indicated in Table IV. We shall consider them in a two-step approach, by examining first the unhydrated complexes and, in a second step, the significance of the conservation of partial hydration of the ion in the calcium complex.

The term ΔE_1 measures the intramolecular repulsions in the A^- moieties due to the repulsive interactions between the liganding subunits of the A^-, namely their benzoxazole and ketopyrrole rings. These repulsions are different in the two complexes because of their slightly different conformations. They are stronger in

TABLE IV. *Interaction energies (kcal/mole) in the A^{-}-M^{2+}-A^{-} complexes*

	Mg^{2+}	Ca^{2+}
ΔE_i	+ 14.2	+ 10.2
ΔE (A^-, M^{2+}, A^-)	—682.5	—573.1
ΔE_T	—668.3	—562.9
ΔE (A^-, M^{2+}, W_a, A^-)	—	—585.6
ΔE (A^-, M^{2+}, W_a, W_b, W_c, A^-)	—	—612.8
ΔE_H	477.6	398.8
δE	—190.7	—164.1
δE (W_a)	—	—186.8
δE (W_a, W_b, W_c)	—	—214.0

ΔE_i = intramolecular repulsions in the A^- moieties.
ΔE = intermolecular interaction energies between the entities in parentheses treated together in the complex.
ΔE_T = $\Delta E_i + \Delta E$ (A^-, M^{2+}, A^-).
ΔE_H = dehydration energies of the cations.
δE = energy balance for unhydrated complexes = $\Delta E_T - \Delta E_H$;
δE (W_a) = corresponding balance including W_a;
δE (W_a, W_b, W_c) = corresponding balance including W_a, W_b, W_c.

the Mg^{2+} complex and their difference between the two structures favours the Ca^{2+} complex by 4.0 kcal/mole.

The term ΔE (A^-, M^{2+}, A^-) comprises the intermolecular interaction energy of the cation with the two ionophores and the mutual interactions of the ionophores with each other in the presence of the cation. This value is substantially greater for the Mg^{2+} complex than for the Ca^{2+} complex.

The term ΔE_T comprises the sum of the intra- and intermolecular contributions. It remains largely, by 105 kcal/mole, in favour of the Mg^{2+} complex.

ΔE_T indicates the experimental dehydration enthalpies of the two cations (Goldman and Bates 1972). The value is greater for Mg^{2+} than for Ca^{2+} but only by 78 kcal/mole.

Hence, an enthalpy balance δE in which the dehydration enthalpy of the cation is subtracted from the interaction energy ΔE_T in A^- - M^{2+} - A^- would, at this stage, result in an appreciable advantage (27 kcal/mole) in favour of Mg^{2+}.

As a next step we consider the influence on the energy-balance of the presence of the water molecule visible in the crystal of the Ca^{2+} cation (W_a). The computations were carried out by positioning W_a in the vicinity of the cation, starting from the data provided by the crystal structure and performing an energy-minimization on the six variables (three translations and three rotations) defining its location with respect to the complex A^- - Ca^{2+} - A^-. This optimized interaction energy of W_a with the A^- - Ca^{2+} - A^- complex, ΔE (W_a), is - 27.4 kcal/mole.

The total stabilization energy, ΔE, of the resulting complex W_a - A^- - Ca^{2+} - A^- amounts to - 584.8 kcal/mole. [This value is smaller than the summed value (- 588.3 kcal/mole) of ΔE and ΔE (W_a), a feature which is due to the non-additivity of the polarization contribution of the ionophores by the complexing entities, whether Ca^{2+} alone or $Ca^{2+} + W_a$. When the entity is $Ca^{2+} + W_a$, the polarizing field exerted by Ca^{2+} is opposed by the one, albeit much weaker, exerted by the water molecule, which approaches the complex through its oxygen atom, hence a weaker field and a reduced magnitude of the polarization energy of the ionophores].

We would like to stress at this point the fact that we have verified by computation that there is no possibility of binding a water molecule as a seventh ligand to the cation in the crystal structure of the Mg^{2+} complex.

If now we compare the stabilization energies of the Mg^{2+} and the Ca^{2+} complexes, including W_a in the latter, δE (W_a), we find that the difference in favour of Mg^{2+} has decreased to 4.7 kcal/mole. In fact one should also take into account the energy expenditure necessary to extract the water molecule from the bulk of the solvent and transpose it to the Ca^{2+} complex. If we *tentatively* assume (e.g. Gresh and B. Pullman 1980, B. Pullman et al. 1979) that this energy expenditure corresponds, for the extracted molecule, to the water-water dimerization energy, namely - 5.3 kcal/mole as computed in the framework of the present procedure, the difference between the two complexes becomes 10 kcal/mole in favour of Mg^{2+}.

It thus appears, at this point, that although the presence of a water molecule as a seventh ligand to calcium in the complex brings about a large increment in the global energy, the balance between Ca^{2+} and Mg^{2+} remains nevertheless somewhat in favour of the Mg^{2+} complex.

It must be noted, however, that divalent cations are known to be able to strongly bind and structure a second solvation shell (Veillard 1977, Berthod et al. 1978). This incited us to investigate the possibility of binding two additional water molecules, W_b and W_c, through the hydrogen atoms of W_a in a second solvation shell of Ca^{2+}. This was performed by means of energy-minimization of the set of the variables defining the orientation of W_b and W_c and simultaneous energy-reminimization of the orientation of W_a in the presence of W_b and W_c.

The binding energies of W_b and W_c amount to - 17 kcal/mole for each, and the stabilization energy of the whole complex W_a-W_b-W_c-A^--Ca^{2+}-A^- reaches then - 612.4 kcal/mole. If now we compute the energy balance with the solvation energy we obtain a value of - 214.0, thus 24.7 kcal/mole larger than in the Mg^{2+} complex. Even after subtraction of the energy admittedly necessary to detach three water molecules from the water bulk (15.9), the balance remains in favour of the calcium complex by 9 kcal/mole.

Our results demonstrate thus that the relative stabilization of the Ca^{2+} complex with respect to the Mg^{2+} complex is due essentially to the presence, in the former, of water molecules coordinated with the cation. Following our computations this stabilization requires the intervention of not only the water molecule clearly distinguished in the crystal but of one or two supplementary such molecules bound to the former. That such a situation is a reasonable possibility in solution is indicated by the observation that, from the Fourier map of the calcium complex, one may deduce the presence of a second water molecule, with high thermal motion and probable partial occupancy, hydrogen-bonded to the well-defined calcium-bound water molecule and to the carboxylate oxygen of an adjacent complex (Smith and Duax, 1976).

4. Conclusion

Altogether, considering the three types of ionophore studies, it appears that although the energy balances reported cannot be claimed as completely accurate (see discussion in references Gresh *et al.*, 1979; Gresh and Pullman 1980 a), these calculations represent the first successful attempt at a reliable evaluation of the various terms involved in the selectivity of ionophores. The results obtained on the three examples reported show the intricacy of the interplay of the different energy components: in valinomycin the values of the complexation energies to the cavity and those of the desolvation energies of the cation rank in the same order from Na^+ to Cs^+ but their differences modify the overall ordering, yielding the experimental one. In nonactin the same factors play similarly for K^+/Na^+, but their interplay is not sufficient to account for the specificity order observed, and it is the different repulsions between the carbonyl groups in the two different cavities which are the decisive element. For the couple NH_4/K^{++} the selectivity is already present in the cation-ionophore binding energies and desolvation does not change the order. In the A 23187 ionophore, it is the water molecule(s) bound to the Ca^{2+} but absent in the Mg^{2+} complex which ensure the comparable affinity of the carrier for the two ions (with possibly a small preference for Ca^{2+}).

I would like to add that for all these ionophores we have explored also a number of other structural properties (accessibilities, electrostatic molecular potentials and fields) which may be related to different aspects of their activity as ion carriers: mechanism, in particular the initiation, of the complexation (Etchebest *et al.*, 1982), the possibility of co-transport of water (Gresh and Pullman, 1983) or of an anion (Etchebest *et al.*, 1982) etc.

Finally, I would like to mention that very recently we have extended the same type of computations to a representative of trans-membrane channels: gramicidine A (A. Pullman and C. Etchebest, 1983). We have computed, in particular, the energy profile for single and double cation occupancy in a head-to-head $\beta^{6.3}_{3,3}$ - helical dimer backbone. So far, the computations have been carried out only for Na^+ and do not relate thus, as yet, to the problem of specificity. They provide, however, useful information on the general aspect of these types of ion transport.

REFERENCES

Alleaume, M. and Barrans, Y., 1983, Acta Crystallographica, in press.

Anteunis, M., 1977, Bioorg. Chem. *6*, 1.

Berthod, H., Pullman, A. and Pullman, B., 1978, Int. Quant. Chem., Quantum Biol. Symp. *5*, 79.

Chaney, M., Demarco, P., Jones, N. and Occolowitz, J., 1974, J. Am. Chem. Soc. *96*, 1932.

Deber, C. and Pfeiffer, D., 1976, Biochemistry *15*, 132.

Debono, M., Molloy, R., Dorman, D., Paschal, J., Babcok, D., Deber, C. and Pfeiffer, D., 1981, Biochemistry *20*, 6866.

Eisenmann, G. and Krasne, S., 1975, MTP Int. Rev. Sci. Biochem. Ser. *2*, 27.

Eisenmann, G., Krasne, S. and Ciani, S., 1975, Ann. N.Y. Acad. Sci. *264*, 34.

Etchebest, C., Lavery, R. and Pullman, A., 1982, Studia Biophysica *90*, 7.

Goldman, S. and Bates, R., 1972, J. Am. Chem. Soc. *94*, 1476.

Grell, E., Funck, Th. and Eggers, F., 1972, in Molecular Mechanisms of Antibiotic Action on Protein Biosynthesis and Membranes, Proceeding of a Symposium in Granada, Munoz, E., Garcia-Ferrandiz, F. and Vasquez, D. (Eds), Elsevier, Amsterdam, p. 646.

Gresh, N., Claverie, P. and Pullman, A., 1979, Int. J. Quantum Chem. Symp. *13*, 243.

Gresh, N., Etchebest, C., de la Luz Rojas, O. and Pullman, A., 1981, Int. J. Quantum Chem., Quantum Biol. Symp. *8*, 109.

Gresh, N. and Pullman, A., 1978, Theoret. Chim. Acta *49*, 283.

Gresh, N. and Pullman, A., 1982, Intern. J. Quantum Chem. *22*, 709.

Gresh, N. and Pullman, A., 1983, Intern. J. Quantum Chem., Quantum Biol. Symp. 10, 215.

Gresh, N. and Pullman, B., 1980 a, Biochim. Biophys. Acta *608*, 47.

Gresh, N. and Pullman, B., 1980 b, Biochim. Biophys, Acta *625*, 356.

Gresh, N. and Pullman, B., 1983, Intern. J. Quantum Chem., *24*, 491.

Gresh, N. and Pullman, B., 1984, Theoret. Chim. Acta *64*, 383.

Hamilton, J.A., Sabesan, M.N. and Steinrauf, L.K., 1981, J. Am. Chem. Soc. *103*, 5880.

Hedwig, G. and Parker, A., 1974, J. Am. Chem. Soc. *96*, 6573.

Kaufmann, E., Lehn, J.M. and Sauvage, J.P., 1976, Helv. Chim. Acta *59*, 1099.

Krause, G., Grell, E., Albrecht-Gary, A.M., Boyd, D. and Schwing, J.P., 1983, in: Physical Chemistry of Transmembrane Ion Motions, G. Spach (Ed.) Elsevier.

Morf, W. and Simon, W., 1971 a, Helv. Chim. Acta *54*, 794.

Morf, W. and Simon, W., 1971 b, Helv. Chim. Acta *54*, 2683.

Morf, W., Züst, Ch. and Simon, W., 1972, in: Molecular Mechanisms of Antibiotic Action on Protein Synthesis and Membranes, Munoz, Z., Garcia-Ferrandiz, F. and Vasquez, D. (Eds), Elsevier, Amsterdam, p. 523.

Nakashima, R., Dordick, R. and Garlid, K., 1982, J. Biol. Chem. *257*, 12540.

Neupert-Laves, K. and Dobler, M., 1975, Helv. Chim. Acta *58*, 432.

Ovchinnikov, Y.A., Ivanov, I.V. and Shkrob, A.M., 1972, in: Molecular Mechanisms of Antibiotic Action on Protein Synthesis and Membranes, Munoz, E., Garcia-Ferrandiz, F. and Vasquez, D., Eds., Elsevier, Amsterdam, 459.

Pfeiffer, D. and Lardy, H., 1976, Biochemistry *15*, 935.

Pfeiffer, D., Reed, P. and Lardy, H., 1974, Biochemistry, *13*, 4007.

Pohl, G., Kreikenbohm, R. and Seuwen, K., 1980, Z. Naturforsch. Biosci. *23 C*, 562.

Pullman, A., 1983 a, in: Physical Chemistry of Transmembrane Ion Motions, (G. Spach Ed.), Elsevier, Amsterdam, p. 153.

Pullman, A., 1983 b, in: Ions and Molecules in Solution (Tanaka, N., Ohtaki, H. and Tamamushi, R. Eds.), Elsevier, Amsterdam, p. 373.

Pullman, A. and Etchebest, C. 1983, FEBS Letters, *163*, 199.

Pullman, A., Zakrzewska, K. and Perahia, D., 1979, Intern. J. Quantum Chem. *16*, 395.

Pullman, B., Miertus, S. and Perahia, D., 1979, Theoret. Chim. Acta *50*, 317.

Puskin, J. and Gunter, T., 1975, Biochemistry *14*, 187.

Reed, P. and Lardy, H., 1972, J. Biol. Chem. *247*, 6970.

Shemyakin, M.M., Ovchinnikov, Yu. A., Ivanov, V.T., Antonov, V.K., Vinogradova, E.I., Skhrob, A.M., Malenkov, G.G., Evstratov, A.V., Melnik, E.I. and Ryabova, I.D., 1969, J. Membr. Biol. *1*, 402.
Steinrauf, L.K., Hamilton, J.A. and Sabesan, H.N., 1982, J. Am. Chem. Soc. *104*, 4085.
Smith, G. and Duax, W., 1976, J. Am. Chem. Soc. *98*, 1578.
Veillard, H., 1977, J. Am. Chem. Soc. *99*, 7194.
Züst, C., Fruh, P. and Simon, W., 1973, Helv. Chim. Acta *56*, 495.

GPSR Compliance

The European Union's (EU) General Product Safety Regulation (GPSR) is a set of rules that requires consumer products to be safe and our obligations to ensure this.

If you have any concerns about our products, you can contact us on ProductSafety@springernature.com

In case Publisher is established outside the EU, the EU authorized representative is:

Springer Nature Customer Service Center GmbH
Europaplatz 3
69115 Heidelberg, Germany

Batch number: 10809439

Printed by Printforce, the Netherlands